Computational Protein-Protein Interactions

Computational
Protein-Protein
Interactions

Edited by
Ruth Nussinov
Gideon Schreiber

CRC Press
Taylor & Francis Group
Boca Raton London New York

CRC Press is an imprint of the
Taylor & Francis Group, an **informa** business

CRC Press
Taylor & Francis Group
6000 Broken Sound Parkway NW, Suite 300
Boca Raton, FL 33487-2742

First issued in paperback 2017

© 2009 by Taylor and Francis Group, LLC
CRC Press is an imprint of Taylor & Francis Group, an Informa business

No claim to original U.S. Government works

ISBN-13: 978-1-4200-7005-7 (hbk)
ISBN-13: 978-1-138-11335-0 (pbk)

Library of Congress Cataloging-in-Publication Data

Computational protein-protein interactions / editors, Ruth Nussinov and Gideon
    Schreiber.
        p. ; cm.
    Includes bibliographical references and index.
    ISBN 978-1-4200-7005-7 (hardcover : alk. paper)
    1. Protein-protein interactions--Computer simulation. 2. Protein-protein
interactions--Mathematical models. I. Nussinov, Ruth. II. Schreiber, Gideon.
    [DNLM: 1. Protein Interaction Mapping. 2. Proteins--physiology. 3. Computational
Biology--methods. 4. Models, Molecular. QU 55 C7378 2009]

    QP551.5.C66 2009
    572'.64--dc22   2009012638

Visit the Taylor & Francis Web site at
http://www.taylorandfrancis.com

and the CRC Press Web site at
http://www.crcpress.com

Contents

v

Preface

Proteins are the working horse of the cellular machinery. They are responsible for diverse functions ranging from molecular motors to signaling. They catalyze reactions, transport, form the building blocks of viral capsids, traverse the membranes to yield regulated channels, and transmit information from the DNA to the RNA. They synthesize new molecules, and they are responsible for their degradation. Proteins are the vehicles of the immune response and of viral entry into cells.

Perhaps the most common dominator of all proteins is their ability to interact with one another and with many other types of molecules, whether small or large. Not only do proteins interact with most known chemical components, but they do so specifically. That is, they interact at a specific location, with a specified affinity and kinetics. This is the result of the varied chemistry of the amino acids and cofactors, and the specific three-dimensional shapes of proteins. Protein–protein interactions are divided (somewhat artificially) into permanent and transient interactions, but even the transient complexes can bind with picomol or nanomol affinity, and with rate constants of association and dissociation ranging six orders of magnitude. Another line of division is between homo- and heterocomplexes, which can be further distinguished by the number of proteins involved in the complex (dimers, trimers, large multiprotein complexes, etc.). Thus, protein interactions can be discussed from the point of view of a biophysicist or bioinformatician: The first is interested in understanding how the various forms of interactions work and assemble, and the latter is focused on the analysis of the interactions; both aiming toward prediction of the interaction. However, within cells, proteins often function as parts of large networks of interactions, also called interactome. In recent years, many aspects of biology have been likened to these networks, in which distinct nodes (e.g., individual proteins) can be defined that interact with one another within a system to perform various biological functions. Network maps have been constructed to depict all of the possible protein–protein interactions within a cell (i.e., the interactome), essentially providing a low-resolution view of molecular recognition. The distinct view of protein–protein interactions, from the atomic detail to the cellular interactome arrangement, has to be investigated at different levels (structure, function, organization, energetics, dynamics) with each of these levels being investigated experimentally as well as computationally. To obtain a more complete understanding of cellular processes, a combination of all of these will be needed.

To be able to predict protein–protein interactions, there is a need to figure out various aspects of their associations. These range from shape complementarity to the organization and the relative contributions of the physical components to their stability. Proteins interact through their surfaces. Thus, to analyze protein–protein interactions, residues (or atoms) that are in contact across the two-chain interface are studied. In addition, residues in their vicinity are also inspected to explore their supporting matrix. At the same time, it behooves us to remember that proteins that are free in solution exist in ensembles of native, though distinct, conformers. In viewing proteins as static structures, the properties of a particular population are explored. Yet, different populations may preferentially associate with different partners. The overriding reasons for the heightened interest in protein–protein interactions are that better understanding and better quantization of the key features controlling the interactions should lead to higher success in the prediction of protein associations. This would assist in the elucidation of cellular pathways and in drug design.

This book covers a broad range of aspects related to prediction of protein–protein interactions, to interfering with protein–protein interactions, and to their design. These relate to cellular pathways with the goals of understanding and predicting function, and to strategies in drug targeting. With the increasing accumulation of experimental data, the need for computational approaches is rapidly increasing. This can also be gauged from the rapid growth in the literature in this direction and the creation of new computational/bioinformatics journals. Accordingly, this book provides an overview, with the chapters carefully selected and written by leaders in the field.

Although it is important to predict protein associations, it is a daunting task. Some associations are obligatory, whereas others are transient, continuously forming, and dissociating. From the physical standpoint, any two proteins can interact. The question is under what conditions and at what strength. Protein–protein interactions are largely driven by the hydrophobic effect. In addition, hydrogen bonds and electrostatic interactions play important roles. The physical principles of protein–protein interactions are general, and many of the interactions observed *in vitro* are the result of experimental overexpression or of crystal effects, complicating functional prediction.

Joël Janin, a pioneer in the field of protein science, provides in the first chapter an overview of the basic principles of protein–protein interactions and the biophysical forces driving them. Two basic models are suggested to explain the energetic composition of protein–protein binding sites: the buried surface model and the hot spot model. However, this is an oversimplification, as protein interactions have diverse solutions to accommodate the differences in binding, with rates and affinities spanning 10 orders of magnitude. With this diversity of interactions, which relate to the diversity of life, it is clear that no one model of interaction fits all. Moreover, many interactions are feasible only after the protein undergoes chemical modification, such as phosphorylation, allowing for the interactions to be controlled. Still, the basic biophysical principles of the interactions are the same, independent of the type or lifetime of the complex.

In the second chapter, "Low-Resolution Recognition Factors Determine Major Characteristics of the Energy Landscape in Protein–Protein Interaction," Ilya Vakser seeks the general shapes and features that make a binding site. The first part of the chapter provides a detailed description of databases of protein–protein interactions, which are used for computational analysis. This "technical" part is of great importance, as the results are strongly influenced by the quality and coverage of the database used. Detailed analysis of binding clearly shows the similarities between binding and folding of proteins. We are used to examining binding sites at high resolution. However, Vakser points out that the general shape of proteins already dictates their binding. The existence of large-scale structural recognition features in protein association explains the funnel-like intermolecular energy landscape. As in protein folding, this concept is necessary to explain the kinetics data for protein–protein association and is useful for docking simulations.

Eric Sundberg takes the investigation of the architecture of binding sites one step further in the third chapter, and provides a detailed molecular description of its composition and how this drives binding. In Chapter 3, "The Molecular Architecture of Protein–Protein Binding Sites," he describes the binding site as a dense network of interacting amino acids, with most of them forming well-defined local clusters. Within this network of interactions, one can find both negative and positive cooperativity. It is important that a number of amino acids interact over long distances providing long-range communication within binding sites.

High-resolution methods to decipher the architecture and energetics of binding sites can be slow and are limited to a small set of protein complexes. In recent years, alternative approaches, based on selection from large libraries, have become more popular to provide more comprehensive answers. The most commonly used vehicle for such analysis is phage display. Particularly interesting is the development of the "shotgun" approach by Sachdev Sidhu and co-workers. In Chapter 4, "Mapping Protein Function by Combinatorial Mutagenesis," Gábor Pál and Sidhu describe the technical aspects of this method and provide different applications for it. In addition to mapping the contribution of residues to binding, this method provided a more global picture of the relation between sequence, structure, and conservation of binding sites. Moreover, the general concept has been extended beyond phage display by the use of other combinatorial methods, which hold much promise for the future. It is clear that combinatorial analysis with well-defined libraries and selections can be used to explore diverse protein functions in a rapid manner and serve as a more complete set for computational biology than provided by traditional mutation studies.

The structure of a protein–protein interaction, its affinity and thermodynamic characteristics, depict a "frozen" state of a complex. This picture ignores the kinetic nature of complex formation and dissociation, which are of major biological and biophysical interest. In the next two chapters, Gideon Schreiber and Rebecca Wade provide a summary of the pathway for protein–protein association. In Chapter 5, Schreiber analyzes the pathway of association as a three-step reaction. After collision, the proteins form an encounter complex, which develops into the final complex through a transition state. Electrostatic forces are the main determinants of this reaction. The structures of the encounter complex and transition state are discussed. In Chapter 6, Pachov, Gabdoulline, and Wade discusses Brownian dynamics simulation methods, which are used to simulate the reaction coordinates and rates.

One of the most exciting fronts in computational protein–protein interactions is the use of the existing knowledge on protein–protein interactions for interface design. In Chapter 7, "Computational Design of Protein–Protein Interactions," Julia Shifman provides numerous examples of successes and failures, yielding an up-to-date picture of where we are, the main problems facing us, and what we can expect in the near future. She very elegantly divides the design problem into subareas (including affinity design; how to achieve specificity; de novo interface design; association versus dissociation; and protein–protein, protein–peptide, and protein–DNA interactions) and discusses the various tools used to achieve success in each area.

The number of files in the Protein Data Bank is rapidly growing, now exceeding 50,000. However, structural information is often not available and even if available it is often not straightforward to use to predict the protein function. Yet, the involvement of protein–protein interactions in all cellular processes and the consequent crucial need to figure out their functions has led to focused efforts to predict functions from sequences and, if available, from their structures. A practical way to predict protein function is through identification of the binding partners. Since the vast majority of protein chores in living cells are mediated by protein–protein interactions, if the function of at least one of the components with which the protein interacts is identified, it is expected to facilitate its functional and pathway assignment. Through the network of protein–protein interactions, we can map cellular pathways and their intricate cross-connectivity. Because two protein partners cannot simultaneously bind at the same (or overlapping) site, discovery of the ways in which the proteins associate should assist in inferring their dynamic regulation. Identification of protein–protein interactions is at the heart of functional genomics. Prediction of protein–protein interactions is also crucial for drug discovery. Knowledge of the pathway, its topology, length, and dynamics should provide useful information for forecasting side effects. Six chapters of this book address different computational approaches to map binding.

Methods to map binding consist of a number of layers of information and resolution. At the high end is protein docking. To be able to dock proteins, the information on the partners as well as their structure has to be available (or at least the structure of a close homologue). Howook Hwang, Brian Pierce, and Zhiping Weng (Chapter 8) provide a detailed description of how protein–protein docking works and the criteria of success.

However, producing an interactome using high-resolution docking algorithms is limited by the lack of available structures and knowledge of protein partners. A complementary method, for which less or no structural information is required, is mapping protein binding sites. Yanay Ofran, in Chapter 9, "Prediction of Protein Interaction Sites," provides a very detailed description and analysis of different methods for the prediction of protein–protein binding sites, with the plusses and minuses of the different methods. In Chapter 10, "Predicting Molecular Interactions in Structural Proteomics," Irina Kufareva and Ruben Abagyan provide an exciting outlook on how to use the structural information to map function as embedded in the subcellular structural organization of the proteins; that is, the relationship between binding and function, and how we can build a cellwide dynamic and structural interaction map. The basic questions to solve are which protein is interacting with which, where the interaction takes place, and what the different (in structure and composition) complexes look like.

Most eukaryotic proteins are composed of multiple domains, with each being an independent folding unit. Multidomain proteins allow the acquisition of new properties without disrupting the ones they already have. One of the most important properties a protein can acquire is the ability to interact with other proteins, and thus defining its interactome. In Chapter 11, Inbar Cohen-Gihon, Roded Sharan, and Ruth

Nussinov describe the mechanisms by which domain rearrangements occur in the genome and highlight the role of co-occurring domains in protein–protein interactions. Due to the modularity of the protein domain world, it is straightforward to use graph theoretical tools to explore domain composition of proteins. Although the numbers of domains across several genomes are similar, the sizes of highly connected domain subgraphs grow with evolution, and thus the complexity of the organism.

A large fraction of cellular proteins are estimated to be "natively disordered," that is, unstable in solution. The structures of disordered proteins are not random. Rather, the disordered state has a significant residual structure. In the disordered state, a protein exists in an ensemble of conformers. Disordered proteins are believed to account for a large fraction of all cellular proteins and to play roles in cell-cycle control, signal transduction, transcriptional and translational regulation, and large macromolecular complexes. Although disordered on their own, their native conformation is stabilized upon binding. Vladimir Uversky and colleagues discuss these proteins in Chapter 12, "Intrinsically Disordered Proteins and Their Recognition Functions." It was suggested that the increasing abundance of intrinsically disordered proteins in higher organisms is likely due to the change in the cellular requirements for certain protein functions, particularly regulatory functions/cellular signaling. Many "hub" domains, such as SH2, SH3, and PDZ, bind to disordered regions, apparently because disordered regions can bind partners fast, with both high specificity and low affinity. In this chapter, the authors describe functions and molecular mechanisms of these disordered peptides with specific focus on recognition and attempt to create links with the structural properties of these proteins.

As protein–protein interactions play a crucial role in many biological processes, their disruption can lead to a disease state or cure. Therefore, it is of great interest to consider them as potential drug targets. In Chapter 13, "Identification of Druggable Hot Spots on Proteins and in Protein–Protein Interfaces," Dmitri Beglov and co-workers describe a powerful approach to the identification of druggable regions on the protein surfaces by computational mapping, using small molecular probes such as small organic molecules. Computational mapping places the molecular probes, whether small molecules or functional groups, on the surface of the protein to identify the most favorable binding positions. Although x-ray crystallography and nuclear magnetic resonance (NMR) indicate that organic solvents bind to a limited number of sites on a protein, computational mapping methods can result in hundreds of energy minima and do not reveal why some sites bind molecules with different sizes and polarities, thus presenting a problem in the prediction of these regions. The authors review the mapping algorithms in the literature and the difficulties that are involved. Next they describe their mapping based on the fast Fourier transform (FFT) correlation approach, which samples possible configurations on a dense translational and rotational grid. The positions are scored using an energy function that includes attractive and repulsive van der Waals terms, electrostatic interaction energy based on Poisson–Boltzmann calculations, a cavity term to represent the effect of nonpolar enclosures, and a structure-based pairwise interaction potential. Finally, they provide two interesting applications.

Finally, in Chapter 14, "Designing Protein–Protein Interaction Inhibitors," Matthieu Montes reviews the various methods available today for virtual compound screening of protein–protein interactions inhibitors. Such methods are on the way to replacing the more traditional blind, high throughput, fragment-based screening, reducing cost and increasing coverage. In particular, protocols using a wise combination of structure-based virtual ligand screening and ligand-based virtual ligand screening methods have led to very interesting inhibitors displaying original scaffolds, which can be used as a basis to develop new compounds with therapeutical interest on challenging targets. However, these methods suffer from similar problems as other computational methods, such as the need to improve scoring functions, better account for electrostatics and solvation, and the fundamental problem of how small molecules can compete with the binding of large proteins on the same binding site.

Overall, although the chapters span the broad area of computational protein–protein interactions, the area is very extensive, and to keep the size of the book manageable it is not possible to include all aspects. In particular, areas that are not addressed in this book relate to membrane proteins and molecular dynamic simulations of protein–protein interactions, with the goal of obtaining deeper insights into how the function is performed. Nonetheless, it is hoped that this book provides a basic outline of major directions in computational protein–protein interactions.

Ruth Nussinov
Gideon Schreiber

Editors

Ruth Nussinov received her Ph.D. in 1977 from the biochemistry department at Rutgers University, New Brunswick, NJ. She did her postdoctoral work in the structural chemistry department of the Weizmann Institute of Science, Rehovot, Israel. Subsequently, Dr. Nussinov worked in the chemistry department at the University of California, Berkeley, the biochemistry department at Harvard University, Cambridge, MA, and the National Institutes of Health (NIH), Bethesda, MD. She joined Tel Aviv University, Israel, in 1984, and in 1990, she became a professor in the department of human genetics at Tel Aviv University's medical school, the Sackler School of Medicine. In 1985, Dr. Nussinov accepted a concurrent position at the National Cancer Institute of the NIH, where she is a Senior Principal Investigator heading the Computational Structural Biology Group. She has authored around 350 scientific papers, edited numerous journals, and speaks frequently at both national and international meetings. Her interests largely focus on protein folding and dynamics, protein–protein interactions, amyloid conformations and toxicity, and large multi-molecular associations with the goal of understanding the protein structure–function relationship.

Gideon Schreiber received his doctorate in biochemistry at the Hebrew University, Jerusalem, in 1992. After a postdoctoral period at the Medical Research Council's (MRC) Laboratory of Molecular Biology in Cambridge, U.K., working in the laboratory of Professor Alan Fersht, he joined the Weizmann Institute of Science, Rehovot, Israel, as Senior Scientist. Presently, he is an associate professor at the department of biological chemistry at the same institute. His research interests include the investigation of protein–protein interactions, from the basic understanding of the mechanism to protein–design. His work spans bioinformatics and algorithm development, biophysical bench work, protein-design and engineering to applied biology. In addition, Dr. Schreiber is a director of the Israel Structural Proteomic Center, located at the Institute, and which aims to provide structures of proteins and protein–complexes to the biological community.

Contributors

Ruben Abagyan
The Scripps Research Institute
La Jolla, California

Dmitri Beglov
Structural Bioinformatics Lab
Boston University
Boston, Massachusetts

Ryan Brenke
Department of Biomedical
 Engineering
Boston University
Boston, Massachusetts

Gwo-Yu Chuang
Department of Biomedical
 Engineering
Boston University
Boston, Massachusetts

Inbar Cohen-Gihon
Sackler Faculty of Medicine
Tel Aviv University
Tel Aviv, Israel

A. Keith Dunker
Center for Computational Biology and
 Bioinformatics
Indiana University School of Medicine
Indianapolis, Indiana

Monika Fuxreiter
Biological Research Center
Hungarian Academy of Sciences
Budapest, Hungary

Razif R. Gabdoulline
Molecular and Cellular Modeling Group
EML Research gGmbH
Heidelberg, Germany
and
BioQuant
University of Heidelberg
Heidelberg, Germany

David Hall
Department of Biomedical
 Engineering
Boston University
Boston, Massachusetts

Howook Hwang
Bioinformatics Program
Boston University
Boston, Massachusetts

Joël Janin
Yeast Structural Genomics
Université Paris-Sud
Orsay, France

Dima Kozakov
Department of Biomedical
 Engineering
Boston University
Boston, Massachusetts

Irina Kufareva
The Scripps Research Institute
La Jolla, California

Melissa Landon
Department of Biochemistry
Rosenstiel Basic Medical Sciences
   Research Center
Brandeis University
Waltham, Massachusetts

Matthieu Montes
Chaire de Bioinformatique
Conservatoire National des Arts et
   Métiers
Paris, France

Chi Ho Ngan
Department of Biomedical
   Engineering
Boston University
Boston, Massachusetts

Yanay Ofran
The Goodman Faculty of Life Science
Bar Ilan University
Ramat Gan, Israel

Christopher J. Oldfield
Center for Computational Biology and
   Bioinformatics
Indiana University School of Medicine
Indianapolis, Indiana

Georgi V. Pachov
Molecular and Cellular Modeling Group
EML Research gGmbH
Heidelberg, Germany

Gábor Pál
Department of Biochemistry
Eötvös Loránd University
Budapest, Hungary

Brian Pierce
Bioinformatics Program
Boston University
Boston, Massachusetts

Roded Sharan
School of Computer Science
Tel Aviv University
Tel Aviv, Israel

Yang Shen
Computer Science and Artificial
   Intelligence Laboratory
Department of Biological
   Engineering
Massachusetts Institute of
   Technology
The Stata Center
Cambridge, Massachusetts

Julia M. Shifman
Department of Biological Chemistry
The Hebrew University of Jerusalem
Jerusalem, Israel

Sachdev S. Sidhu
Banting and Best Department of
   Medical Research
Terrence Donnelly Centre for Cellular
   and Biomolecular Research
University of Toronto
Toronto, Ontario

Eric J. Sundberg
Boston Biomedical Research Institute
Watertown, Massachusetts

Spencer Thiel
Structural Bioinformatics Lab
Boston University
Boston, Massachusetts

Peter Tompa
Biological Research Center
Hungarian Academy of Sciences
Budapest, Hungary

Vladimir N. Uversky
Center for Computational Biology and
    Bioinformatics
Indiana University School of Medicine
Indianapolis, Indiana
and
Institute for Biological Instrumentation
Russian Academy of Sciences
Pushchino, Moscow Region, Russia
and
Institute for Intrinsically Disordered
    Protein Research
Indiana University School of Medicine
Indianapolis, Indiana

Sandor Vajda
Department of Biomedical Engineering
Boston University
Boston, Massachusetts

Ilya A. Vakser
Center for Bioinformatics and
    Department of Molecular Biosciences
University of Kansas
Lawrence, Kansas

Rebecca C. Wade
Molecular and Cellular Modeling
    Group
EML Research gGmbH
Heidelberg, Germany

Zhiping Weng
Bioinformatics Program and
    Department of Biomedical
    Engineering
Boston University
Boston, Massachusetts
and
Program in Bioinformatics and
    Integrative Biology
University of Massachusetts Medical
    School
Worcester, Massachusetts

Brandon Zerbe
Department of Biomedical Engineering
Boston University
Boston, Massachusetts

1 Basic Principles of Protein–Protein Interaction

Joël Janin

CONTENTS

Proteins are the major players in molecular recognition at the heart of all processes of life. They interact with the other components of the cell, small molecules, nucleic acids, membranes, and other proteins to build supramolecular assemblies and elaborate molecular machines that perform all sorts of functions, from chemical catalysis and mechanical work to signaling and regulation (Alberts, 1998). Protein–protein recognition is the mechanism by which the specific interaction between polypeptide chains creates functional units. Its study has been part of biochemistry, structural biology, and computational biology for more than 30 years, and it has now spread to all domains of biology and medical science (Eisenberg et al., 2000). Protein–protein recognition must be given a chemical and physical basis, which in practice requires high-resolution three-dimensional structures. The Protein Data Bank (PDB; Berman et al., 2000) contains that information for several hundreds of protein assemblies, mostly transient binary complexes and oligomeric proteins. Cells contain plenty of larger assemblies, still poorly represented in the PDB, with the exception of the icosahedral viruses, ribosomes, and a few others (Dutta & Berman, 2005). Their analysis is the next frontier in our understanding of molecular recognition in biology.

The structures of binary complexes and oligomeric proteins present in the PDB form only a small sample of what exists in nature, yet they have stimulated a rich body of biochemical studies by site-directed mutagenesis, supported by biophysical studies of their thermodynamics and kinetics. The results have been extensively analyzed and they are the topics of several reviews and collective books (Jones &

Thornton, 1996, 2000; Larsen et al., 1998; Kleanthous, 2000; Janin & Wodak, 2003; Fu, 2004; Russell et al., 2004; Ponstingl et al., 2005; Janin et al., 2007). Two models of protein–protein interaction have emerged over the years from these studies: the buried surface model and the hot spot model. The first is geometric and defines the interface as the protein surface that is solvent accessible in the isolated components, but not in the complex (Chothia & Janin, 1975); it implies that the interaction is distributed more or less evenly over that surface. On the other hand, the hot spot model states that the significant interactions are highly localized. The model was inspired by the site-directed mutagenesis study of the human growth hormone/growth hormone receptor system (Clackson & Wells, 1995) and subsequent alanine scanning experiments performed on other systems. In alanine scanning, the residues of one component in contact with the other are systematically mutated to Ala, and the affinity of the mutants is compared to that of the wild type. Many of the mutations cause little or no change in affinity, and those that do define the hot spots (Bogan & Thorn, 1998; DeLano, 2002; Wells & McClendon, 2007).

I believe that the two views can be reconciled, and that the nonuniform nature of protein–protein interfaces can be accounted for by splitting them into a core and a rim depending on the solvent accessibility of the interface atoms. The rim, which has an amino acid composition and other properties similar to the solvent accessible surface, contains very few hot spots. The core differs in its composition, it contains most of the hot spots, and it is better conserved in evolution than the rim and the rest of the protein surface, which suggests that it is the main target of the selection exerted by protein–protein recognition on the protein sequence.

DIVERSITY OF PROTEIN–PROTEIN INTERACTION

In spite of its limited size, the sample of protein assemblies for which structural data are available shows a diversity that reflects the diversity of life itself (Nooren & Thornton, 2003a). A broad classification may be based on the time scale on which the assembly process takes place. At one end of the scale, the collisions that occur at every instance within the crowded space of the cell create short-lived (submicroseconds) contacts of no biological significance, except that they compete with functional interactions. Their equivalent in the PDB are the crystal packing contacts, which are mostly nonspecific and yield stable assemblies only because each molecule is in contact with many neighbors. The interactions seen in crystal packing may thus be compared to those in complexes and oligomeric proteins to give a structural basis to specificity (Janin, 1997; Bahadur et al., 2004).

At the other end of the scale, oligomeric proteins have a long-lived quaternary structure that self-assembles at the time the subunits are synthesized. Many oligomers dissociate *in vitro* only when they are made to unfold, and *in vivo* only when they enter a degradation pathway; thus, they can be considered as permanent. In between, protein–protein complexes are made of polypeptide chains that fold independently and associate only when they happen to meet. Most are transient, but the range of affinities and lifetimes covers at least eight orders of magnitude. Examples of long-lived associations are the trypsin/pancreatic inhibitor complex, with a half-life of months (Vincent & Ladzunski, 1972), and the complex of barnase, a bacterial ribonuclease, with its intracellular inhibitor barstar, which has a $K_d \approx 10^{-14}$ M and a half-life of days (Schreiber & Fersht, 1993). Antigen–antibody complexes are not quite as stable; they have a K_d in the range 10^{-8}–10^{-10} M and a half-life of minutes to hours (Foote & Eisen, 1995; Braden & Poljak, 2000; Sundberg & Mariuzza, 2002). In the immune system, there are much weaker interactions that play an equally important role: the T cell receptor interacts with its different partners to form complexes that have $K_d \approx 10^{-6}$ M (Foote & Eisen, 2000). Weak, short-lived interactions are fully functional in many other processes. In general, the complex between an enzyme, a protein kinase for instance, and its substrate cannot be long lived because its dissociation would limit the reaction turnover. Similarly, a fraction of a second is sufficient for redox proteins to carry out an electron transfer reaction after they come to be in contact (Crowley & Carrondo, 2004). Cell signaling relies on both short-lived and stable protein–protein interactions. The response of a cell to an external stimulus frequently involves forming a loose initial complex that may become a stable assembly when it recruits new partners, undergoes phosphorylation or other chemical changes, and translocates to a different cell compartment. The timescale may be minutes, or milliseconds in the case of the visual signal.

Irrespective of their stability, all these interactions are biologically significant, they play major roles in essential processes, and thus are subject to a Darwinian selection that affects the sequence of the polypeptide chains and the physical chemical properties of their interfaces.

ACCESSIBLE SURFACE AREA VERSUS FREE ENERGY: THE HYDROPHOBIC EFFECT REVISITED

In the buried surface model, the interface between two macromolecules is the set of atoms and residues that lose solvent accessibility in the assembly (Chothia & Janin, 1975; Janin & Chothia, 1990). This geometric definition has a thermodynamical counterpart due to the relationship between the free enthalpy of a nonpolar organic solute in water (ΔG_{np}) and its solvent accessible surface area (ASA; Lee & Richards, 1971). The following relation is verified when hydrocarbons are transferred from a nonpolar solvent to water (Hermann, 1972):

$$\Delta G_{np} = \gamma \, ASA \qquad (1.1)$$

Chothia (1974, 1975) used the hydrocarbon solubility data to place the coefficient γ in the range 20–25 cal.mol^{-1}.Å^{-2}. Later estimates yield $\gamma = 29$ cal.mol^{-1}.Å^{-2} for aromatic hydrocarbons and 31 cal.mol^{-1}.Å^{-2} for aliphatic compounds (Vajda et al., 1995). A still higher value, 50 cal.mol^{-1}.Å^{-2}, has been derived from a comparison with the macroscopic process, γ being the microscopic equivalent of a surface tension coefficient (Sharp et al., 1991). In addition, analytical models of the hydration of hard spheres suggest that linearity is achieved only above a certain size of the spheres (Lum et al., 1999), which implies that γ ought to be larger for proteins than for small molecules.

Equation 1.1 is a quantitative expression of the hydrophobic effect. Because a nonpolar solute cannot give or receive H-bonds, the water molecules in contact with it lose part of their H-bond energy and/or their freedom of orientation (Kauzmann, 1959; Tanford, 1997). This costs free energy, and Equation 1.1 states that, within a family of similar molecules, the cost is proportional to the number of water molecules concerned. It can just as well be written with the number of carbon atoms in the hydrocarbon molecule, or the volume it occupies, in place of the ASA, but the ASA is a more suitable parameter when dealing with a folded protein that is in contact with the solvent only through its surface atoms.

The free enthalpy of a nonpolar solute contains terms other than ΔG_{np}, but they take similar values in water and organic solvents. The transfer experiment lumps together with the hydrophobic effect the balance of the van der Waals interactions with made water vs. the organic solvent, also likely to increase linearly with the ASA. However, the reasoning does not apply to polar groups that make H-bonds, less numerous but much more energetic than van der Waals interactions. Their contribution is a balance between the free enthalpy of water–solute and water–water H-bonds. It can be positive or negative depending on the nature of the polar groups and the details of their geometry, and is generally difficult to assess. With large molecules that make many H-bonds, one may attempt to average the contributions of individual bonds and use Equation 1.1 for polar as well as nonpolar groups, with appropriate values of γ. However, there are no families of compounds with variable numbers of polar groups on which to calibrate the coefficients, and the sets that have been proposed over the years show large discrepancies (Eisenberg & McLachlan, 1986; Ooi et al., 1987; Makhatadze & Privalov, 1994; Xie & Freire, 1994).

PROTEIN–PROTEIN INTERFACES IN THE BURIED SURFACE MODEL

Given the atomic coordinates of the complex between a receptor protein (R) and a ligand (L; we make this distinction only for convenience, and L may be also a protein), the size of the RL interface is measured by the buried surface area:

$$BSA = ASA_L + ASA_R - ASA_{RL} \qquad (1.2)$$

where ASA_L, ASA_R, and ASA_{RL} are the solvent-accessible surface areas of free R, free L, and the RL complex, respectively. When RL dissociates, nonpolar groups in R and L move from a protein environment to water. The relevant free enthalpy term can be calculated from Equation 1.1 and the nonpolar contribution to the BSA, but the value of γ derived from hydrocarbon transfer experiments may not be appropriate, due to the discrepancies noted earlier between microscopic and macroscopic approaches and because the protein environment is more dense and better packed than an organic solvent.

Nevertheless, the BSA has proved to be a very useful parameter to evaluate the interaction between two proteins. Its estimation from atomic coordinates is robust, and it distinguishes between different categories of interactions. Protein–protein complexes have an average BSA of 1910 Å², and 58% of that BSA belongs to nonpolar groups (Table 1.1). Lo Conte et al. (1999) noted that, in a sample of 75 complexes,

TABLE 1.1
Properties of Protein–Protein Interfaces

Interface Parameter	Protein–Protein Complexes[a]	Homodimers[b]	Weak Dimers[c]	Crystal Packing[d]
Number in data set	70	122	19	188
BSA (Å²)	1910	3900	1620	570/1510
(SD)	(760)	(2200)	(670)	(520)
Number of amino acids	57	104	50	48
% in the interface core[e]	55	60	51	40
Chemical composition (%)[f]				
Nonpolar	58	65	62	58
Neutral polar	28	23	25	25
Charged	14	12	13	17
Atomic packing[g]				
Buried atoms f_{bu} (%)	34	36	28	21
L_D packing index	42	45	34	32
S_c complementarity score	0.69	0.70	—	0.63
Number per 1000 Å² BSA				
H-bonds	5.3	4.8	4.3	3.6
Hydration waters[h]	10	11	—	15
Residue conservation[i] s core/rim ratio	0.82	0.87		—

[a] Data of Chakrabarti and Janin (2002) on a subset of the complexes of Lo Conte et al. (1999).

[b] Data of Bahadur et al. (2003).

[c] Homodimers in equilibrium with the monomer according to the literature (Lévy, 2007).

[d] Pairwise interfaces in crystals of monomeric proteins. The first mean BSA value is for the 1320 interfaces in the 152 crystal forms analyzed by Janin and Rodier (1995). All other numbers are for the 188 interfaces with BSA >800 Å2 in Bahadur et al. (2004).

[e] Core residues contain interface atoms with zero ASA in the assembly.

[f] Fraction of the BSA contributed by nonpolar (carbon-containing) chemical groups; groups that contain N, O, or S are counted as neutral polar, or charged in Asp, Glu, Arg, and Lys side chains.

[g] f_{bu} is the fraction of interface atoms with zero ASA in the assembly; L_D is defined in Bahadur et al. (2004), S_c is defined in Lawrence and Colman (1993).

[h] Data from Rodier et al. (2005).

[i] Ratio of the mean values of the Shannon entropy (s) of the residues of the interface core and rim in the aligned sequences of homologous proteins: 52 protein components of the complexes (excluding antigen–antibody complexes), 121 homodimers, and 102 monomeric proteins in crystal contacts (Guharoy & Chakrabarti, 2005).

many enzyme–inhibitor complexes and nearly all the complexes between a protein antigen and a cognate antibody have an interface that buries 1200–2000 Å², which they called "standard size." Figure 1.1 represents the BSA distribution in a larger set of complexes recently assembled by Hwang et al. (2008). The average BSA in that set is the same as in Table 1.1, but the range of the values (800–5800 Å²) is broader than in earlier studies. Nevertheless, all but 2 of the 25 antigen–antibody complexes

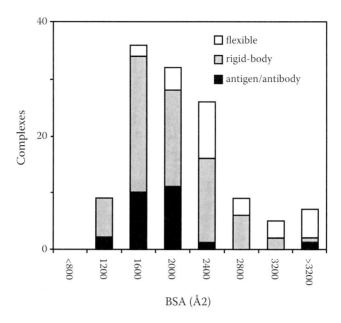

FIGURE 1.1 BSA and the mode of protein–protein recognition in complexes. Histogram of the buried surface area in the nonredundant set of 123 protein–protein complexes assembled by Hwang et al. (2008). This set includes the structures of the components as well as of the complexes. After least-square superimposition of the components and the complexes, the RMS distance between Cα atoms is less than 1.8 Å in 23 of the 25 antigen–antibody complexes. The other complexes are marked "rigid body" when the RMS distance is less than 1.8 Å and marked "flexible" when it is larger.

and 55% of the other complexes have standard-size interfaces; 7% have a BSA less than 1200 Å², and 38% a BSA larger than 2000 Å².

The set of Hwang et al. (2008) was assembled to benchmark protein docking algorithms, and it includes the structures of the free components as well as the complexes. When the free and bound structures are compared by least-square superposition, the root mean squared (RMS) distance between the Cα atoms ranges between 0.2 and 8 Å. Small RMS distances imply that the components of the complex associate as rigid bodies to a good approximation; they only undergo side-chain rotations and small main-chain movements. Large RMS distances point to major conformation changes and to a mechanism of induced fit or flexible recognition. In Figure 1.1, we set the limit between the two categories at 1.8 Å. With that cutoff, all the complexes with a BSA <1200 Å² and 92% of those with a standard-size interface are in the rigid body category, which includes all but two of the antigen–antibody complexes and 70% of the other complexes. The induced fit category contains only 8% of the complexes with a BSA <2000 Å² and 47% of those with a larger BSA. Thus, the new sample supports the remark made by Lo Conte et al. (1999) that large interfaces correlate with large conformation changes in protein–protein complexes.

This remark can be extended to oligomeric proteins, most of which contain large interfaces (Janin et al., 1988; Jones & Thornton, 1995; Bahadur et al., 2003). Their subunits usually fold as they assemble, undergoing disorder-to-order transitions that are extreme cases of conformation changes. In contrast, crystal packing contacts tend to create small interfaces and induce only minor conformation changes. Table 1.1 reports mean values of the BSA in homodimers and crystal contacts for comparison with the complexes. In crystals of monomeric proteins, the average pairwise contact buries 570 Å^2 (Janin & Rodier, 1995), less than the minimum value of 800 Å^2 observed in complexes or homodimers. Nevertheless, a significant fraction of the crystal packing interfaces have a BSA >800 Å^2, and thus are comparable in size to the interfaces of complexes. Table 1.1 cites properties of these large nonspecific interfaces along with those of the biologically significant assemblies. The crystal packing interfaces have the same nonpolar fraction as in the complexes, but they contain fewer buried atoms (f_{bu} = 21% vs. 28%), fewer H-bonds (3.6 vs. 5.3 per 1000 Å^2), and more water molecules (15 vs. 10 per 1000 Å^2) in proportion of their size. Moreover, their S_c complementarity score (Lawrence & Colman, 1993) and L_D packing index (Bahadur et al., 2004) are low, which suggests that the nonspecific interfaces are less tightly packed than specific ones.

The set of homodimeric proteins assembled by Bahadur et al. (2003) has a mean BSA of 3900 Å^2; all the interfaces are at least standard size, and some bury as much as 10,000 Å^2, a surface equivalent to the one buried when a small protein folds. With 65% of the BSA coming from nonpolar groups, the homodimer interfaces are more hydrophobic than in complexes. They bury a greater proportion of their atoms and may also be better packed, but the differences indicated by the L_D and S_c parameters are marginal, and the density of polar interactions (H-bonds and hydration waters) is not significantly different from that in complexes. In addition, Table 1.1 also mentions "weak dimers," a set of homodimers known to be in equilibrium with the monomers, assembled by Dey et al. (in preparation) with the help of the PiQSi database (http://www.supfam.org/elevy/piqsi/; Lévy, 2007). In this set, the BSA range is 750–3000 Å^2 and the mean is 1620 Å^2, close to the value reported by Noreen and Thornton (2003b) in an earlier set of the same type. The interfaces of the weak dimers are comparable in size to those of the complexes, but they tend to be less polar and bury fewer atoms; moreover, their low L_D index suggests that they are poorly packed like the crystal packing interfaces.

BURIED SURFACE AREAS AND BINDING FREE ENERGIES

The stability of a complex RL and the affinity of R for L are characterized by the equilibrium constant (K_d) or by the standard state free enthalpy of dissociation per mol of complex:

$$\Delta G_d = -RT \ln K_d/c° \tag{1.3}$$

where c° is the standard state concentration (1M by convention); RT $\approx$ 0.6 kcal.mol^{-1} at 300K. For short, we shall call ΔG_d a "binding free energy," but not a "binding energy" as the literature often does. Writing *energy* for *enthalpy* ignores the pressure dependence of the equilibrium, of no significance for most applications. Omitting *free* overlooks the crucial role of entropy, and masks the point that, for a bimolecular reaction, the value of ΔG_d depends on the choice of c°.

The data in Table 1.1 suggest a broad correlation between the stability of a protein assembly, and the size and nonpolar character of its interface. To make it quantitative, Horton and Lewis (1992) established a linear relationship between the binding free energy of a set of complexes and their BSA appropriately weighted for the polar and nonpolar components. The set comprised mostly enzyme–inhibitor complexes that have standard-size interfaces and assemble as a rigid body, and the correlation is unlikely to extend to other systems. It would predict very large binding free energies for all the complexes that have a BSA >2000 Å², and thus make them much more stable than they are in reality. In general, conformation changes and other energy terms not directly related to the interface size mask any correlation of the BSA with stability. For instance, most of the complexes involved in signal transduction have larger interfaces than protease-inhibitor complexes, yet they are often short lived and display large conformation changes (Lo Conte et al., 1999).

On the other hand, the existence of a correlation between ΔG_d and BSA is supported by experiments that make small changes in carefully chosen systems. When the dissociation constant K'_d of a mutant complex is compared to the K_d of the wild type, Equation 1.3 (from which c° is eliminated) yields the change in the binding free energy:

$$\Delta\Delta G = RT \ln K'_d / K_d \qquad (1.4)$$

In the simple case where the mutation affects neither the conformation of the components nor the polar interactions, the nonpolar contribution dominates:

$$\Delta\Delta G \approx \Delta\Delta G_{np} = \gamma \,\Delta BSA \qquad (1.5)$$

where ΔBSA is the change of the buried surface area caused by the mutation.

Mariuzza and collaborators (Sundberg et al., 2000; Li et al., 2005) have observed such a relation in experiments where they introduced side chains of different sizes at given positions of the antigen-combining site of two antilysozyme monoclonal antibodies. They measured the dissociation constant of each mutant complex with the antigen, then determined x-ray structures to check that there was no conformation change. ΔG_d was a linear function of the BSA in both series of mutants, but the slope was nearly three times as large for the H63 antibody mutated on a tyrosine placed at the center of the interface than for the D1.3 antibody where the mutation site was a tryptophan at the periphery.

Another example of linear relationship between binding free energy and BSA concerns complexes with nonprotein ligands. Wells and McClendon (2007) compared the potency of a series of small molecules that bind to protein targets of pharmaceutical interest. Expressed as a binding free energy, the potency is linearly related to the number of nonhydrogen atoms in the ligands. In Figure 1.2, that number was converted into a BSA by assuming that all the ligand atoms are in contact

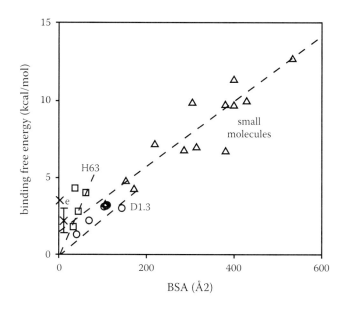

FIGURE 1.2   Binding free energies and buried surface area. (○): Mutant data for antibody D1.3 (Sundberg et al., 2000); ΔΔG is plotted against the BSA change in six mutants of residue $TrpV_L92$, located at the periphery of the interface with the antigen hen egg lysozyme; the slope of the regression line is 23 cal.mol^{-1}.Å^{-2}. (□): Mutant data for antibody H63 (Li et al., 2005); ΔΔG is plotted against the BSA change in four mutants of residue Tyr V_H33 at the center of the interface; the slope of the line is 66 cal.mol^{-1}.Å^{-2}. (△): ΔG_d of small molecules bound to proteins of pharmacological interest (Wells & McClendon, 2007); each nonhydrogen atom is assumed to contribute 9.5 Å^2 to the BSA; the slope of the line is 21 cal.mol^{-1}.Å^{-2}. (×): Mean and standard deviation of ΔG_d for xenon bound to three proteins (myoglobin, lipid transfer protein, and T4 lysozyme; Desvaux et al., 2005).

with the protein and that the average BSA per atom is the same as in protein–protein complexes (9.5 Å^2; Chakrabarti & Janin, 2002). With these assumptions, the slope of the regression line is in the range of the values of γ for hydrocarbons cited earlier. The figure also includes the antibody mutation data of Sundberg et al. (2000) and Li et al. (2005) for comparison. The antibodies have about 80 atoms in contact with the antigen, and as the mutations remove only a few, the lines cannot be extrapolated to the whole interface. On the other hand, the small molecules of Wells and McClendon (2007) have up to 56 atoms, and it is not unreasonable to extend that line to 80. This predicts $\Delta G_d \approx 17.5$ kcal.mol^{-1}, much more than the 11 kcal.mol^{-1} reported for the D1.3 or H63/lysozyme complexes, but other protein–protein complexes with interfaces of the same size, barnase/barstar for instance, have binding free energies of that order or greater. Thus, there is a qualitative, but not quantitative, agreement between the observed binding free energies and the values expected from the line in Figure 1.2. Even that must be qualified: the small molecules contain polar atoms, and we do not know how much of their surface is buried when they bind; and the plot altogether ignores what happens on the target protein or on the antigen side of the interface.

HOT SPOTS AND THE NONADDITIVITY OF BINDING FREE ENERGIES

In the D1.3 antibody–lysozyme complex, about 19 residues of each partner contribute to the BSA, and, except for the glycines, nearly all have been mutated to alanine (Ala; Dall'Acqua et al., 1996, 1998). In these experiments, a factor of 2 in K_d equivalent to $\Delta\Delta G = 0.4$ kcal.mol^{-1} is considered as significant; occasionally, the mutation improves affinity and $\Delta\Delta G$ is negative, but rarely less than -0.4 kcal.mol^{-1}. The hot spots are positions where $\Delta\Delta G$ exceeds 1.5 kcal.mol^{-1} (Clackson & Wells, 1995) or 2 kcal.mol^{-1} (Bogan & Thorn, 1998), which represent a factor of 12 and 30 in K_d, respectively. In antibody D1.3 and lysozyme, 6 of 26 mutations yield a $\Delta\Delta G$ below 0.4 kcal.mol^{-1}, and another 6 a $\Delta\Delta G$ above 1.5 kcal.mol^{-1}. Thus, about one-quarter of the interface residues are silent, and another quarter are hot spots.

The ASEdb database (http://nic.ucsf.edu/asedb; Thorn & Bogan, 2001) reports results of alanine-scanning studies in several systems that show a similar fraction of hot spots, but more silent residues, sometimes up to 50%. It should be noted that site-directed mutagenesis is blind to interactions involving the protein main chain, which contributes one-fifth of the BSA and an even larger proportion of the H-bonds in protein–protein complexes (Lo Conte et al., 1999). In addition, Gly and Ala residues are often not mutated. On the other hand, the residues that make large contributions to the BSA are almost always hot spots. More than half of the residues with $\Delta\Delta G > 4$ kcal.mol^{-1} in ASEdb are tryptophans, tyrosines, or arginines, with side chains that often bury over 100 Å^2 in a complex. Thus, mutating to Ala the peripheral tryptophan of antibody D1.3 removes 145 Å^2, or one-sixth of the ASA lost in contacts with lysozyme, while decreasing its affinity for the antigen by 4 kcal.mol^{-1} (Sundberg et al., 2000). In the trypsin–soybean trypsin inhibitor complex (PDB entry 1avw), the arginine residue in P1 position loses 245 Å^2, over one-quarter of the ASA lost by the inhibitor. The effect on affinity of substituting that arginine is not known, but in the pancreatic inhibitor, the equivalent lysine-to-Ala mutation raises K_d by eight orders of magnitude ($\Delta\Delta G = 10$ kcal.mol^{-1}; Castro & Anderson, 1996; Krowarsch et al., 1999), possibly the greatest affinity drop ever measured for a point mutant.

It has sometimes been claimed that hot spots quantitatively account for the observed binding free energies. If we assume that the effects of mutations are additive, a "shaved" complex in which n hot spot residues have been converted to Ala is expected to have:

$$\Delta G'_d = \Delta G_d - \Sigma_n \Delta\Delta G \qquad (1.6)$$

On occasion, the summation yields $\Delta G'_d \approx 0$, which is the basis for the claim. However, this only predicts for the shaved complex a $K_d \approx 1$ M, a value of no particular significance; $\Delta G'_d$ could just as well be negative and $K'_d > 1$M. Thus, adding up $\Delta\Delta G$ values says little about the nature of the interaction or the role of the hot spots. For the same reason, the small molecule line in Figure 1.2 does not have to pass through the origin (the mutant lines do, since the origin corresponds to the wild-type complex). Instead, the point where it crosses the vertical axis yields a predicted ΔG_d value for a zero-atom ligand. With L just a point in space, the energy of the system does not change when RL dissociates, but its entropy does, because L gains freedom to move by translation. The relevant term in ΔG_d can be approximated as:

$$\Delta G_{trans} \approx RT \ln V/V° \qquad (1.7)$$

where V and V° are the volumes available to L before and after dissociation. V° is fixed by the standard state concentration (recall that ΔG_d is standard state dependent): $V° \approx 1600 \text{ Å}^3$ in the conventional 1M standard state. V depends on how rigidly bound the ligand is to the protein; if we assume that it can move by $\delta x = 0.3–1 \text{ Å}$ while remaining bound, $V \approx 4\pi/3 \ \delta x^3$, and Equation 1.7 yields $\Delta G_{trans} \approx -3.5–6 \text{ kcal.mol}^{-1}$. This term is the only one in ΔG_d for a zero-atom ligand, and for a one-atom ligand, it represents the contribution of the ligand entropy, purely translational. Whereas the zero-atom ligand is just a convention, the one-atom ligand can be real. Xenon binds to a number of proteins, which makes it useful as a heavy atom for determining phases in protein crystallography. The observed K_d values are in the range 5–100 mM (Desvaux et al., 2005), which happens to place the rare gas right on the small molecule line of Figure 1.2.

The translational contribution to ΔG_d is by nature nonadditive: with more atoms in L, the added degrees of freedom are rotational and vibrational, and their free enthalpy is not volume dependent. Thus, the range of ΔG_{trans} values cited earlier for the zero-atom ligand may be valid for a protein molecule (Finkelstein & Janin, 1989; Ruvinsky, 2007). Other nonadditive terms derive from the interactions that are made in the complex. If a pair of chemical groups in R and L interact with an energy (ε), deleting one in R′ and the other in L′ should cause affinity to drop by $\Delta\Delta G = \varepsilon$ in the R′L and the RL′ complexes, and also in the R′L′ complex, whereas additivity would predict 2ε. On the other hand, $\Delta\Delta G_{np}$ is additive because the change induced by the mutations concerns the ASA of free R or free L, not the complex.

In reality, $\Delta\Delta G$ almost never represents the energy of the interactions made by the deleted atoms. This is most obvious for polar interactions. If a polar side chain in R is engaged in a buried H-bond at the RL interface, the R′L mutant complex will contain an unpaired polar group in L, and $\Delta\Delta G$ will reflect the net loss of the water–protein H-bond that this group makes in free L. The contribution of the protein–protein H-bond to ΔG_d, probably much smaller, can be recovered by deleting the unpaired polar group in L and preparing the R′L′ complex. This reasoning is at the basis of the double mutant cycle (Fersht, 1988). Many studies show that the free energy contribution of an H-bond between neutral side chains is less than 1 kcal.mol^{-1}, which implies that the two-alanine mutation in the double mutant can be silent even though each of the mutated residues is a hot spot. The method can be extended to cycles of more than two mutants to test whether the contributions of the pairwise interactions are themselves additive. Neighboring pairs of interacting residues often display non-additive cooperative effects that point to a modular architecture of the binding sites (Reichmann et al., 2005, 2007).

THE CORE/RIM MODEL OF PROTEIN–PROTEIN INTERFACES

Based on alanine-scanning data, Bogan and Thorn (1998) proposed an O-ring model of protein–protein interfaces, in which the hot spots are surrounded by energetically unimportant residues that occlude solvent from them. In parallel, Lo Conte et al. (1999) noted that only about one-third of the interface atoms are actually buried (zero ASA) in protein–protein complexes (34% in Table 1.1); another third are in contact with immobilized water, the remainder with bulk solvent. These authors represented the interfaces as having a core of buried atoms surrounded by rings of atoms accessible to the solvent. This led Chakrabarti and Janin (2002) to split protein–protein interfaces into two regions: the core, made of the residues that contain the buried interface atoms, and the rim, made of the residues in which all the interface atoms remain solvent accessible. In the average protein–protein complex, the core contains 55% of the interface residues, the rim 45%. The amino acid composition of the rim is very similar to that of the solvent accessible surface, except that the core is depleted by a factor of nearly two in the charged residues Asp/Glu/Lys (but not Arg), and enriched by the same factor in the aromatic residues Phe/Tyr/Trp (Chakrabarti & Janin, 2002). The core represents a larger fraction of the interface residues in homodimer proteins, and its composition is enriched in aliphatic as well as in aromatic residues (Bahadur et al., 2003). Weak dimers have interfaces that bury fewer atoms and have only 51% of their residues in the core; crystal packing contacts, which bury only one-fifth of their atoms, have only 40% (Table 1.1).

The core/rim model of the interfaces is structure based, but it has a counterpart in evolution. Although commonly used to identify binding sites (Arnon et al., 2001; Lichtarge & Sowa, 2002; Ma et al., 2003; Caffrey et al., 2004), the conservation of the interfaces is far from obvious in many systems. One reason is that the evolutionary pressure is not homogeneous within an interface. Guharoy and Chakrabarti (2005) calculate the Shannon entropy (s) of the interface residues in sets of homologous protein sequences; s measures the sequence variability at individual positions of the sequence, and it is zero at fully conserved positions. Table 1.1 shows that, in the average homodimeric protein or component of a protein–protein complex, the Shannon entropy takes lower values for residues of the interface core than the rim. With no such effect being found at crystal contacts, one may conclude that the specific interaction between two proteins exerts a stronger selection pressure on the core than the rim of their interface.

Figure 1.3 illustrates the core and rim and their relationship to sequence conservation in the Gα/Gβγ interface of transducin, a heterotrimeric G-protein that interacts with rhodopsin to initiate the visual signal in the retina (Lambright et al., 1996). The Gα subunit of transducin has 45 residues in contact with Gβγ. In the left panel of Figure 1.3, its molecular surface is colored red for the core residues, blue for the rim residues; in the right panel, it is colored according to their Shannon entropy. The Gα/Gβγ interface is in two patches. The minor patch implicates the N-terminal helix of Gα that points out of the subunit on the top; the helix, which comprises 6 core and 11 rim residues, is disordered in the free subunit. The major patch involves the main body of the subunit, close to the GTP binding site. It has a well-defined core of 14 residues surrounded by an equivalent number of rim residues, and resembles

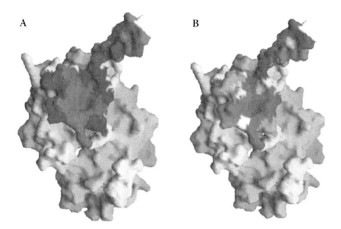

FIGURE 1.3 (SEE COLOR INSERT FOLLOWING PAGE 174.) The core/rim model and the conservation of interface residues. The surface of the Gα subunit of transducin (PDB entry 1got; Lambright et al., 1996) is rendered in gray except for the region in contact with Gβγ. The feature protruding on the top right is the N-terminal helix. (A) The interface core, made of residues containing atoms buried at the interface, is in red; the rim, made of residues in which all interface atoms remain solvent accessible, is in blue. (B) The interface is colored according to the Shannon entropy that measures the divergence of each position in aligned sequences, ranging from 0 (red) to 0.4 (pink) to 1.4 (dark blue). Figure made by M. Guharoy (Bose Institute, Calcutta) with GRASP (Nicholls et al., 1991).

the standard-size interface of a protease–inhibitor complex both in size and in the O-ring like arrangement of its core and rim. A comparison of the two panels indicates that whereas the sequence of the N-terminal helix is poorly conserved ($s > 1$ in blue), the major patch is fully conserved ($s = 0$ in red) in the core, and moderately (pink) to highly (red) conserved in the rim.

The description of the interface conservation given by the Shannon entropy is generally consistent with the data from alanine-scanning experiments. Guharoy and Chakrabarti (2005) report a correlation between $\Delta\Delta G$ and the contribution to the BSA of the residues of interface core, but not the rim. The correlation yields a slope $\gamma = 26$–38 cal.mol^{-1}.Å^{-2}, close to the values derived from the solubility of hydrocarbons. Other approaches, for instance, the "residue depth" of Chakravarty and Varadarajan (1999) or the "hot regions" of Keskin et al. (2005), give a similar picture of the way hot spots are distributed within an interface. In Figure 1.4, based on the data on five complexes reported in ASEdb, nearly all of the mutations with large effects on affinity ($\Delta\Delta G > 2$ kcal.mol^{-1}) are seen to concern residues of the interface core. Mutations of the interface rim are silent ($\Delta\Delta G < 0.4$ kcal.mol^{-1}) or have a moderate effect (0.4–2 kcal.mol^{-1}). This does not imply that the rim plays no part in the interaction, only that its contribution to ΔG_d does not depend heavily on the nature of the side chains. The mutations that affect residues outside the interface are silent, with a few exceptions that may be due to conformation changes and other indirect effects. A few silent mutants belong to the interface core; some may represent interactions of main

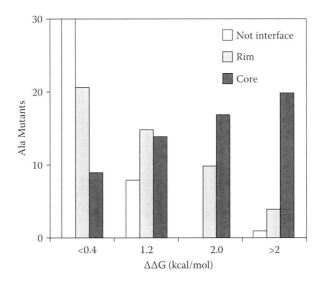

FIGURE 1.4 Alanine scanning and the core/rim model. The alanine-scanning data on five complexes are taken from ASEdb: barnase/barstar (1brs), Factor VII/Tissue factor (1dan), RNase inhibitor/RNase A (1dvf), and the two antigen–antibody complexes D1.3/lysozyme (1vfb) and D1.3/E5.2 (1vfb).

chain atoms; others result from compensating effects that cannot be assessed with the present data.

DESIGNING INTERACTIONS

In a complex with a standard-size interface, the core typically comprises 26 residues, 13 on each component, and mutating a few suffices to destroy affinity. The remark can be turned around to state that mutating a few properly chosen residues should enable us to create stable complexes. This, of course, is what the immune system does when it makes antibodies, but the same result has to be reached by selecting sequences in a rationally designed combinatorial library created by introducing degenerate codons in a synthetic gene. In Nygren's affibodies, the three-α-helix scaffold of protein Z is made variable at a dozen surface sites, and the selection is made by phage display (Nord et al., 1995; Nygren & Uhlen, 1997; Nygren, 2008). In Plückthun's DARPins, the scaffold contains a variable number of ankyrin repeats, and the selection tool is ribosome display (Binz et al., 2003, 2005). X-ray structures are available for an affibody/protein Z complex (PDB entry 1lp1; Högbom et al., 2003), and a DARPin/caspase 2 complex (1p2c; Schweizer et al., 2007). Both display interfaces with a BSA ≈ 1600 Å^2 that implicate mostly, but not exclusively, the randomized residues. In the affibody complex, 11 of the 13 randomized residues lose ASA, and they contribute 70% of the BSA; the remainder comes from seven framework positions. The DARPin contains four ankyrin repeats mutated at a total of 22 positions, 14 of which are part of the interface and contribute 86% of the BSA. The interface size, the total number of residues involved, and the abundance of aromatic residues that are selected at the randomized positions all resemble monoclonal antibodies, and the affinity is at least as good.

The number of mutation sites can be much reduced, and the selection step eliminated altogether, with the help of symmetry. Grueninger et al. (2008) have introduced nonpolar side chains at selected surface sites in four bacterial proteins that include Uro, a homodimer, and Rua, a cyclic tetramer, aiming to create stable assemblies with twice the original number of subunits. Crystal structures show that a Uro variant with three substitutions has acquired the designed tetrameric structure, and that a single mutation changes Rua from a *C4* tetramer to a *D4* octamer. Another variant with two substitutions yields a nonsymmetrical octamer, and a third forms fibers. In these systems, one to three point mutations suffice to generate new assemblies instead of 11 to 13 in affibodies and DARPins, but each mutation creates several (presumably) favorable contacts as a result of the symmetry, and the new interfaces include many nonmutated residues. In the *D4* Rua octamer, for instance, the tetramer/tetramer interface does implicate the eight symmetry-related tyrosines that replace alanines in the wild type, but they contribute only 9% of the BSA; the remainder comes from 21 other residues and their symmetry counterparts.

CONCLUSION

Most of the site-directed mutagenesis and biophysical data discussed concern systems in which rigid-body recognition is a valid approximation. In the Rua octamer or the complexes with affibodies and DARPins, the components retain their structure, and the designed assemblies obey the rules that we observe in natural assemblies. The Rua octamer was designed ab initio, the affibodies and DARPins were obtained by a combination of rational design and *in vitro* selection. The success of the two approaches proves that our understanding of the rigid-body mode of protein–protein interaction has reached the point where we can predict novel modes of interaction and build protein molecules that use them. The CAPRI (Critical Assessment of Predicted Interactions) experiment, designed to test protein docking methods (Janin et al., 2003), confirms that view. In seven years, CAPRI has demonstrated conclusively that the structure of a complex can be accurately predicted from that of its components as long as the conformation changes are small; the prediction becomes inaccurate or fails altogether when they are large (Schueler-Forman et al., 2005; Janin & Wodak, 2007; Lensink et al., 2007). Flexible recognition and induced fit often involve major changes in the partner proteins, including disorder-to-order transitions in which protein folding is coupled to ligand binding, as for the N-terminal helix of transducin Gα. As they play a major role in many processes, uncovering their mechanism will be of great interest in the years to come.

ACKNOWLEDGMENTS

I acknowledge support of the 3D-Repertoire and SPINE2-Complexes programs of the European Union, and the very productive collaboration of Dr. R. Bahadur

(Jacobs University, Bremen) and Pr. P. Chakrabarti (Bose Institute, Calcutta). Their colleague Dr. M. Guharoy is thanked for Figure 1.3.

REFERENCES

Alberts B. The cell as a collection of protein machines: Preparing the next generation of molecular biologists. *Cell* 1998 92:291–294.

Armon A, Graur D, Ben-Tal N. ConSurf: An algorithmic tool for the identification of functional regions in proteins by surface-mapping of phylogenetic information. *J. Mol. Biol.* 2001 307:447–463.

Bahadur RP, Chakrabarti P, Rodier F, Janin J. Dissecting subunit interfaces in homodimeric proteins. *Proteins* 2003 53:708–719.

Bahadur RP, Chakrabarti P, Rodier F, Janin J. A dissection of specific and non-specific protein–protein interfaces. *J. Mol. Biol.* 2004 336:943–955.

Berman HM, Westbrook J, Feng Z, Gilliland G, Bhat TN, Weissig H, Shindyalov IN, Bourne PE. The Protein Data Bank. *Nucleic Acids Res.* 2000 28:235–242.

Binz HK, Amstutz P, Plückthun A. Engineering novel binding proteins from nonimmunoglobulin domains. *Nat. Biotechnol.* 2005 23:1257–1268.

Binz HK, Stumpp MT, Forrer P, Amstutz P, Plückthun A. Designing repeat proteins: Well-expressed, soluble and stable proteins from combinatorial libraries of consensus ankyrin repeat proteins. *J. Mol. Biol.* 2003 332:489–503.

Bogan AA, Thorn KS. Anatomy of hot spots in protein interfaces. *J. Mol. Biol.* 1998 280:1–9.

Braden BC, Poljak RJ. Structure and energetics of anti-lyzosome antibodies, in *Protein–Protein Recognition*, C. Kleanthous, ed., Oxford University Press, UK, 2000, pp. 126–161.

Caffrey D, Somaroo S, Hughes J, Mintseris J, Huang E. Are protein–protein interfaces more conserved in sequence than the rest of the protein surface? *Protein Sci.* 2004 13:190–202.

Castro MJ, Anderson S. Alanine point-mutations in the reactive region of bovine pancreatic trypsin inhibitor: Effects on the kinetics and thermodynamics of binding to beta-trypsin and alpha-chymotrypsin. *Biochemistry* 1996 35:11435–11446.

Chakrabarti P, Janin J. Dissecting protein–protein recognition sites. *Proteins* 2002 47:334–343.

Chakravarty S, Varadarajan R. Residue depth: A novel parameter for the analysis of protein structure and stability. *Structure* 1999 7:723–732.

Chothia C. Hydrophobic bonding and accessible surface area in proteins. *Nature* 1974 248:338–339.

Chothia C. Structural invariants in protein folding. *Nature* 1975 254:304–308.

Chothia C, Janin J. Principles of protein–protein recognition. *Nature* 1975 256:705–708.

Clackson T, Wells JA. A hot spot of binding energy in a hormone-receptor interface. *Science* 1995 267:383–386.

Crowley PB, Carrondo MA. The architecture of the binding site in redox protein complexes: Implications for the fast dissociation. *Proteins* 2004 55:603–612.

Dall'Acqua W, Goldman ER, Eisenstein E, Mariuzza RA. A mutational analysis of the binding of two different proteins to the same antibody. *Biochemistry* 1996 35:9667–9676.

Dall'Acqua W, Goldman ER, Lin W, Teng C, Tsuchiya D, Li H, Ysern X, Braden BC, Li Y, Smith-Gill SJ, Mariuzza RA. A mutational analysis of binding interactions in an antigen-antibody protein–protein complex. *Biochemistry* 1998 Jun 37:7981–7991.

DeLano WL. Unraveling hot spots in binding interfaces: Progress and challenges. *Curr. Opin. Struct. Biol.* 2002 12:14–20.

Desvaux H, Dubois L, Huber G, Quillin ML, Berthault P, Matthews BW. Dynamics of xenon binding inside the hydrophobic cavity of pseudo-wild-type bacteriophage T4 lysozyme explored through xenon-based NMR spectroscopy. *J. Am. Chem. Soc.* 2005 127:11676–11683.

Dutta S, Berman HM. Large macromolecular complexes in the Protein Data Bank: A status report. *Structure* 2005 13:381–388.

Eisenberg D, Marcotte EM, Xenarios, I, Yeates TO. Protein function in the post-genomic era. *Nature* 2000 405:823–826.

Eisenberg D, McLachlan AD. Solvation energy in protein folding and binding. *Nature* 1986 319:199–203.

Fersht AR. Relationships between apparent binding energies measured in site-directed mutagenesis experiments and energetics of binding and catalysis. *Biochemistry* 1988 27:1577–1580.

Finkelstein AV, Janin J. The price of lost freedom: entropy of biomolecular complex formation. *Protein Eng.* 1989 3:1–3.

Foote J, Eisen HN. Kinetic and affinity limits on antibodies produced during immune responses. *Proc. Nat. Acad. Sci. USA* 1995 92:1254–1256.

Foote J, Eisen HN. Breaking the affinity ceiling for antibodies and T cell receptors. *Proc. Nat. Acad. Sci. USA* 2000 97:10679–10681.

Fu H, ed. *Protein–Protein Interactions: Methods and Applications* (Methods in Molecular Biology, Vol. 261). Humana Press, Totowa, NJ, 2004.

Grueninger D, Treiber N, Ziegler MO, Koetter JW, Schulze MS, Schulz GE. Designed protein–protein association. *Science* 2008 319:206–209.

Guharoy M, Chakrabarti P. Conservation and relative importance of residues across protein–protein interfaces. *Proc. Nat. Acad. Sci. USA* 2005 102:15447–15452.

Högbom M, Ecklund M, Nygren PA, Nordlund P. Structural basis for recognition by an in vitro evolved affibody. *Proc. Nat. Acad. Sci. USA* 2003 100:3191–3196.

Hermann RB. Theory of hydrophobic bonding. II. Correlation of hydrocarbon solubility in water with solvent cavity surface-area. *J. Phys. Chem.* 1972 76:2754–2759.

Horton N, Lewis M. Calculation of the free energy of association for protein complexes. *Protein Sci.* 1992 1:169–181.

Hwang H, Pierce B, Mintseris J, Janin J, Weng Z. Protein–protein docking benchmark version 3.0. *Proteins* 2008 73:705–709.

Janin J. Specific versus non-specific contacts in protein crystals. *Nature Struct. Biol.* 1997 4:973–974.

Janin J, Chothia C. The structure of protein–protein recognition sites. *J. Biol. Chem.* 1990 265:16027–16030.

Janin J, Henrick K, Moult J, Eyck LT, Sternberg MJ, Vajda S, Vakser I, Wodak SJ. CAPRI: A Critical Assessment of Predicted Interactions. *Proteins* 2003 52:2–9.

Janin J, Miller S, Chothia C. Surface, subunit interfaces and interior of oligomeric proteins. *J. Mol. Biol.* 1988 204:155–164.

Janin J, Rodier F. Protein–protein interaction at crystal contacts. *Proteins* 1995 23:580–587.

Janin J, Rodier F, Chakrabarti P, Bahadur RP. Macromolecular recognition in the Protein Data Bank. *Acta Crystallogr. D Biol. Crystallogr.* 2007 63:1–8.

Janin J, Wodak S. The third CAPRI assessment meeting. *Structure* 2007 15:755–759.

Janin J, Wodak SJ, eds. *Protein Modules and Protein–Protein Interaction* (Advances in Protein Chemistry, Vol. 61). Academic Press, San Diego, CA, 2003.

Jones S, Thornton JM. Protein–protein interactions: A review of protein dimer structures. *Prog. Biophys. Mol. Biol.* 1995 63:31–65.

Jones S, Thornton JM. Principles of protein–protein interactions. *Proc. Natl. Acad. Sci. USA* 1996 93:13–20.

Jones S, Thornton JM. Analysis and classification of protein–protein interactions from a structural perspective, in *Protein–Protein Recognition*, C. Kleanthous, ed., Oxford University Press, UK, 2000, pp. 33–59.

Kauzmann W. Some factors in the interpretation of protein denaturation. *Adv. Protein. Chem.* 1959 14:1–63.

Keskin O, Ma B, Nussinov R. Hot regions in protein–protein interactions: The organization and contribution of structurally conserved hot spot residues. *J. Mol. Biol.* 2005 345:1281–1294.

Kleanthous C, ed. *Protein–Protein Recognition* (Frontiers in Molecular Biology). Oxford University Press, UK, 2000.

Krowarsch D, Dadlez M, Buczek O, Krokoszynska I, Smalas AO, Otlewski J. Interscaffolding additivity: Binding of P1 variants of bovine pancreatic trypsin inhibitor to four serine proteases. *J. Mol. Biol.* 1999 289:175–186.

Lambright DG, Sondek J, Bohm A, Skiba NP, Hamm HE, Sigler PB. The 2.0 Å crystal structure of a heterotrimeric G protein. *Nature* 1996 379:311–319.

Larsen TA, Olson AJ, Goodsell DS. Morphology of protein–protein interfaces. *Structure* 1998 6:421–427.

Lawrence MC, Colman PM. Shape complementarity at protein/protein interfaces. *J. Mol. Biol.* 1993 234:946–950.

Lee BK, Richards FM. The interpretation of protein structures: Estimation of static accessibility. *J. Mol. Biol.* 1971 55:379–400.

Lensink MF, Méndez R, Wodak SJ. Docking and scoring protein complexes: CAPRI 3rd Edition. *Proteins* 2007 69:704–718.

Lévy ED. PiQSi: Protein quaternary structure investigation. *Structure* 2007 15:1364–1367.

Li Y, Huang Y, Swaminathan CP, Smith-Gill SJ, Mariuzza RA. Magnitude of the hydrophobic effect at central versus peripheral sites in protein–protein interfaces. *Structure* 2005 13:297–307.

Lichtarge O, Sowa M. Evolutionary predictions of binding surfaces and interactions. *Curr. Opin. Struct. Biol.* 2002 12:21–27.

Lo Conte L, Chothia C, Janin J. The atomic structure of protein–protein recognition sites. *J. Mol. Biol.* 1999 285:2177–2198.

Lum K, Chandler D, Weeks JD. Hydrophobicity at small and large length scales. *J. Phys. Chem. B* 1999 103:4570–4577.

Ma B, Elkayam T, Wolfson H, Nussinov R. Protein–protein interactions: Structurally conserved residues distinguish between binding sites and exposed protein surfaces. *Proc. Natl. Acad. Sci. USA* 2003 100:5772–5777.

Makhatadze GI, Privalov PL. Hydration effects in protein unfolding. *Biophys. Chem.* 1994 51:291–304.

Nicholls A, Sharp KA, Honig B. Protein folding and association: Insights from the interfacial and thermodynamic properties of hydrocarbons. *Proteins* 1991 11:281–296.

Nooren IM, Thornton JM. Diversity of protein–protein interactions. *EMBO J.* 2003a 22:3486–3492.

Nooren IM, Thornton JM. Structural characterisation and functional significance of transient protein–protein interactions. *J. Mol. Biol.* 2003b 325:991–1018.

Nord K, Nilsson J, Nilsson B, Uhlen M, Nygren PA. A combinatorial library of an α–helical bacterial receptor domain. *Protein Eng.* 1995 8:601–608.

Nygren PA. Alternative binding proteins: Affibody binding proteins developed from a small three-helix bundle scaffold. *FEBS J.* 2008 275:2668–2676.

Nygren PA, Uhlen M. Scaffolds for engineering novel binding sites in proteins. *Curr. Op. Struct. Biol.* 1997 7:463–469.

Ooi T, Oobatake M, Némethy G, Scheraga HA. Accessible surface areas as a measure of the thermodynamic parameters of hydration of peptides. *Proc. Natl. Acad. Sci. USA* 1987 84:3086–3090.

Ponstingl H, Kabir T, Gorse D, Thornton JM. Morphological aspects of oligomeric protein structures. *Prog. Biophys. Mol. Biol.* 2005 89:9–35.

Reichmann D, Rahat O, Albeck S, Meged R, Dym O, Schreiber G. The modular architecture of protein–protein binding interfaces. *Proc. Natl. Acad. Sci. USA* 2005 102:57–62.

Reichmann D, Rahat O, Cohen M, Neuvirth H, Schreiber G. The molecular architecture of protein–protein binding sites. *Curr. Opin. Struct. Biol.* 2007 17:67–76.

Rodier F, Bahadur RP, Chakrabarti P, Janin J. Hydration of protein–protein interfaces. *Proteins* 2005 60:36–45.

Russell RB, Alber F, Aloy P, Davis FP, Korkin D, Pichaud M, Topf M, Sali A. A structural perspective on protein–protein interactions. *Curr. Opin. Struct. Biol.* 2004 14:313–324.

Ruvinsky AM. Calculations of protein-ligand binding entropy of relative and overall molecular motions. *J. Comput. Aided. Mol. Des.* 2007 21:361–370.

Schreiber G, Fersht, AR. Interaction of barnase with its polypeptide inhibitor barstar studied by protein engineering. *Biochemistry* 1993 32:5145–5150.

Schueler-Furman O, Wang C, Bradley P, Misura K, Baker D. Progress in modeling of protein structures and interactions. *Science* 2005 310:638–642.

Schweizer A, Roschitzki-Voser H, Amstutz P, Briand C, Gulotti-Georgieva M, Prenosil E, Binz HK, Capitani G, Baici A, Plückthun A, Grütter MG. Inhibition of caspase-2 by a designed ankyrin repeat protein: Specificity, structure, and inhibition mechanism. *Structure* 2007 15:625–636.

Sharp KA, Nicholls A, Fine RF, Honig B. Reconciling the magnitude of the microscopic and macroscopic hydrophobic effects. *Science* 1991 252:106–109.

Sundberg EJ, Mariuzza RA. Molecular recognition in antibody-antigen complexes. *Adv. Prot. Chem.* 2002 61:119–160.

Sundberg EJ, Urrutia M, Braden BC, Isern J, Tsuchiya D, Fields BA, Malchiodi EL, Tormo J, Schwarz FP, Mariuzza RA. Estimation of the hydrophobic effect in an antigen-antibody protein–protein interface. *Biochemistry.* 2000 39:15375–15387.

Tanford C. How protein chemists learned about the hydrophobic factor. *Protein Sci.* 1997 6:1358–1366.

Thorn KS, Bogan AA. ASEdb: A database of alanine mutations and their effects on the free energy of binding in protein interactions. *Bioinformatics.* 2001 17:284–285.

Vajda S, Weng Z, DeLisi C. Extracting hydrophobicity parameters from solute partition and protein mutation/unfolding experiments. *Protein Eng.* 1995 11:1081–1092.

Vincent JP, Lazdunski M. Trypsin-pancreatic trypsin inhibitor association. Dynamics of the interaction and role of disulfide bridges. *Biochemistry* 1972 11:2967–2977.

Wells JA, McClendon CL. Reaching for high-hanging fruit in drug discovery at protein–protein interfaces. *Nature* 2007 450:1001–1009.

Xie D, Freire E. Structure based prediction of protein folding intermediates. *J. Mol. Biol.* 1994 242:62–80.

2 Low-Resolution Recognition Factors Determine Major Characteristics of the Energy Landscape in Protein–Protein Interaction

Ilya A. Vakser

CONTENTS

OVERVIEW

Protein–protein recognition is a key element of life at the molecular level. Our understanding of the principles of protein recognition is still limited. However, a significant amount of information on the subject has been already accumulated and analyzed. The structure of protein–protein complexes is generally more difficult to determine than the structure of individual proteins. However, the number of experimentally determined complexes is statistically significant. The databases of protein–protein complexes are important for systematic studies of protein interactions and the design of new predictive tools. A number of such databases have been compiled and widely utilized in the research community. The underlying physical principles of protein folding and binding are the same, which translates into the similarity of the recognition factors in folding and docking. The concepts of steric and physicochemical complementarity are the basis for many modeling techniques applicable to both problems. Structural recognition factors relate to energy landscape characteristics that help understand the formation of complexes and create better modeling tools. The multiscale approach to modeling protein interactions reflects the nature of protein recognition, which involves the larger structural factors facilitating complex formation and the smaller local factors responsible for the final lock of the molecules within the complex.

INTRODUCTION

Protein–protein complex formation can be viewed from either a more physical perspective as a minimization of the free energy of the system or from a more empirical point of view as a match between various phenomenological structural and/or physicochemical motifs (so-called recognition factors). In living organisms, proteins recognize their partners among many other proteins and bind in a specific way in short physiological timeframes. Given the complexity of the system, from either the physical or empirical points of view, the formation of a protein–protein complex is a remarkable event, based on the nature's superefficient "energy-minimization protocol" and guided by long-range and short-range recognition factors. Modern methods of protein docking are based on our efforts to simulate and navigate the intermolecular energy landscape, and on our current understanding of the recognition factors governing complex formation.

The three-dimensional (3D) structure of a protein–protein complex, generally, is more difficult to determine experimentally than the structure of an individual protein. Adequate computational techniques to model protein interactions are important because of the growing number of known protein 3D structures, particularly in the context of structural genomics (Russell et al. 2004; Szilagyi et al. 2005; Vakser 2008). The protein docking techniques offer tools for fundamental studies of protein interactions and provide a structural basis for drug design. Since its introduction in the 1970s, the protein–protein docking field has grown substantially through the development of powerful docking algorithms, rapid progress in computer hardware, and significant expansion of available experimental data on structures of protein–protein complexes (Lensink et al. 2007; Vakser and Kundrotas 2008).

Nevertheless, our understanding of the principles of protein recognition is still limited. With the rapid advances in experimental and computational determination of structures of individual proteins, the importance of modeling of protein 3D interactions increases. We now face the challenge of structural modeling of protein-interaction networks on the genome scale, requiring much more powerful docking methodologies, based on the knowledge of protein–protein recognition characteristics.

DATABASES

Although the structure of protein–protein complexes is generally more difficult to determine than the structure of individual proteins, the number of experimentally determined complexes is statistically significant. The databases of protein–protein complexes are indispensable for systematic studies of protein interactions and the design of new predictive tools. A number of databases of co-crystallized protein–protein complexes have been compiled. One early data set of protein–protein complexes was built by Vakser and Sali (unpublished) based on a 1997 release of the Protein Data Bank (PDB) containing 5013 entries. It has been extensively used in studies of knowledge-based potentials (Glaser et al. 2001), intermolecular energy landscapes (Vakser et al. 1999; Tovchigrechko and Vakser 2001; Papoian and Wolynes 2003), docking methodology (Tovchigrechko et al. 2002), and other studies. Other data sets of protein–protein complexes have been compiled and used to address various aspects of physicochemical and structural features of protein–protein interfaces (Dasgupta et al. 1997; Keskin et al. 1998; Larsen et al. 1998; Lo Conte et al. 1999; Ponstingl et al. 2000; Lu et al. 2003; Keskin et al. 2004; Davis and Sali 2005; Gong et al. 2005; Teyra et al. 2006; Jefferson et al. 2007; Kundrotas and Alexov 2007).

The data sets of co-crystallized structures are important for studying protein interfaces. However, their role in validation of docking procedures is limited. The reason is that the bound docking problem (rematching of separated components of a complex in their bound conformation) has been solved by modern docking approaches. The bound docking problem also does not have practical value in the sense that it does not create new structural information (the knowledge of bound conformations assumes that the structure of the complex had been determined). The challenge for the docking techniques is prediction of complexes from the unbound components (experimentally determined and, even more challenging, modeled). For that matter, the databases of unbound protein structures corresponding to complexes of known structure (unbound docking benchmark sets) are important. The selection of crystal structures for such data sets is much more limited than for the bound sets because only a limited number of proteins are crystallized in both bound and unbound form (Mintseris et al. 2005; Gao et al. 2007).

The Dockground resource (http://dockground.bioinformatics.ku.edu) implements a comprehensive database of co-crystallized (bound) protein–protein complexes, providing foundation for the expansion to unbound (experimental and simulated) protein–protein complexes, modeled protein–protein complexes, and systematic sets of docking decoys. The bound part of Dockground is a relational database of annotated structures based on the Biological Unit file (Biounit) provided by the Research Collaboratory for Structural Bioinformatics (RCSB) as a separated file containing a probable biological molecule. Dockground is automatically updated to reflect the growth of the PDB. It contains 102,527 pairwise complexes from 24,596 PDB entries, out of a total 52,263 PDB structures (August 2008). The database includes a dynamic generation of nonredundant data sets of pairwise complexes based either on the structural similarity (Structural Classification of Proteins [SCOP] classification) or on user-defined sequence identity.

The bound part also contains "easy" precompiled complexes sets:

1. Automatically selected representative complexes. The set is regularly updated and is downloadable as an Excel-readable text file. It is built according to the criteria: PDB entry is not obsolete; area of interface buried by each chain >400 Å^2; multimeric state = 2; chains are not tangled, interwoven, or disordered at the interface; no DNA, RNA, or ligand at the interface; no S–S bonds between chains; and chains are not membrane associated. The current release contains 1970 complexes.

2. Manually selected representative complexes. The set is updated less frequently and represents a more sophisticated selection of complexes. The multimeric complexes are considered and monomeric chains clustered with sequence identity 30%, pairwise complexes are reclustered, oligomer complex representatives are selected (based on the best resolution), and the final dimeric representatives are selected. In addition, the selection keeps interfaces containing metal ion, PO4, SO4, and S–S bonds (if those are peripheral to the interface), and membrane associated chains (if the membrane-bound part is not part of the structure), and excludes subunit interaction that may be obligate (according to the reference in the PDB file). If several chains interact with other chains as a whole (judged by visual inspection and analysis of references), they are treated as one entity. If structures with sequence identity >30% have different binding modes, they are considered as different entries. The set contains 523 nonobligate interactions from 508 PDB entries and is downloadable as an Excel-readable text file and as PDB coordinates (separately in four categories: enzyme–inhibitor, antigen–antibody, cytokine or hormone/receptor, and others).

A selection of bound complexes served as the basis for building the unbound data set. The rationale for not using all possible bound complexes was that only the structures relevant to docking should be considered (e.g., structures that are not disordered, have adequate interface area, etc.). The selection criteria were: the structures have to be nonobsolete and have >30 residues. Unbound structures are separate structures that are also co-crystallized in a complex. The web interface allows the user to select the data set based on sequence identity (calculated by BLAST [Basic Local Alignment Search Tool]) between bound and unbound proteins, sequence identity between different bound structures to exclude redundancy, minimal crystallographic resolution, and choice of hetero- or homodimers (or both). As an option, the selection can be done for structures related to a specific protein.

Preselected data sets based on most common criteria are suggested for an "easy" download. These downloadable data sets also include the following important characteristics:

1. Crystal packing (nonbiological interfaces) and obligate complexes (components adopt their folds only within the complex) are excluded. Such complexes were detected for the smaller manually selected representative data set by using related reference in the PDB file and were detected for the

larger automatically selected data set of all complexes by an automated procedure. Weng and co-authors showed that obligate complexes can be distinguished based on properties of the interfaces structure (Mintseris and Weng 2003, 2005). An automatic classification procedure, NOXClass from Lengauer's lab (Zhu et al. 2006), was used to detect obligate and crystal packing interactions based on interface properties.

2. If the unbound structure is not available in the PDB, it is simulated based on the bound structure by assigning side-chain conformations from a rotamer library using SCWRL (Canutescu et al. 2003). Such simulated unbound structures do not involve backbone changes, which is one of the factors limiting their utility. For the future releases, new approaches for unbound structure simulation are designed based on known general differences between bound and unbound conformations that include both side-chain and backbone conformational changes.

The resulting unbound data set, built with >97% sequence identity between bound and unbound structures, contains 4723 nonobligate, biological (not crystal packing) complexes, originating from 1718 PDB entries. Among those, 892 complexes (from 542 PDB entries) have the unbound structures crystallized. The unbound structures for the rest of the complexes were simulated.

From this data set the nonredundant data set is obtained, with <30% sequence identity between the bound complexes. The homomultimers (which are often presumed to be obligate) are excluded by eliminating complexes with >70% sequence identity between units. The resulting set contains 523 complexes (from 508 PDB entries). Among them, 81 complexes are enzyme-inhibitor, 70 are antigen–antibody, 34 are cytokine or hormone/receptor, and 338 are other. Overall, 99 complexes have both components in unbound form crystallized, and 143 had one unbound component crystallized. The rest of the unbound structures were simulated.

The distribution of complexes according to the change from unbound to bound conformations is shown in Figure 2.1. The comparison of bound and unbound crystal structures shows that most changes are <4 Å RMSD (all atoms), with a clear peak in 0.5–2 Å interval. Some complexes have very large RMSD values due to domain movements. The simulated unbound structures are normally distributed with smaller RMSD values because they do not involve the conformational changes of the backbone. More adequate techniques for simulation of unbound conformations that involve backbone conformations are under development.

PARALLELS BETWEEN PROTEIN RECOGNITION AND PROTEIN FOLDING

The underlying physical principles that determine the structure of individual proteins and the structure of protein complexes are identical. Statistically derived residue–residue and atom–atom preferences for protein–protein interfaces were found to be similar to those in protein cores (Tsai et al. 1996; Vajda et al. 1997; Keskin et al. 1998; Glaser et al. 2001; Zhou and Zhou 2002; Liang et al. 2007). A major role

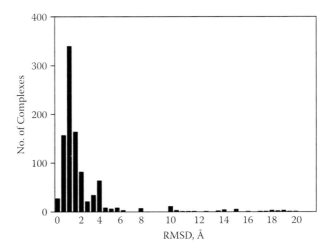

FIGURE 2.1 Distribution of biological, nonobligate complexes according to the bound/unbound all atom RMSD.

of hydrophobicity in protein folding is well established (Richards 1977; Dill 1990). Studies of protein–protein interfaces confirm the importance of hydrophobicity in complex formation as well (Korn and Burnett 1991; Vakser and Aflalo 1994; Young et al. 1994; Tsai et al. 1996; Keskin et al. 2004). The importance of the concept of the energy funnel, first demonstrated for protein folding (Bryngelson et al. 1995; Dill 1999), has been expanded to the intermolecular energy landscape in protein–protein interactions (Tsai et al. 1999; Shoemaker et al. 2000; Tovchigrechko and Vakser 2001; Baker and Lim 2002; Hunjan et al. 2008; O'Toole and Vakser 2008; Ruvinsky and Vakser 2008a, 2008b). Tight packing of structural elements inside proteins is one of the fundamental concepts in our understanding of protein structures (Ponder and Richards 1987; Hubbard and Argos 1994; Jiang et al. 2003). The same concept of compactness applies to protein–protein interfaces as well (Keskin et al. 2004; Douguet et al. 2006).

The interaction of secondary structure elements in protein structures may be formulated in terms of docking, even though docking is traditionally considered to be a problem of matching two separate molecules. The main difference in matching secondary structure elements and matching separate molecules is in the constraints imposed by the environment. A number of studies explored the applicability of docking to secondary structure packing (Ausiello et al. 1997; Yue and Dill 2000; Vakser and Jiang 2002; Inbar et al. 2003; Jiang et al. 2003). A multiplicity of physicochemical factors obviously plays a role in the packing of secondary structure elements in proteins and in the formation of protein complexes. However, the well-known tight packing of structural elements suggests the importance of the geometric fit.

Earlier studies of this subject were primarily focused on helix–helix packing (Richmond and Richards 1978; Cohen et al. 1979; Chothia et al. 1981; Murzin and Finkelstein 1988; Reddy and Blundell 1993; Walther et al. 1996). One reason was the limited number of high-quality crystal structures, mostly containing helices. A traditional biochemical view on interactions of secondary structure elements largely neglected geometric complementarity as an important factor (with the exception of helix–helix interactions). A docking algorithm based on geometric complementarity was applied to a comprehensive database of secondary structure elements derived from the PDB (Jiang et al. 2003). The results show that the steric fit plays an important role in the interaction of all secondary structure elements. Docking procedures have been utilized in protein-structure prediction (Yue and Dill 2000; Haspel et al. 2002; Inbar et al. 2003). In such cases, the secondary structure elements are docked by rigid-body procedures followed by structural refinement.

Docking approaches are popular in modeling the structure of transmembrane (TM) helix bundles in G-protein coupled receptors and other integral membrane proteins. The few existing crystal structures of integral membrane proteins provide useful information on the TM bundle configurations. The TM helices are roughly parallel to each other; they are of similar length (determined by the thickness of the membrane) and are well packed. Thus, it is reasonable to assume that the structure of the bundle is determined primarily by the helix–helix interactions, rather than by the interhelical loops (which, of course, still determine the general topology of the bundle). Most helix–helix interfaces in TM bundles are predominantly binary—if two interfaces overlap, one of them is usually dominant. In that regard, TM bundles are ideal objects for docking predictions. At the same time, helices are simple enough to provide validation ground for new docking concepts (e.g., see Pappu et al. 1999). It has been noted that the side chains at the helix–helix interfaces, on average, are shorter than those at the noninterface helix areas (Jiang and Vakser 2000, 2004). This structural characteristic creates a low-resolution recognition factor that allows one to model the TM bundle at low resolution (Vakser and Jiang 2002). However, a high-resolution model of the TM bundle requires an explicit conformational search of the helix internal coordinates (primarily side chains). Thus, from the practical point, the high-resolution modeling of TM bundles is currently useful only if accompanied by a set of experimentally derived structural constraints.

COMPLEMENTARITY, RECOGNITION MOTIFS, AND HOT SPOTS

The protein–protein binding site architecture has been extensively studied in recent years (Reichmann et al. 2007). Among many factors contributing to protein recognition and the efforts to model it, a tight geometric complementarity between interacting protein surfaces is a cornerstone of protein–protein docking methodology since its inception in 1978 (Wodak and Janin 1978). Systematic database analysis of the rapidly growing number of co-crystallized protein–protein complexes provides an increasing amount of evidence supporting this concept (Keskin et al. 2004; Douguet et al. 2006). A number of investigations of packing and buried surface area at protein–protein interfaces (Lawrence and Colman 1993; Hubbard and Argos 1994; Janin 1995) supported the general conclusion that the interacting proteins have a high degree of surface complementarity, but indicated that there is a significant variation in this regard between different complexes. For example, packing at the antigen–antibody interface is relatively loose (Lawrence and Colman 1993; Mariuzza and Poljak 1993). The contact surface area in protein–protein complexes generally varies from 500 to 5000 $Å^2$ with many complexes having even larger contact areas (Lo Conte et al. 1999; Douguet et al. 2006).

Most protein–protein interfaces are found to be more hydrophobic than exposed areas (Korn and Burnett 1991; Vakser and Aflalo 1994; Young et al. 1994; Tsai and Nussinov 1997). Hydrophobic amino acid residues tend to be enriched in the interface in hydrophobic patches of 200–400 $Å^2$ (Jones and Thornton 1996; Tsai et al. 1996; Lijnzaad and Argos 1997). A high degree of electrostatic and hydrogen-bonding complementarity is also observed for protein–protein interfaces (Janin 1995; McCoy et al. 1997; Tsai et al. 1997; Larsen et al. 1998).

The receptor (the larger protein in the complex) sites are often concave (Ho and Marshall 1990; Peters et al. 1996; Binkowski et al. 2003; Nicola and Vakser 2007). The binding surface is also known to be more conserved than the nonbinding surface. A degree of residues conservation and evolutionary importance is an indicator of the binding and/or functional region (Zhang et al. 1999; Armon et al. 2001; Elcock and McCammon 2001; Cammer et al. 2003; Yao et al. 2003). It has also been determined that entropic properties of the binding site are different from those of the nonbinding surface (Elcock 2001; Rajamani et al. 2004). The binding "hot spots" theory points to existence of a small number (e.g., three) of interface residues that are key to binding. They are usually positioned in the middle of the interface, are inaccessible to solvent, complementary to other hot spot residues across the interface, are evolutionary conserved, and maintain their conformation upon binding (Halperin et al. 2004; Rajamani et al. 2004; Vakser 2004; Keskin et al. 2005; Moreira et al. 2007).

The interface residues with the largest conformational change upon binding were studied by docking techniques (Tovchigrechko and Vakser 2005). The study

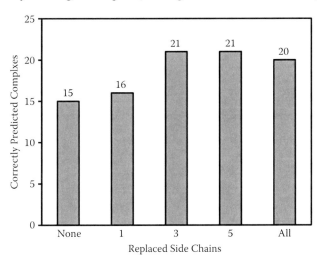

FIGURE 2.2 The significance of the side-chain conformation change in docking. The total number of complexes is 31. The unbound interface side-chains conformers with the largest bound/unbound RMSD were replaced by the bound ones. The criterion for correct prediction is a near-native match (<5 Å ligand interface C^α RMSD) in 10 lowest-energy predictions. See text for details.

determined how the replacement of a limited number of such side chains with their bound conformations affects the performance of a rigid-body docking procedure on the benchmark of the unbound protein structures (Chen et al. 2003). The root mean square deviation (RMSD) of a side chain was calculated after superimposing the N, C, C$^\alpha$, and C$^\beta$ atoms in the same residue taken from bound and unbound protein conformations. All the interface residues were sorted by this side-chain RMSD. N side chains with the largest RMSD were selected and replaced by their bound conformers. The side-chain replacement was done for both receptor and ligand. The value of N was set to 1, 3, 5, and to all interface residues. The benchmarking protocol was run for each pair of generated structures. The results are summarized in Figure 2.2. The number of successfully docked complexes significantly increases when only three side chains are brought into the bound conformation. There is no improvement with a larger number of replaced side chains, and the number of correctly docked complexes still does not reach the one obtained for true bound structures (27 complexes). Therefore, one can conclude that for $N > 3$ the backbone conformational change becomes the limiting factor for the binding.

LARGE-SCALE RECOGNITION FACTORS

One of the most fundamental questions concerning ligand–receptor interaction is whether such a process of intermolecular association is generally determined by local structural elements of the participating molecules or whether there are also large-scale motifs in molecule structures that facilitate complex formation. The local physicochemical and steric factors are responsible for the final "lock" of the molecules when their binding sites are already in close proximity. At the same time, the existing evidence suggests that there are structurally determined factors that contribute to bringing the binding sites to such proximity.

An important insight into the basic rules of protein recognition is provided by the studies of large-scale structural recognition factors, such as correlation of the antigenicity of surface areas with their accessibility to large probes (Novotny et al. 1986), role of the surface clefts (Laskowski et al. 1996), binding-site characterization based on geometric criteria (Ho and Marshall 1990; Peters et al. 1996; Nicola and Vakser 2007), study of the "low-frequency" surface properties (Duncan and Olson 1993), recognition of proteins deprived of atom-size structural features (Vakser 1995; Vakser and Nikiforovich 1995; Vakser 1996b; Vakser et al. 1999), and backbone complementarity in protein recognition (Vakser 1996c). The practical importance of the large-scale recognition factors for docking methodologies is that they often allow one to ignore local structural inaccuracies (e.g., those caused by conformational changes of the partners upon complex formation).

The effect is illustrated in Figure 2.3, showing the lowest energy low-resolution match between unbound hemagglutinin and the BH151 antibody, which is a meaningful approximation of the correct binding mode (Vakser 1997). The binding involves significant conformational changes in the surface side chains. Thus, the high-resolution rigid-body docking mode was unable to produce adequate structures. The match shows the low-resolution surface complementarity between the molecular structures. A closer examination, however, reveals multiple discrepancies in the

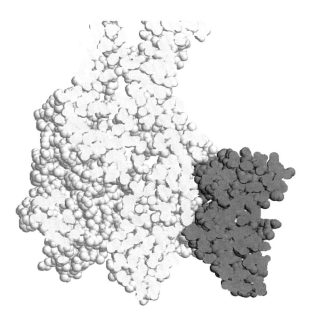

FIGURE 2.3 The lowest-energy low-resolution match between unbound hemagglutinin (light gray) and the BH151 antibody (dark gray).

atom-size details (penetrations, gaps, physicochemical inconsistencies). Both the low-resolution complementarity and the high-resolution mismatches are the direct results of the elimination of small structural details from the docking procedure, which was specifically designed to provide such effects.

The backbone complementarity in protein–protein recognition was studied directly by representing the molecules by C^α atoms only and applying the C^α-centered potentials for intermolecular energy calculations. A systematic six-dimensional search for complementarity between ligand and receptor backbone structures revealed that, in most cases, the low-energy configurations of the complexes are nonrandomly related to their crystal structures.

The computer experiment revealed that all tested backbone structures, except antigen–antibody, in all 10 low-energy configurations (in one case, in 6 of 10), were found within 12 Å from the crystallographically determined position in the complex (Figure 2.4). Taking into account the remarkably nonrandom character of the results, one may conclude that the main-chain fold plays an important role in protein recognition. At the same time, the results showed that the role of the main chain in antigen–antibody complexes is less significant than in the other cases of protein complexes. The reason may be that the antibody molecules, with basically the same main-chain fold, have to recognize different antigens. This means that the backbone cannot be a recognition factor in this case. The conformational differences in the main chain of the recognition loops in the variable domain of Fab may just facilitate the specific arrangement of the side chains, which could reflect certain differences in the principles of complex formation. Thus, the complementarity between the backbones, in general, may facilitate the initial placement of the ligand at the binding site of the

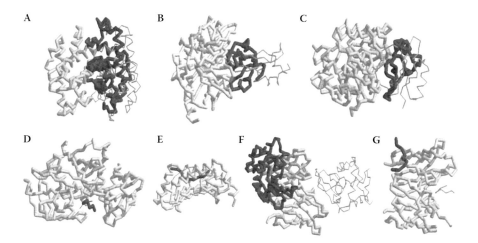

FIGURE 2.4 The lowest-energy complexes of the backbone structures. The molecular pairs are: (A) α and β subunits of human hemoglobin, (B) trypsin and BPTI, (C) subtilisin and chymotrypsin inhibitor, (D) acid proteinase and peptide inhibitor, (E) α1–α2 subunits of MHC I and a peptide, (F) the variable region of Fab and lysozyme, and (G) the variable region of Fab and a peptide. The thick chain represents the receptor (light gray) and the ligand (dark gray). The backbone of the ligand in the crystal structure is shown by the thin chain.

receptor. At the same time, the identity and the specific conformation of the surface side chains play the crucial role at the subsequent stage of the complex formation.

Observation of co-crystallized protein–protein complexes and low-resolution protein–protein docking studies suggests the existence of a binding-related anisotropic shape characteristic of protein–protein complexes. A recent study (Nicola and Vakser 2007) systematically assessed the global shape of proteins in a nonredundant database of co-crystallized protein–protein complexes by measuring the distance of the surface residues to the protein's center of mass. The results showed that on average the binding site residues are closer to the center of mass than the nonbinding surface residues. The data clearly shows a tendency of the interface residues to be closer than average to the center of mass. The effect is not detectable for the small interfaces, but increases dramatically for the large interfaces. The paradigm is illustrated in Figure 2.5. Examples of actual interfaces are shown in Figure 2.6. Arguably, a small interface is geometrically less likely than a large one to have a deep concavity or significant flatness detectable by a simple measure of the average distance to the protein center of mass. On the other hand, a large interface on the larger protein within a complex geometrically can be of any type—concave, convex, or flat (Figure 2.5). The fact that it is by far more likely to be close to the protein center of mass than the rest of the surface does not follow from geometry, but rather is due to free energy aspects of protein binding/folding.

A systematic evaluation of the low-resolution protein–protein recognition was performed on a comprehensive nonredundant database of co-crystallized protein–protein complexes. The docking program GRAMM was used to delete the atom-size structural details and to systematically dock the resulting molecular images. The

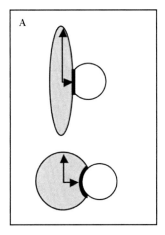
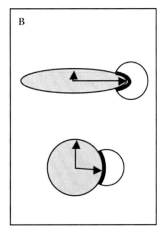

FIGURE 2.5 Schematic illustration of some possible protein–protein complex geometries. The distances from the center of mass (arrows) illustrate (A) more likely geometries (binding site on average is closer to the center of mass than the nonbinding surface) and (B) less likely geometries. For simplicity, the illustration shows proteins of different sizes. However, the same paradigm of binding site close to the center of mass applies to homodimers.

results revealed the existence of the low-resolution recognition in 52% of all complexes in the database and in 76% of the 113 complexes with >4000 Å² interface area. Limitations of the docking and analysis tools used in that study suggested that the actual number of complexes with the low-resolution recognition is higher (Vakser et al. 1999). A more sophisticated approach for the detection of low-resolution recognition was based on different models of random matches (Tovchigrechko and Vakser 2001). The recognition was considered detected if the binding area was more populated by the low-energy docking predictions than by the matches generated in the random models. The number of complexes with detected recognition based on different random models varied significantly. However, the results confirmed that such recognition is likely to be the universal feature in protein–protein association.

The same techniques have been applied to docking of protein models of different accuracies (Tovchigrechko et al. 2002). To simulate the precision of protein models,

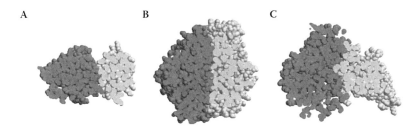

FIGURE 2.6 Examples of protein–protein interfaces. A cross-section through the structure shows (A) small interface with undetectable binding-related shape anisotropy (1138 Å²), (B) large flat interface (7004 Å²), and (C) large concave interface (4055 Å²).

all proteins in the protein–protein database were structurally modified in the range of 1 to 10 Å RMSD, with 1 Å intervals. A sophisticated procedure was specifically designed and implemented for that purpose. All resulting models of the proteins (Figure 2.7 shows an example of models) were docked. The statistical significance of the docking was analyzed, and the results were correlated with the precision of the models. The data showed that even highly imprecise protein models (>6 Å RMSD) still yield structurally meaningful docking results that are accurate enough to predict binding interfaces and to serve as starting points for further structural analysis. An example of docking protein models of low accuracy is shown in Figure 2.8. The study demonstrated the applicability of existing docking techniques to models of various accuracies and, at the same time, the existence of the large recognition factors in protein structures.

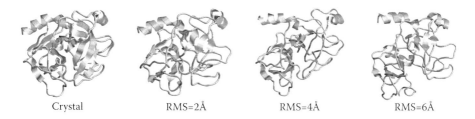

| Crystal | RMS=2Å | RMS=4Å | RMS=6Å |

FIGURE 2.7 The array of trypsin structures, from the x-ray to low-resolution models.

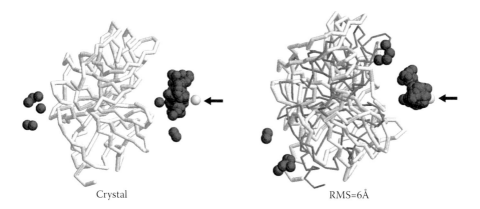

Crystal   RMS=6Å

FIGURE 2.8 Results of the low-resolution docking of trypsin and BPTI. The experimental structures are on the left and the low-resolution models (RMS = 6 Å, both trypsin and BPTI) are on the right. The dark gray spheres are the BPTI center of mass in the 100 lowest energy positions. The light gray sphere (indicated by an arrow) is the BPTI center of mass in the co-crystallized complex. For comparison, the experimental structure of trypsin (thin dark gray chain) is overlapped with the model. The docking of the models clearly preserves the cluster of correct predictions in the area of the binding site.

INTERMOLECULAR ENERGY LANDSCAPE

The existence of the large-scale structural recognition factors in protein association has to do with the funnel-like intermolecular energy landscape. The concept of the funnel-like energy landscapes has had a significant impact on the understanding of protein folding (Dill 1999). The kinetics of the amino acid chain folding into a unique 3D structure are impossible to explain using "flat" energy landscapes, where minima are located on the energy "surface" that do not favor the native structure (so-called golf-course landscapes). The general slope of the energy landscape toward the native structure ("the funnel") explains the kinetics of protein folding. It also provides the basis for protein-structure prediction. The basic physicochemical and structural principles of protein binding are similar, if not identical, to those of protein folding. Thus, the funnel concept can be naturally extended to intermolecular energy (Tsai et al. 1999; Tovchigrechko and Vakser 2001; Wolynes 2005). As in protein folding, this concept is necessary to explain the kinetics data for protein–protein association. The existence of a funnel in protein–protein interactions is supported by considerations regarding long-range electrostatic and/or hydrophobic "steering forces" and the geometry of proteins (Berg and von Hippel 1985; McCammon 1998), energy estimates for near-native complex structures (Camacho et al. 2000), and the binding mechanism that involves protein folding (Shoemaker et al. 2000).

It has been shown that simple energy functions, including coarse-grained (low-resolution) models, reveal major landscape characteristics. The large-scale, systematic studies of protein–protein complexes confirmed the existence of the intermolecular binding funnel (Vakser et al. 1999; Tovchigrechko and Vakser 2001).

A simplified representation of the landscape was used for a systematic study of its large-scale characteristics in a large nonredundant data set of protein complexes. The focus of the study was on the basic features of the low-resolution energy basins and their distribution on the landscape (O'Toole and Vakser 2008). The results clearly show that, in general, the number of such basins is small (Figure 2.9), these basins

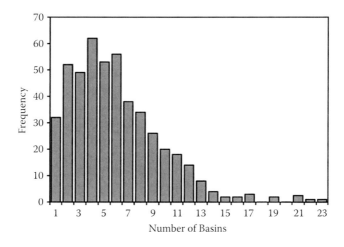

FIGURE 2.9   Distribution of complexes with a certain number of basins.

are well formed, correlated with actual binding modes, and the pattern of basins distribution depends on the type of the complex.

GRAMM-X docking was applied to a comprehensive nonredundant database of nonobligate protein–protein complexes to determine the size of the intermolecular energy funnel (Hunjan et al. 2008). The unbound structures were simulated using a rotamer library. The procedure generated grid-based matches, based on a smoothed Lennard-Jones potential, and minimized them off the grid with the same potential. The minimization generated a distribution of distances, based on a variety of metrics, between the grid-based and the minimized matches. The metric selected for the analysis, ligand interface RMSD, provided three independent estimates of the funnel size: based on the distribution amplitude for the near-native matches (Figure 2.10), deviation from random, and correlation with the energy values. The three methods converge to similar estimates of ~6–8Å ligand interface RMSD. The results indicated dependence of the funnel size on the type of the complex (smaller for antigen–antibody, medium for enzyme–inhibitor, and larger for the rest of the complexes) and the funnel size correlation with the size of the interface.

In a subsequent study, the energy landscapes of 92 protein–protein complexes were described by conformational ensembles of docked protein matches developed for each of the considered resolutions from 1.7 to 5.5 Å (Ruvinsky and Vakser 2008a). The results demonstrated that the ruggedness and the slope are markedly higher for funnels then for other basins at all resolutions. The results also showed that increasing of the potential range decreases the number of multiconformational clusters

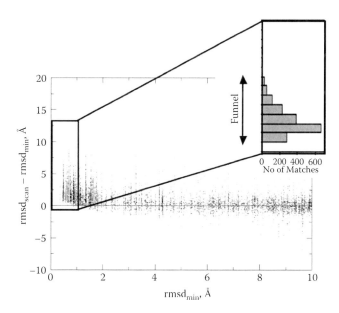

FIGURE 2.10 Distribution of minimization distances (the difference between scan and minimized positions) versus the minimized positions (the end points of the minimization). Each point on the distribution plots corresponds to a match. There are 2000 matches for each of the 399 protein–protein interactions.

substantially, increases the number of near-native docked matches, and keeps the energy gap of the landscape at the same level. The results revealed that averaged characteristics of the energy basins (the ruggedness and the energy slope) may be a more stable indicator of the native funnel than the basin depth and its occupancy.

Since the protein–protein intermolecular energy landscape is based on distance-dependent potentials, it is grounded in the shape of the interacting molecules. This connection is especially transparent in the case of van der Waals–only force fields, including the digitized Lennard-Jones potentials, used in a number of surface complementarity docking algorithms including GRAMM. This geometry–energy dichotomy was addressed in earlier docking studies (Vakser 1996a) and allows one to conveniently use the energetic and geometric considerations interchangeably in exploring protein recognition factors.

The potential smoothing approach in GRAMM is based on extending the range of the potential, at the same time lowering the resolution of the geometric molecular image (Vakser 1995, 1996a). This allows direct tracking of major landscape charac-teristics to the structural details of protein shape. The atomic size details correspond to "high-frequency" landscape fluctuation. The individual residues side chains and larger structural fragments (e.g., secondary structure elements) produce progres-sively larger (lower frequency) fluctuations. The largest (lowest frequency) fluctua-tions (funnels/basins) correspond to macrostructural recognition factors (binding sites and similar size shape characteristics; Tovchigrechko and Vakser 2001).

Beyond these general considerations, a limited number of basin distributions can be specifically tracked back to a known class of shape characteristics. One example is enzyme–inhibitor complexes, which often have a dominant basin (typically, the fun-nel) that corresponds to the geometrically pronounced binding site on the enzyme. However, in most other cases the connection is not easily made and requires detailed structure-function investigation of the interacting proteins.

IMPLICATIONS FOR DOCKING

The knowledge of the major characteristics of the binding sites, such as hydrophobic clusters, hot spot residues, and so forth, helps to narrow the global search for the binding mode, by providing the opportunity to focus on specific sites on the protein surfaces, as well as on limited areas of the intramolecular conformational space.

A number of protein docking approaches implement a multistage/multiscale approach, where the initial global search is performed at lower resolution, followed by the local refinement to a higher resolution (Pappu et al. 1999; Gray et al. 2003; Li et al. 2003; Carter et al. 2005; Tovchigrechko and Vakser 2005). The change of resolution is an essential part of the refinement. A major impediment to the refine-ment protocols is the uncertainties in the landscape transformation (Pappu et al. 1999). Such uncertainties led to the loss of the refinement trajectories, unneces-sary oversampling, and so forth. Thus, the quantitative description of the landscape change according to the resolution is important for designing refinement procedures for docking.

The use of the simpler energy functions at the first (scan) stage of docking has two related aspects. First, it is to make the procedure computationally feasible, since it involves global search in the docking coordinates space. Second, for some algorithms explicitly, and for many implicitly, simpler functions result in a simpler landscape that still serves as a meaningful approximation of the "real" landscape. The value of such simpler landscapes is that they reduce the inherent "high-frequency" energy fluctuations from more detailed force fields, which allows a sparse sampling procedure to detect the binding funnel area.

A typical second docking stage often involves scoring (reevaluation of the same matches with a more accurate, but computationally expensive, energy function) and sometimes refinement (minimization, often more detailed than at the first stage potential, involving structure movement; e.g., off the grid, on a finer grid, etc., depending on the algorithm).

In such a scheme, it is critical that the set of the matches submitted to the second stage contains at least one match within the funnel. Otherwise the refinement, which is local by design, will not produce a near-native structure. This requirement is by far not trivial, given the scope of the global search, and the fact that the scan stage of most global search procedures is the rigid-body one. This results in the need for the scan stage to generate huge numbers of matches (often in the hundreds of thousands) for the subsequent refinement. Such numbers of starting points for the refinement reduces its reliability, because the accuracy has to be compromised for the sake of computational feasibility.

Obviously improving the methods of the funnel detection at the scan stage would drastically reduce the number of starting points for the second stage, thus allowing it to perform a better refinement. For such a task, the knowledge of the average number of funnel-like basins per complex and the size of the funnel is important. Specifically, the knowledge of the average funnel size suggests the maximal distance of a scan match from the native structure in order for the refinement to succeed. The funnel sizes detected by geometry-only procedures may underestimate the "real" size of the funnel, due to the absence of electrostatics and desolvation components and the rigid-body approximation. However, they are adequate to docking algorithms where the global search scan stage is based on the rigid-body approximation and the steric fit is the principal component of the force field.

Another important implication for practical docking directly relates to the ruggedness and slope characteristics. The basin depth-related ruggedness and slope have not been utilized in protein docking. Among various approaches to funnel detection, docking procedures typically use the energy/score of the top-ranked match and, optionally, the cluster occupancy, in part related to the density-related ruggedness. Thus, the existing docking methods do not properly account for the basin shape. As a distinct property of the funnel, the depth-related ruggedness and the slope should complement the energy and the cluster occupancy in the docking funnel detection.

ACKNOWLEDGMENTS

The author is grateful to many colleagues who helped him understand the principles and characteristics of protein recognition. A significant part of the presented material is based on the contributions of Andrey Tovchigrechko, Dominique Douguet, Ying Gao, George Nicola, Nicholas O'Toole, Anatoly Ruvinsky, Jagtar Hunjan, and others who worked over the years in the Vakser lab. The author gratefully acknowledges the support of National Institutes of Health through the R01 GM074255 grant.

REFERENCES

Armon, A., Graur, D., and Ben-Tal, N. 2001. ConSurf: An algorithmic tool for the identification of functional regions in proteins by surface mapping of phylogenetic information. *J. Mol. Biol.* **307:** 447–463.

Ausiello, G., Cesareni, G., and Helmer-Citterich, M. 1997. ESCHER: A new docking procedure applied to the reconstruction of protein tertiary structure. *Proteins* **28:** 556–567.

Baker, D., and Lim, W. A. 2002. Folding and binding. From folding towards function. *Curr. Opin. Struct. Biol.* **12:** 11–13.

Berg, O. G., and von Hippel, P. H. 1985. Diffusion-controlled macromolecular interactions. *Ann. Rev. Biophys. Biophys. Chem.* **14:** 131–160.

Binkowski, T. A., Naghibzadeh, S., and Liang, J. 2003. CASTp: Computed atlas of surface topography of proteins. *Nucleic Acids Res.* **31:** 3352–3355.

Bryngelson, J. D., Onuchic, J. N., Socci, N. D., and Wolynes, P. G. 1995. Funnels, pathways, and the energy landscape of protein folding: A synthesis. *Proteins* **21:** 167–195.

Camacho, C. J., Gatchell, D.W., Kimura, S.R., and Vajda, S. 2000. Scoring docked conformations generated by rigid-body protein–protein docking. *Proteins* **40:** 525–537.

Cammer, S. A., Hoffman, B. T., Speir, J. A., Canady, M. A., Nelson, M. R., Knutson, S., Gallina, M., Baxter, S. M., and Fetrow, J. S. 2003. Structure-based active site profiles for genome analysis and functional family subclassification. *J. Mol. Biol.* **334:** 387–401.

Canutescu, A. A., Shelenkov, A. A., and Dunbrack, R. L. 2003. A graph-theory algorithm for rapid protein side-chain prediction. *Protein Sci.* **12:** 2001–2014.

Carter, P., Lesk, V. I., Islam, S.A., and Sternberg, M. J. E. 2005. Protein–protein docking using 3D-Dock in rounds 3, 4, and 5 of CAPRI. *Proteins* **60:** 281–288.

Chen, R., Mintseris, J., Janin, J., and Weng, Z. 2003. A protein–protein docking benchmark. *Proteins* **52:** 88–91.

Chothia, C., Levitt, M., and Richardson, D. 1981. Helix to helix packing in proteins. *J. Mol. Biol.* **145:** 215–250.

Cohen, F. E., Richmond, T. J., and Richards, F. M. 1979. Protein folding: Evaluation of some simple rules for the assembly of helices into tertiary structures with myoglobin as an example. *J. Mol. Biol.* **132:** 275–288.

Dasgupta, S., Iyer, G. H., Bryant, S. H., Lawrence, C. E., and Bell, J. A. 1997. Extent and nature of contacts between protein molecules in crystal lattices and between subunits of protein oligomers. *Proteins* **28:** 494–514.

Davis, F. P., and Sali, A. 2005. PIBASE: A comprehensive database of structurally defined protein interfaces *Bioinformatics* **21:** 1901–1907.

Dill, K. A. 1990. Dominant forces in protein folding. *Biochemistry* **29:** 7133–7155.

Dill, K. A. 1999. Polymer principles and protein folding. *Protein Sci.* **8:** 1166–1180.

Douguet, D., Chen, H. C., Tovchigrechko, A., and Vakser, I. A. 2006. Dockground resource for studying protein–protein interfaces. *Bioinformatics* **22:** 2612–2618.

Duncan, B. S., and Olson, A. J. 1993. Approximation and characterization of molecular surfaces. *Biopolymers* **33:** 219–229.

Elcock, A. H. 2001. Prediction of functionally important residues based solely on the computed energetics of protein structure. *J. Mol. Biol.* **312:** 885–896.

Elcock, A. H., and McCammon, J. A. 2001. Identification of protein oligomerization states by analysis of interface conservation. *Proc. Natl. Acad. Sci. USA* **98:** 2990–2994.

Gao, Y., Douguet, D., Tovchigrechko, A., and Vakser, I. A. 2007. Dockground system of databases for protein recognition studies: Unbound structures for docking. *Proteins* **69:** 845–851.

Glaser, F., Steinberg, D., Vakser, I. A., and Ben-Tal, N. 2001. Residue frequencies and pairing preferences at protein–protein interfaces. *Proteins* **43:** 89–102.

Gong, S., Park, C., Choi, H., Ko, J., Jang, I., Lee, J., Bolser, D. M., Oh, D., Kim, D. S., and Bhak, J. 2005. A protein domain interaction interface database: InterPare. *BMC Bioinformatics* **6:** 207.

Gray, J. J., Moughon, S., Wang, C., Schueler-Furman, O., Kuhlman, B., Rohl, C. A., and Baker, D. 2003. Protein–protein docking with simultaneous optimization of rigid-body displacement and side-chain conformations. *J. Mol. Biol.* **331:** 281–299.

Halperin, I., Wolfson, H. J., and Nussinov, R. 2004. Protein interactions: Coupling of structurally conserved residues and of hot spots across protein–protein interfaces. Implications for docking. *Structure* **12:** 1027–1038.

Haspel, N., Tsai, C. J., Wolfson, H., and Nussinov, R. 2002. Reducing the computational complexity of protein folding via fragment folding and assembly. *Protein Sci.* **12:** 1177–1187.

Ho, C. M. W., and Marshall, G. R. 1990. Cavity search: An algorithm for the isolation and display of cavity-like binding regions. *J. Comput. Aided Mol. Des.* **4:** 337–354.

Hubbard, S. J., and Argos, P. 1994. Cavities and packing at protein interfaces. *Protein Sci.* **3:** 2194–2206.

Hunjan, J., Tovchigrechko, A., Gao, Y., and Vakser, I. A. 2008. The size of the intermolecular energy funnel in protein–protein interactions. *Proteins* **72:** 344–352.

Inbar, Y., Benyamini, H., Nussinov, R., and Wolfson, H. J. 2003. Protein structure prediction via combinatorial assembly of sub-structural units. *Bioinformatics* **19:** i158–i168.

Janin, J. 1995. Principles of protein–protein recognition from structure to thermodynamics. *Biochimie* **77:** 497–505.

Jefferson, E. R., Walsh, T. P., Roberts, T. J., and Barton, G. J. 2007. SNAPPI-DB: A database and API of structures, interfaces and alignments for protein–protein interactions. *Nucleic Acids Res.* **35:** D580–D589.

Jiang, S., Tovchigrechko, A., and Vakser, I. A. 2003. The role of geometric complementarity in secondary structure packing: A systematic docking study. *Protein Sci.* **12:** 1646–1651.

Jiang, S., and Vakser, I. A. 2000. Side chains in transmembrane helices are shorter at helix-helix interfaces. *Proteins* **40:** 429–435.

Jiang, S., and Vakser, I. A. 2004. Shorter side chains optimize helix–helix packing. *Protein Sci.* **13:** 1426–1429.

Jones, S., and Thornton, J. M. 1996. Principles of protein–protein interactions. *Proc. Natl. Acad. Sci. USA* **93:** 13–20.

Keskin, O., Bahar, I., Badretdinov, A. Y., Ptitsyn, O. B., and Jernigan, R. L. 1998. Empirical solvent-mediated potentials hold for both intra-molecular and inter-molecular inter-residue interactions. *Protein Sci.* **7:** 2578–2586.

Keskin, O., Ma, B., and Nussinov, R. 2005. Hot regions in protein–protein interactions: The organization and contribution of structurally conserved hot spot residues *J. Mol. Biol.* **345:** 1281–1294.

Keskin, O., Tsai, C. J., Wolfson, H., and Nussinov, R. 2004. A new, structurally nonredundant, diverse data set of protein–protein interfaces and its implications. *Protein Sci.* **13:** 1043–1055.

Korn, A. P., and Burnett, R. M. 1991. Distribution and complementarity of hydropathy in multisubunit proteins. *Proteins* **9:** 37–55.

Kundrotas, P. J., and Alexov, E. 2007. PROTCOM: Searchable database of protein complexes enhanced with domain–domain structures. *Nucleic Acids Res.* **35:** D575–D579.

Larsen, T. A., Olson, A. J., and Goodsell, D. S. 1998. Morphology of protein–protein interfaces. *Structure* **6**: 421–427.

Laskowski, R. A., Luscombe, N. M., Swindells, M. B., and Thornton, J. M. 1996. Protein clefts in molecular recognition and function. *Protein Sci.* **5**: 2438–2452.

Lawrence, M. C., and Colman, P. M. 1993. Shape complementarity at protein/protein interfaces. *J. Mol. Biol.* **234**: 946–950.

Lensink, M. F., Mendez, R., and Wodak, S. J. 2007. Docking and scoring protein complexes: CAPRI 3rd Edition. *Proteins* **69**: 704–718.

Li, L., Chen, R., and Weng, Z. 2003. RDOCK: Refinement of rigid-body protein docking predictions. *Proteins* **53**: 693–707.

Liang, S., Liu, S., Zhang, C., and Zhou, Y. 2007. A simple reference state makes a significant improvement in near-native selections from structurally refined docking decoys. *Proteins* **69**: 244–253.

Lijnzaad, P., and Argos, P. 1997. Hydrophobic patches on protein subunit interfaces: Characteristics and prediction. *Proteins* **28**: 333–343.

Lo Conte, L., Chothia, C., and Janin, J. 1999. The atomic structure of protein–protein recognition sites. *J. Mol. Biol.* **285**: 2177–2198.

Lu, H., Lu, L., and Skolnick, J. 2003. Development of unified statistical potentials describing protein–protein interactions. *Biophys. J.* **84**: 1895–1901.

Mariuzza, R. A., and Poljak, R. J. 1993. The basics of binding: Mechanisms of antigen recognition and mimicry by antibodies. *Curr. Opin. Immunol.* **5**: 50–55.

McCammon, J. A. 1998. Theory of biomolecular recognition. *Curr. Opin. Struct. Biol.* **8**: 245–249.

McCoy, A. J., Epa, V. C., and Colman, P. M. 1997. Electrostatic complementarity at protein/protein interfaces. *J. Mol. Biol.* **268**: 570–584.

Mintseris, J., and Weng, Z. 2003. Atomic contact vectors in protein–protein recognition. *Proteins* **53**: 629–639.

Mintseris, J., and Weng, Z. 2005. Structure, function, and evolution of transient and obligate protein–protein interactions. *Proc. Natl. Acad. Sci. USA* **102**: 10930–10935.

Mintseris, J., Wiehe, K., Pierce, B., Anderson, R., Chen, R., Janin, J., and Weng, Z. 2005. Protein–protein docking benchmark 2.0: An update. *Proteins* **60**: 214–216.

Moreira, I. S., Fernandes, P. A., and Ramos, M. J. 2007. Hot spots—A review of the protein–protein interface determinant amino-acid residues. *Proteins* **68**: 803–812.

Murzin, A. G., and Finkelstein, A. V. 1988. General architecture of the alpha-helical globule. *J. Mol. Biol.* **204**: 749–769.

Nicola, G., and Vakser, I. A. 2007. A simple shape characteristic of protein–protein recognition. *Bioinformatics* **23**: 789–792.

Novotny, J., Handschumacher, M., Haber, E., Bruccoleri, R. E., Carlson, W. B., Fanning, D. W., Smith, J. A., and Rose, G. D. 1986. Antigenic determinants in proteins coincide with surface regions accessible to large probes (antibody domains). *Proc. Natl. Acad. Sci. USA* **83**: 226–230.

O'Toole, N., and Vakser, I. A. 2008. Large-scale characteristics of the energy landscape in protein–protein interactions. *Proteins* **71**: 144–152.

Papoian, G. A., and Wolynes, P. G. 2003. The physics and bioinformatics of binding and folding—An energy landscape perspective. *Biopolymers* **68**: 333–349.

Pappu, R. V., Marshall, G. R., and Ponder, J. W. 1999. A potential smoothing algorithm accurately predicts transmembrane helix packing. *Nature Struct. Biol.* **6**: 50–55.

Peters, K. P., Fauck, J., and Frommel, C. 1996. The automatic search for ligand binding sites in proteins of known three-dimensional structure using only geometric criteria. *J. Mol. Biol.* **256**: 201–213.

Ponder, J. W., and Richards, F. M. 1987. Internal packing and protein structural classes. In *Cold Spring Harbor Symposia on Quantitative Biology*, pp. 421–428. Cold Spring Harbor Laboratory, New York.

Ponstingl, H., Henrick, K., and Thornton, J. M. 2000. Discriminating between homodimeric and monomeric proteins in the crystalline state. *Proteins* **41:** 47–57.

Rajamani, D., Thiel, S., Vajda, S., and Camacho, C. J. 2004. Anchor residues in protein–protein interactions. *Proc. Natl. Acad. Sci. USA* **101:** 11287–11292.

Reddy, B. V. B., and Blundell, T. L. 1993. Packing of secondary structure elements in proteins. Analysis and prediction of inter-helix distances. *J. Mol. Biol.* **233:** 464–479.

Reichmann, D., Rahat, O., Cohen, M., Neuvirth, H., and Schreiber, G. 2007. The molecular architecture of protein–protein binding sites. *Curr. Opin. Struct. Biol.* **17:** 67–76.

Richards, F. M. 1977. Areas, volumes, packing, and protein structure. *Ann. Rev. Biophys. Bioeng.* **6:** 151–176.

Richmond, T. J., and Richards, F. M. 1978. Packing of alpha-helices: Geometrical constraints and contact areas. *J. Mol. Biol.* **119:** 537–555.

Russell, R. B., Alber, F., Aloy, P., Davis, F. P., Korkin, D., Pichaud, M., Topf, M., and Sali, A. 2004. A structural perspective on protein–protein interactions. *Curr. Opin. Struct. Biol.* **14:** 313–324.

Ruvinsky, A. M., and Vakser, I. A. 2008a. Chasing funnels on protein–protein energy landscapes at different resolutions. *Biophys. J.* **95:** 2150–2159.

Ruvinsky, A. M., and Vakser, I. A. 2008b. Interaction cutoff effect on ruggedness of protein–protein energy landscape. *Proteins* **70:** 1498–1505.

Shoemaker, B. A., Portman, J. J., and Wolynes, P. G. 2000. Speeding molecular recognition by using the folding funnel: The fly-casting mechanism. *Proc. Natl. Acad. Sci. USA* **97:** 8868–8873.

Szilagyi, A., Grimm, V., Arakaki, A. K., and Skolnick, J. 2005. Prediction of physical protein–protein interactions. *Phys. Biol.* **2:** S1–S16.

Teyra, J., Doms, A., Schroeder, M., and Pisabarro, M. T. 2006. SCOWLP: A web-based database for detailed characterization and visualization of protein interfaces. *BMC Bioinformatics* **7:** 104.

Tovchigrechko, A., and Vakser, I. A. 2001. How common is the funnel-like energy landscape in protein–protein interactions? *Protein Sci.* **10:** 1572–1583.

Tovchigrechko, A., and Vakser, I. A. 2005. Development and testing of an automated approach to protein docking *Proteins* **60:** 296–301.

Tovchigrechko, A., Wells, C. A., and Vakser, I. A. 2002. Docking of protein models. *Protein Sci.* **11:** 1888–1896.

Tsai, C.-J., Kumar, S., Ma, B., and Nussinov, R. 1999. Folding funnels, binding funnels, and protein function. *Protein Sci.* **8:** 1181–1190.

Tsai, C.-J., Lin, S. L., Wolfson, H. J., and Nussinov, R. 1996. Protein–protein interfaces: Architectures and interactions in protein–protein interfaces and in protein cores. Their similarities and differences. *Crit. Rev. Biochem. Mol. Biol.* **31:** 127–152.

Tsai, C.-J., Lin, S. L., Wolfson, H., and Nussinov, R. 1997. Studies of protein–protein interfaces: A statistical analysis of the hydrophobic effect. *Protein Sci.* **6:** 53–64.

Tsai, C.-J., and Nussinov, R. 1997. Hydrophobic folding units at protein–protein interfaces: Implications to protein folding and to protein–protein association. *Protein Sci.* **6:** 1426–1437.

Vajda, S., Sippl, M., and Novotny, J. 1997. Empirical potentials and functions for protein folding and binding. *Curr. Opin. Struct. Biol.* **7:** 222–228.

Vakser, I. A. 1995. Protein docking for low-resolution structures. *Protein Eng.* **8:** 371–377.

Vakser, I. A. 1996a. Long-distance potentials: An approach to the multiple-minima problem in ligand-receptor interaction. *Protein Eng.* **9:** 37–41.

Vakser, I. A. 1996b. Low-resolution docking: Prediction of complexes for underdetermined structures. *Biopolymers* **39:** 455–464.

Vakser, I. A. 1996c. Main-chain complementarity in protein–protein recognition. *Protein Eng.* **9:** 741–744.

Vakser, I. A. 1997. Evaluation of GRAMM low-resolution docking methodology on the hemagglutinin-antibody complex. *Proteins* **Suppl. 1:** 226–230.

Vakser, I. A. 2004. Protein–protein interfaces are special. *Structure* **12:** 910–912.

Vakser, I. A. 2008. PSI has to live and become PCI: Protein complex initiative. *Structure* **16:** 1–3.

Vakser, I. A., and Aflalo, C. 1994. Hydrophobic docking: A proposed enhancement to molecular recognition techniques. *Proteins* **20:** 320–329.

Vakser, I. A., and Jiang, S. 2002. Strategies for modeling the interactions of the transmembrane helices of G-protein coupled receptors by geometric complementarity using the GRAMM computer algorithm. *Methods Enzym.* **343:** 313–328.

Vakser, I. A., and Kundrotas, P. 2008. Predicting 3D structures of protein–protein complexes. *Curr. Pharm. Biotech.* **9:** 57–66.

Vakser, I. A., Matar, O. G., and Lam, C. F. 1999. A systematic study of low-resolution recognition in protein–protein complexes. *Proc. Natl. Acad. Sci. USA* **96:** 8477–8482.

Vakser, I. A., and Nikiforovich, G. V. 1995. Protein docking in the absence of detailed molecular structures. In *Methods in Protein Structure Analysis*, M. Z. Atassi and E. Appella, eds., pp. 505–514. New York: Plenum Press.

Walther, D., Eisenhaber, F., and Argos, P. 1996. Principles of helix–helix packing in proteins: The helical lattice superposition model. *J. Mol. Biol.* **255:** 536–553.

Wodak, S. J., and Janin, J. 1978. Computer analysis of protein–protein interactions. *J. Mol. Biol.* **124:** 323–342.

Wolynes, P. G. 2005. Recent successes of the energy landscape theory of protein folding and function. *Quart. Rev. Biophys.* **38:** 405–410.

Yao, H., Kristensen, D. M., Mihalek, I., Sowa, M. E., Shaw, C., Kimmel, M., Kavraki, L., and Lichtarge, O. 2003. An accurate, sensitive, and scalable method to identify functional sites in protein structures. *J. Mol. Biol.* **326:** 255–261.

Young, L., Jernigan, R. L., and Covell, D. G. 1994. A role for surface hydrophobicity in protein–protein recognition. *Protein Sci.* **3:** 717–729.

Yue, K., and Dill, K. A. 2000. Constraint-based assembly of tertiary protein structures from secondary structure elements. *Protein Sci.* **9:** 1935–1946.

Zhang, B., Rychlewski, L., Pawlowski, K., Fetrow, J. S., Skolnick, J., and Godzik, A. 1999. From fold predictions to function predictions: Automation of functional site conservation analysis for functional genome predictions. *Protein Sci.* **8:** 1104–1115.

Zhou, H., and Zhou, Y. 2002. Distance-scaled, finite ideal-gas reference state improves structure-derived potentials of mean force for structure selection and stability prediction. *Protein Sci.* **11:** 2714–2726.

Zhu, H., Domingues, F. S., Sommer, I., and Lengauer, T. 2006. NOXclass: Prediction of protein–protein interaction types. *BMC Bioinformatics* **7:** 27.

3 The Molecular Architecture of Protein–Protein Binding Sites

Eric J. Sundberg

CONTENTS

OVERVIEW

In recent years, many aspects of biology have been likened to networks, in which distinct nodes (e.g., cells or molecules) can be defined that interact with one another within a system to perform various biological functions. While networks have most commonly been invoked to describe large, organismal level systems, they have also found some traction in illustrating the ways in which proteins interact with one another. Network maps have been constructed to depict all of the possible protein–protein interactions within a cell (e.g., the interactome), essentially a low-resolution view of molecular recognition. At higher resolution, thinking of protein–protein binding sites as networks of amino acid residues that communicate with one another both structurally and energetically has begun to reveal how the modular architecture of protein interfaces and the networked communications within them serve as driving forces for protein complex specificity and affinity. Studies aimed at defining the biophysical basis of these communication events within protein–protein binding sites may serve as an experimental foundation for improving algorithms designed to predict protein–protein interactions.

INTRODUCTION

Interactions between proteins are essential for nearly all cellular processes (Gascoigne and Zal, 2004; Pawson and Nash, 2000; Warren, 2002) and aberrant protein–protein interactions contribute to the pathogenesis of numerous human diseases (Rual et al., 2005). As the genomewide mapping of protein–protein interactions has identified many of the molecular components of numerous physiological and pathological processes (Bouwmeester et al., 2004; Giot et al., 2003; Ito et al., 2001; Li et al., 2004; Uetz et al., 2000) and structural genomics efforts have determined structures of many of the constituent protein domains involved in these interactions, the ability to predict the binding specificities and energies of protein complexes from protein structures alone has reached paramount importance.

With the postgenomic emergence of systems biology, many biological events have been likened to networks, in which numerous distinct nodes are described that interact with one another within a system resulting in various functions. The description of biological events as networks and the application of network analysis tools to these structures have provided novel insights into biological systems that had been overlooked prior to defining these events in a networked manner (Bader et al., 2008).

Networks can also be defined for subsystems within the context of a broader organismal system, such as for a protein interactome. For instance, the entire set of proteins (e.g., the proteome) within a particular system, such as a cell, that bind to one another can be defined as a network. Interactome maps of proteins have now been assembled for the proteomes of numerous organisms and cell types, to varying degrees of completeness. These protein interaction maps have, for instance, provided insight into the modularity of the proteome, in which a relatively small number of protein core complexes, or machines, carry out a large number of cellular processes and these core complexes are functionally modified by changes in the attachment of proteins to them, rather than by dissolution and reconstitution of wholly new protein core complexes.

At even greater detail, the molecular interfaces formed when two proteins interact with one another can also be defined as networks. In this way, individual amino acids (e.g., nodes) within a binding site (e.g., the system) are interconnected in myriad ways, and this very interconnectedness is essential for protein–protein binding (e.g., function). As such, any protein–protein interaction can be described as a networked system. Structural and energetic connections between the individual amino acids in an interface exist and defining them quantitatively is currently a major focus in the structural biology community. These networked connections within protein–protein interfaces may prove to be one of the key driving forces in the development of improved algorithms for protein–protein interaction prediction in the future.

What follows is a description of how our view of protein binding interfaces has evolved and where it may be going, as well as how it could affect computational efforts in modeling protein–protein interactions and, ultimately, drug discovery and design.

STRUCTURAL HETEROGENEITY IN PROTEIN–PROTEIN INTERFACES

Our evolving understanding of protein–protein interactions indicates that there is a distinct physical organization to protein binding sites. This molecular architecture of protein–protein interfaces has direct consequences for the specificities and affinities of protein complexes. Describing the architectural details that allow for productive associations is essential for our understanding of, and for our ability to predict, protein–protein interactions.

Protein domains are subsets of protein sequences that can fold stably and function independently from the rest of the protein chain from which it derives. Domains can also evolve independently of larger proteins, as patterns of domain interactions are commonly observed within organisms and across taxa (Pereira-Leal and Teichmann, 2005). Most proteins are constructed of multiple domains (Pawson and Nash, 2003), which can mediate interactions with other proteins, often through associations with other domains. An overrepresentation of domain pairs in large data sets of experimentally determined protein–protein interactions has been observed (Deng et al., 2002; Han et al., 2004; Liu et al., 2005; Riley et al., 2005; Sprinzak and Margalit, 2001). Also, structurally based domain–domain interaction databases (Finn et al., 2005; Stein et al., 2005) include many domain–domain interactions that are shared between diverse protein–protein complexes. Thus, protein–protein interactions are thought to be mediated by a limited set of domain–domain interactions and, as such, domains act as primary recognition elements for protein–protein interactions. Accordingly, it has been argued that this constitutes a "protein recognition code" (Sudol, 1998) and that cell regulatory and signaling systems are assembled largely through protein domain interactions (Pawson and Nash, 2003).

Just as domains, being subsets of whole proteins, are generally responsible for driving interactions, only some fraction of the residues on the molecular surface of a domain are involved in binding. That there is chemical heterogeneity within a protein binding site has been obvious since the first structures of proteins were determined, and it was clear that molecular interfaces would necessarily be populated by mixtures of different amino acids that contributed distinct chemical groups to the interface. Once numerous structures of protein complexes had been determined to high resolution, it was observed that although there was a great deal of heterogeneity in the chemical and structural makeup of interfaces, they were also similarities. In general, protein–protein interfaces are on the order of 1500–2000 Å^2 in total buried surface area, relatively planar in shape, exhibit a hydrophobicity intermediate between the protein core and the entirety of the protein molecular surface, and contain approximately one hydrogen bond per 100 Å^2 of buried surface area from each protein (Janin et al., 2007; Jones and Thornton, 1996).

Which of these general characteristics that are inherent to protein interfaces is most critical for specific and high affinity interactions? Certainly the hydrophobic effect is the main driving force for protein association. That protein binding and protein folding share the same critical determinant is not entirely surprising, as numerous investigators have made the argument that protein binding can be considered, in actuality, a subset of protein folding events.

To understand any system, a series of perturbations and observations is required. For the interrogation of protein–protein interactions, this generally means that proteins, or protein complexes, are altered by mutagenesis and quantitative measurements of the changes in the energetics of binding and/or structural modifications to the protein interface are measured. The correlations between energetic and structural changes in perturbed protein–protein interactions can form the basis of predictive algorithms.

In this way, quantitative estimations of the hydrophobic effect in protein–protein interactions by mutating large hydrophobic residues within an interface to various residues with exceedingly smaller and less hydrophobic side chains have been made. The thermodynamic and structural changes associated with these mutations are then measured by isothermal titration calorimetry and x-ray crystallography, respectively. In this way, an estimated energetic contribution to binding due to hydrophobicity at the center of an interface has been measured to be 46 cal/mol/$Å^2$ (Li et al., 2005), while at the periphery it was determined to be 21 cal/mol/$Å^2$ (Sundberg et al., 2000). Despite the fact that protein–protein interfaces are generally relatively planar, they still exhibit some degree of curvature, especially at the edge of the interface, and the significant energetic difference between positionally distinct sites is expected due to this curvature of the binding site, as well as greater solvent accessibility at the periphery of the interface.

If hydrophobicity was the only important parameter for protein binding, however, the exquisite specificity of protein–protein interactions would be dramatically curtailed and the tendency to aggregate heightened. Indeed, many other structural and/or chemical properties of protein interfaces have been correlated with binding. A far from exhaustive list of these important parameters includes properties such as: the shape and chemical complementarity of the two binding surfaces; the amount of surface area buried upon complex formation; the number and distribution of hydrogen bonds and side chains formed across the interface; whether water molecules are excluded upon binding or remain within the binding site and, if so, whether they mediate intermolecular electrostatic interactions; the amino acid composition within the interface; and the degree of conservation of particular residues across species.

Indeed, numerous protein interface prediction algorithms have been developed that rely on the various attributes that, to some extent at least, distinguish protein binding sites from other portions of the protein molecular surface. Since initial efforts to predict surface patches that coincide with protein–protein interfaces (Jones and Thornton, 1997), several dozen methods to predict interface residues have been published (reviewed in Zhou and Qin, 2007). These interface prediction algorithms generally rely on distinguishing characteristics of protein interface residues, including:

1. Sequence conservation of interface residues are generally more conserved evolutionarily than noninterface residues.
2. Amino acid type—Hydrophobic and aromatic residues, as well as arginine, are more abundant, while charged residues are reduced in frequency in interfaces.
3. Secondary structure—Depending on the data set used, β-strands may be found to be more common than α-helices in protein interfaces (Neuvirth et al., 2004), while the opposite can also be found, and nonregular secondary structures are even more common (Guharoy and Chakrabarti, 2007).
4. Solvent accessibility—Interface residues tend to be more solvent accessible in comparison to noninterface residues.
5. Conformational entropy of side chains—Interface residues are more rigid, most likely in order to reduce the entropic cost of binding.

Each of these characteristics alone, however, are relatively weak signifiers of whether an amino acid will reside in an interface or not, and thus numerous data streams are required to provide confidence in these predictions.

Despite situations in which perhaps each of these parameters critically contribute to the association of two proteins, even such a comprehensive sequential and structural analysis of an interface amounts to an insufficient description of a protein–protein interface that is required for accurate predictions of the specificity and affinity of interactions. The situation is further complicated because similar structural and/or chemical entities are not necessarily equally important energetically across all protein complexes. Clearly, something is encoded in protein–protein interfaces that transcends structure, at least static structure, a functional component that is at least one of the keys to understanding protein–protein interactions.

ENERGETIC MOSAICITY IN PROTEIN–PROTEIN INTERFACES

It was not until the mid-1990s that the energetic mosaicity of protein binding sites began to be appreciated. When the structure of a complex of proteins is determined, all of the atoms that make intermolecular contacts are readily identified. The totality of these contact atoms, and the amino acid residues from which they come, constitute what is referred to as the "structural epitope." That all, or at least most, of these contact residues are energetically favorable for binding would have seemed a reasonable assumption. This turns out, in general, not to be the case, as was first determined when Clackson and Wells (1995) adopted a strategy of alanine scanning mutagenesis to assess the energetic contributions of individual amino acids in a hormone–receptor complex. In this type of analysis, each interface residue is systematically mutated to alanine (in effect, paring its side-chain moiety back to a single methyl group) and the change in binding energy upon complex formation relative to the wild-type complex is measured. Certain amino acid residues within this hormone–receptor interface contribute significantly to the binding energy and were thus termed "hot spots," while other residues were energetically silent with respect to the interaction was evident. This subset of energetically significant residues within the interface is often referred to as the "functional epitope."

Within a given protein–protein interface, hot spot residues are more likely to be found in the central portion of the binding site, often surrounded by a ring of less energetically important residues (Bogan and Thorn, 1998). This distribution of binding energy within the interface mirrors the construction of folded globular proteins and likely serves a similar purpose, that is, to exclude water from the sites of energetic importance. Indeed, occlusion of solvent from the center of the interface is a requirement for high affinity interactions. Additionally, even in a situation where residues from the interface core and periphery can make equal energetic contributions to binding, those at the periphery may be more easily replaced by energetically stabilizing water molecules when their side chains are pared back (Janin, 1999). Despite this general topological arrangement of hot spots concentrated in the protein interface core, there are numerous and notable exceptions to this energetically important core/silent ring architecture including protein complexes in which no hot spots can be identified (Roisman et al., 2005; Svensson et al., 2004) or interactions in which hot spot residues extend to the periphery of the interface (Buonpane et al., 2005).

In addition to the generalized topology of interface core hot spot residency, there is no significant correlation between the surface accessibility and binding energy contribution of any particular interface residue (Bogan and Thorn, 1998; Lo Conte et al., 1999), although residues in the core of the interface tend to exhibit a correlation between changes in accessible surface area and binding free energies that is not seen for peripheral interface residues (Guharoy and Chakrabarti, 2005). The difficulty in identifying energetically important interface residues from examination of the three-dimensional structure of a protein complex alone is exacerbated by the fact that many types of amino acids can serve as hot spots, but no one amino acid type always does, and that some residues that appear to make few contacts within an interface can contribute significantly to binding energetics, sometimes due to destabilization of the unbound proteins (DeLano, 2002).

Still, a number of notable attempts have been made to predict hot spot residues within protein–protein interfaces. Kortemme and Baker (2002) developed a quantitative model for binding energies based on an all-atom rotamer description of side chains with an energy function dominated by Lennard-Jones interactions, solvent interactions, and hydrogen bonding. Using this algorithm, 79% of the hot spots in 19 protein–protein interactions with a total of 233 mutations were correctly predicted with an average error of 1.06 kcal/mol. While this suggests that the underlying physical principles incorporated into this model are in fact important drivers of protein associations, several aspects were not well predicted. In particular, the magnitude of electrostatic effects and the effects of replacing water-mediated hydrogen bonds with direct protein-to-protein hydrogen bonds across interfaces were underpredicted. In a similar effort, Serrano and co-workers developed an energy function of a physical description of protein–protein interactions that was informed by considering a training set of nine protein complexes with 339 mutations in order to optimize the set of parameters and weighting factors that best accounted for changes in the stability of the mutant proteins (Guerois et al., 2002). When applied to a set of four protein complexes with 82 mutations, the correlation between the experimental and theoretical changes in binding free energy was 0.64 with a standard deviation of 0.8 kcal/mol.

Even in the absence of experimentally determined three-dimensional structures, hot spots can be identified with a reasonable degree of confidence. Ofran and Rost (2007a) applied ISIS, an algorithm for predicting all interface residues, to predict only hotspots by training the method on: (1) the sequence environment of each residue, including four residues on each side; (2) the evolutionary profile of this nine-residue window; (3) the predicted solvent accessibility of the residue; (4) the solvent accessibility of the immediate sequence environment, including one residue on each side; (5) the predicted secondary structure state of the residue and its immediate sequence environment; and (6) the evolutionary conservation of the residue (Ofran and Rost, 2007b). When applied to a set of experimental mutations with binding free energy changes of greater than 2.5 kcal/mol, this prediction method using only sequence, evolutionary conservation, and predicted structure was able to identify roughly half of the hot spots that an in silico alanine scanning model such as those described earlier were able to predict. Despite these encouraging advances in predicting which residues within a protein–protein interface will act as hot spots, the necessity of performing extensive experimental mutagenesis and binding analysis to quantitatively describe the energetic contributions of individual amino acids with such an interface persists.

ARCHITECTURAL MODULARITY IN PROTEIN INTERFACES AND COOPERATIVE BINDING ENERGETICS

Whether hot spot residues in a protein–protein interface adopt positions within the conventional core–ring archetype or are more dispersed throughout the binding site, they are not simply distributed in a random fashion throughout the interface. Instead, hot spots tend to be clustered within discrete groups, or "hot regions" (Keskin et al., 2005; Reichmann et al., 2005). The resulting decomposition of protein interfaces into modules, which has been shown both computationally and experimentally, has significant energetic consequences for protein–protein interactions.

Further contributing to the heterogeneity of protein–protein interfaces is the frequent presence of cooperativity, in that the energetic contribution to binding of a protein that has been simultaneously mutated at multiple residues is significantly different than the summation of the changes in binding energy of the single-site mutants (Albeck et al., 2000; Bernat et al., 2004; Yang et al., 2003). That is, not only can hot spots be of varying energetic significance in and of themselves, but also their energetic contributions to binding can vary depending on whether and where other hot spot residues are located in the interface. In many protein–protein interactions, such site-to-site energetic communication is a major contributor to protein binding. Compelling evidence has been mounting of late that the modular architecture that is structurally imprinted on protein binding sites not only results in a certain roughness to the energetic landscape of the interface, but serves as the driver of networked energetic communication in protein–protein interactions.

A recent analysis (Keskin et al., 2005) of a structurally nonredundant database of all hot regions (Keskin et al., 2004) in the Protein Data Bank (Berman et al., 2000) at the time has suggested that hot spots are both preorganized in the unbound state of the protein and that they are clustered into densely packed hot regions. Energetic contributions from hot spots within a single hot region were, in general, cooperative, while those residing in separate hot regions were energetically additive. This type of networking of interactions within a protein–protein interface may be a general strategy by which energetic cooperativity between residues can be utilized to dictate the stability of protein–protein complexes.

Indeed, this modular architecture of protein–protein binding sites has been rigorously investigated experimentally. In the TEM1-β-lactamase/β-lactamase inhibitor protein (TEM1-BLIP) complex, Schreiber and co-workers constructed contact maps of the interface, taking into account physical interactions including hydrogen bonds and van der Waals interactions, by which the interface was divided into five individual clusters or modules, each with numerous interacting residues and few interactions between (Reichmann et al., 2005). Using a combination of alanine scanning mutagenesis, surface plasmon resonance (SPR) analysis, and x-ray crystallography, it was shown that mutations residing in distinct modules do not affect one another energetically, and thus, entire modules could be deleted (i.e., by paring back all side chains within that module by mutation to alanine) with negligible structural or energetic consequences on the remainder of the interface. Conversely, mutations within a single module were responsible for cooperative energetic and structural changes within that module.

Another way in which to perturb an interface to affect the affinity of an interaction is to subject one of the proteins in a complex to directed evolution, such as by phage or yeast display. This iterative process of mutation and selection (in this case for tighter binding to an unmodified target protein) describes an affinity maturation pathway of protein variants that, in total, can span many orders of magnitude in affinity. Because numerous mutations are made that together increase the affinity, the dissection of these affinity maturation pathways by interrogating the structural and energetic changes associated with different combinations of mutations makes this is an especially powerful method for investigating biophysical parameters that are combinatorial by definition, such as energetic cooperativity.

Following this strategy, we recently presented detailed structural and energetic analyses of additive versus cooperative effects within a protein–protein interaction. Using a model system consisting of a yeast display affinity-matured T cell receptor (TCR) protein that exhibited a ~1500-fold affinity increase for the bacterial superantigen SEC3 (Kieke et al., 2001), group and individual TCR maturation and reversion pathway mutations were analyzed for binding to SEC3 by surface plasmon resonance analysis (Yang et al., 2003). As in the TEM1-BLIP complexes, energetic cooperativity was observed within a single hot region, in this case defined by the second complementarity determining region (CDR2) loop, while combinations of mutations from distinct hot regions were found to be energetically additive. Even though this is one of the most highly affinity-matured complexes characterized to date, the ultimate high affinity variant was found to be restricted by negative cooperativity (i.e., the summation of the changes in the binding free energies of the individual mutations exceed the change in binding free energy of the final, fully evolved variant). Two maturation mutations in particular accounted quantitatively for the entirety of this negative cooperativity. By determining the x-ray crystal structures of several of these variant TCR proteins that define this affinity maturation pathway, it was observed that the mutations at these two positions exerted opposing conformational changes on the CDR2 loop, providing a structural basis for short-range negative cooperativity (Cho et al., 2005).

In a similar study involving another affinity-matured TCR–superantigen model protein–protein interaction system, we investigated whether amino acids separated by long distances and residing at the peripheral extremes of the interface could act in an energetically cooperative manner (Moza et al., 2006). The hVβ2.1 TCR had been previously affinity-matured by yeast display to bind the superantigen TSST-1 with an increased affinity of greater than 3000-fold relative to the wild-type TCR (Buonpane et al., 2005). Analysis of each of the individual residue changes revealed that there were four mutations within the interface that were energetically significant in the affinity maturation process. Three of these positions are located within the CDR2 loop of the TCR and form one hot region, while the fourth is located in the third framework region (FR3) loop and forms a distinct hot region. From the x-ray crystal structure of this TCR–superantigen complex (Moza et al., 2007), it is evident that these two hot regions are separated by more than 20 Å and each lies at the periphery of the interface. TCR variants in which every possible combination of these four amino acids as either their wild type or affinity-matured residue were tested for binding to the superantigen, and the binding free energy of the combinatorial variants were compared to the summation of binding free energies of their corresponding single-site mutants to ascertain the extent of cooperativity. As expected, several of the amino acids within the CDR2 hot region exhibited cooperative energetics. Surprisingly, though, combinations of mutations involving residues from each of the CDR2 and FR3 hot regions were also found to be energetically cooperative, and furthermore, the magnitude of this inter-hot regional cooperativity was significantly greater than the observed intra-hot regional cooperativity (Moza et al., 2006).

If, in all protein complexes, cooperative energetics existed only within hot regions, and not between them, the quantitative prediction of protein–protein interactions may be considerably simplified. The aforementioned example suggests that this may not be the case. However, the jury remains out on this question as a recent bioinformatics analysis of the hVβ2.1 TCR–superantigen complex in question (del Sol and Carbonell, 2007) has suggested that the CDR2 and FR3 hot regions form a single, albeit large, module in which one might reasonably expect energetically cooperative residues at any distance. In such a rapidly evolving field such as the analysis of cooperativity in protein–protein interactions, this may be more indicative of a mere semantic discrepancy than an actual biophysical rule.

CAPITALIZING ON UNDERSTANDING: PREDICTING INTERACTIONS AND DESIGNING DRUGS

Although progress in developing computational methods for the quantitative predictions of protein–protein interactions has been made recently (Guerois et al., 2002; Huo et al., 2002; Kortemme and Baker, 2002; Massova and Kollman, 1999; Sharp, 1998), the current robustness of these algorithms is not such that the laborious task of determining the structure of a given protein complex can be circumvented. It is clear that these methods are unable to account for aspects of molecular recognition that are important in determining complex formation, but for which we currently have a fundamental lack of understanding.

A fundamental lack of understanding of cooperative binding energetics may be one of the major impediments to formulating with greater accuracy algorithms for protein–protein interaction prediction. If cooperativity existed only within hot regions, and not between them, the task of accurately predicting the binding parameters for protein complexes would be greatly simplified. Some recent results suggest that this may be an overly generalized representation of macromolecular interfaces and that a broader consideration of cooperativity within protein–protein interactions, while more technically and computationally demanding, may ultimately lead to more accurate predictive algorithms. It also appears from recent results that only a subset of hot regions may need to be considered as potentially cooperative. The recent advances in defining the molecular architectures of protein–protein interfaces as networks of individual amino acids residues provide an experimental avenue by which such predictive algorithms can be built.

Because protein–protein interactions are pervasive in biological processes, they are also important therapeutic targets, and thus the prediction of protein–protein interactions is critical for drug design. The development of small molecule inhibitors of such interactions has proven difficult (Arkin and Wells, 2004), largely due to the relatively planar nature of these interfaces, which tend not to present well-defined binding pockets. The presence of hot spots and hot regions within protein interfaces provides possible sites at which potent small molecule inhibitors may bind to effectively block the association of much larger molecules. Indeed, small peptides selected by phage display generally bind their protein binding partners at hot spots (Sidhu et al., 2003), and the discovery of small molecules that inhibit the interaction of B7-1 with CD28 and modulate T cell activation, and in which the drug binds at a hot spot, has been reported (Erbe et al., 2002; Green et al., 2003).

If certain distinct hot regions may be linked energetically, the potency of a small-molecule inhibitor that targets a cooperative hot region may be amplified relative to a small molecule that targets a hot region that is strictly additive. This could have important ramifications for the choice of which hot region within a protein–protein interaction to target for small molecule inhibition, for instance, by structure-based drug design.

REFERENCES

Albeck, S., Unger, R., and Schreiber, G. (2000). Evaluation of direct and cooperative contributions towards the strength of buried hydrogen bonds and salt bridges. *J Mol Biol 298*, 503–520.

Arkin, M. R., and Wells, J. A. (2004). Small-molecule inhibitors of protein–protein interactions: Progressing towards the dream. *Nat Rev Drug Discov 3*, 301–317.

Bader, S., Kuhner, S., and Gavin, A.C. (2008). Interaction networks for systems biology. *FEBS Lett 582*, 1220–1224.

Berman, H. M., Westbrook, J., Feng, Z., Gilliland, G., Bhat, T. N., Weissig, H., Shindyalov, I. N., and Bourne, P. E. (2000). The Protein Data Bank. *Nucleic Acids Res 28*, 235–242.

Bernat, B., Sun, M., Dwyer, M., Feldkamp, M., and Kossiakoff, A. A. (2004). Dissecting the binding energy epitope of a high-affinity variant of human growth hormone: Cooperative and additive effects from combining mutations from independently selected phage display mutagenesis libraries. *Biochemistry 43*, 6076–6084.

Bogan, A. A., and Thorn, K. S. (1998). Anatomy of hot spots in protein interfaces. *J Mol Biol 280*, 1–9.

Bouwmeester, T., Bauch, A., Ruffner, H., Angrand, P. O., Bergamini, G., Croughton, K., Cruciat, C., Eberhard, D., Gagneur, J., Ghidelli, S., et al. (2004). A physical and functional map of the human TNF-alpha/NF-kappa B signal transduction pathway. *Nat Cell Biol 6*, 97–105.

Buonpane, R. A., Moza, B., Sundberg, E. J., and Kranz, D. M. (2005). Characterization of T cell receptors engineered for high affinity against toxic shock syndrome toxin-1. *J Mol Biol 353*, 308–321.

Cho, S., Swaminathan, C. P., Yang, J., Kerzic, M. C., Guan, R., Kieke, M. C., Kranz, D. M., Mariuzza, R. A., and Sundberg, E. J. (2005). Structural basis of affinity maturation and intramolecular cooperativity in a protein–protein interaction. *Structure (Camb) 13*, 1775–1787.

Clackson, T., and Wells, J. A. (1995). A hot spot of binding energy in a hormone-receptor interface. *Science 267*, 383–386.

del Sol, A., and Carbonell, P. (2007). The modular organization of domain structures: Insights into protein–protein binding. *PLoS Computational Biol 3*, e239.

DeLano, W. L. (2002). Unraveling hot spots in binding interfaces: Progress and challenges. *Curr Opin Struct Biol 12*, 14–20.

Deng, M., Mehta, S., Sun, F., and Chen, T. (2002). Inferring domain–domain interactions from protein–protein interactions. *Genome Res 12*, 1540–1548.

Erbe, D. V., Wang, S., Xing, Y., and Tobin, J. F. (2002). Small molecule ligands define a binding site on the immune regulatory protein B7.1. *J Biol Chem 277*, 7363–7368.

Finn, R. D., Marshall, M., and Bateman, A. (2005). iPfam: Visualization of protein–protein interactions in PDB at domain and amino acid resolutions. *Bioinformatics (Oxford, England) 21*, 410–412.

Gascoigne, N. R., and Zal, T. (2004). Molecular interactions at the T cell-antigen-presenting cell interface. *Curr Opin Immunol 16*, 114–119.

Giot, L., Bader, J. S., Brouwer, C., Chaudhuri, A., Kuang, B., Li, Y., Hao, Y. L., Ooi, C. E., Godwin, B., Vitols, E., et al. (2003). A protein interaction map of Drosophila melanogaster. *Science 302*, 1727–1736.

Green, N. J., Xiang, J., Chen, J., Chen, L., Davies, A. M., Erbe, D., Tam, S., and Tobin, J. F. (2003). Structure-activity studies of a series of dipyrazolo[3,4-b:3′,4′-d]pyridin-3-ones binding to the immune regulatory protein B7.1. *Bioorg Med Chem 11*, 2991–3013.

Guerois, R., Nielsen, J. E., and Serrano, L. (2002). Predicting changes in the stability of proteins and protein complexes: A study of more than 1000 mutations. *J Mol Biol 320*, 369–387.

Guharoy, M., and Chakrabarti, P. (2005). Conservation and relative importance of residues across protein–protein interfaces. *Proc Natl Acad Sci USA 102*, 15447–15452.

Guharoy, M., and Chakrabarti, P. (2007). Secondary structure based analysis and classification of biological interfaces: Identification of binding motifs in protein–protein interactions. *Bioinformatics (Oxford, England) 23*, 1909–1918.

Han, D. S., Kim, H. S., Jang, W. H., Lee, S. D., and Suh, J. K. (2004). PreSPI: A domain combination based prediction system for protein–protein interaction. *Nucleic Acids Res 32*, 6312–6320.

Huo, S., Massova, I., and Kollman, P. A. (2002). Computational alanine scanning of the 1:1 human growth hormone-receptor complex. *J Comput Chem 23*, 15–27.

Ito, T., Chiba, T., Ozawa, R., Yoshida, M., Hattori, M., and Sakaki, Y. (2001). A comprehensive two-hybrid analysis to explore the yeast protein interactome. *Proc Natl Acad Sci USA 98*, 4569–4574.

Janin, J. (1999). Wet and dry interfaces: The role of solvent in protein–protein and protein-DNA recognition. *Structure 7*, R277–R279.

Janin, J., Rodier, F., Chakrabarti, P., and Bahadur, R. P. (2007). Macromolecular recognition in the Protein Data Bank. *Acta Crystallogr D Biol Crystallogr 63*, 1–8.

Jones, S., and Thornton, J. M. (1996). Principles of protein–protein interactions. *Proc Natl Acad Sci USA 93*, 13–20.

Jones, S., and Thornton, J. M. (1997). Prediction of protein–protein interaction sites using patch analysis. *J Mol Biol 272*, 133–143.

Keskin, O., Ma, B., and Nussinov, R. (2005). Hot regions in protein–protein interactions: The organization and contribution of structurally conserved hot spot residues. *J Mol Biol 345*, 1281–1294.

Keskin, O., Tsai, C. J., Wolfson, H., and Nussinov, R. (2004). A new, structurally nonredundant, diverse data set of protein–protein interfaces and its implications. *Protein Sci 13*, 1043–1055.

Kieke, M. C., Sundberg, E., Shusta, E. V., Mariuzza, R. A., Wittrup, K. D., and Kranz, D. M. (2001). High affinity T cell receptors from yeast display libraries block T cell activation by superantigens. *J Mol Biol 307*, 1305–1315.

Kortemme, T., and Baker, D. (2002). A simple physical model for binding energy hot spots in protein–protein complexes. *Proc Natl Acad Sci USA 99*, 14116–14121.

Li, S., Armstrong, C. M., Bertin, N., Ge, H., Milstein, S., Boxem, M., Vidalain, P. O., Han, J. D., Chesneau, A., Hao, T., et al. (2004). A map of the interactome network of the metazoan C. elegans. *Science 303*, 540–543.

Li, Y., Huang, Y., Swaminathan, C. P., Smith-Gill, S. J., and Mariuzza, R. A. (2005). Magnitude of the hydrophobic effect at central versus peripheral sites in protein–protein interfaces. *Structure (Camb) 13*, 297–307.

Liu, Y., Liu, N., and Zhao, H. (2005). Inferring protein–protein interactions through high-throughput interaction data from diverse organisms. *Bioinformatics (Oxford, England) 21*, 3279–3285.

Lo Conte, L., Chothia, C., and Janin, J. (1999). The atomic structure of protein–protein recognition sites. *J Mol Biol 285*, 2177–2198.

Massova, I., and Kollman, P. A. (1999). Computational alanine scanning to probe protein–protein interactions: A novel approach to evaluate binding free energies. *J Am Chem Soc 121*, 8133–8143.

Moza, B., Buonpane, R. A., Zhu, P., Herfst, C. A., Rahman, A. K., McCormick, J. K., Kranz, D. M., and Sundberg, E. J. (2006). Long-range cooperative binding effects in a T cell receptor variable domain. *Proc Natl Acad Sci USA 103*, 9867–9872.

Moza, B., Varma, A. K., Buonpane, R. A., Zhu, P., Herfst, C. A., Nicholson, M. J., Wilbuer, A. K., Seth, N. P., Wucherpfennig, K. W., McCormick, J. K., et al. (2007). Structural basis of T-cell specificity and activation by the bacterial superantigen TSST-1. *Embo J 26*, 1187–1197.

Neuvirth, H., Raz, R., and Schreiber, G. (2004). ProMate: A structure based prediction program to identify the location of protein–protein binding sites. *J Mol Biol 338*, 181–199.

Ofran, Y., and Rost, B. (2007a). ISIS: Interaction sites identified from sequence. *Bioinformatics (Oxford, England) 23*, e13–e16.

Ofran, Y., and Rost, B. (2007b). Protein–protein interaction hotspots carved into sequences. *PLoS Computational Biology 3*, e119.

Pawson, T., and Nash, P. (2000). Protein–protein interactions define specificity in signal transduction. *Genes Dev 14*, 1027–1047.

Pawson, T., and Nash, P. (2003). Assembly of cell regulatory systems through protein interaction domains. *Science 300*, 445–452.

Pereira-Leal, J. B., and Teichmann, S. A. (2005). Novel specificities emerge by stepwise duplication of functional modules. *Genome Research 15*, 552–559.

Reichmann, D., Rahat, O., Albeck, S., Meged, R., Dym, O., and Schreiber, G. (2005). The modular architecture of protein–protein binding interfaces. *Proc Natl Acad Sci USA 102*, 57–62.

Riley, R., Lee, C., Sabatti, C., and Eisenberg, D. (2005). Inferring protein domain interactions from databases of interacting proteins. *Genome Biol 6*, R89.

Roisman, L. C., Jaitin, D. A., Baker, D. P., and Schreiber, G. (2005). Mutational analysis of the IFNAR1 binding site on IFNalpha2 reveals the architecture of a weak ligand-receptor binding-site. *J Mol Biol 353*, 271–281.

Rual, J. F., Venkatesan, K., Hao, T., Hirozane-Kishikawa, T., Dricot, A., Li, N., Berriz, G. F., Gibbons, F. D., Dreze, M., Ayivi-Guedehoussou, N., et al. (2005). Towards a proteome-scale map of the human protein–protein interaction network. *Nature 437*, 1173–1178.

Sharp, K. A. (1998). Calculation of HyHel10-lysozyme binding free energy changes: Effect of ten point mutations. *Proteins 33*, 39–48.

Sidhu, S. S., Fairbrother, W. J., and Deshayes, K. (2003). Exploring protein–protein interactions with phage display. *Chembiochem 4*, 14–25.

Sprinzak, E., and Margalit, H. (2001). Correlated sequence-signatures as markers of protein–protein interaction. *J Mol Biol 311*, 681–692.

Stein, A., Russell, R. B., and Aloy, P. (2005). 3did: Interacting protein domains of known three-dimensional structure. *Nucleic Acids Res 33*, D413–D417.

Sudol, M. (1998). From Src homology domains to other signaling modules: Proposal of the "protein recognition code." *Oncogene 17*, 1469–1474.

Sundberg, E. J., Urrutia, M., Braden, B. C., Isern, J., Tsuchiya, D., Fields, B. A., Malchiodi, E. L., Tormo, J., Schwarz, F. P., and Mariuzza, R. A. (2000). Estimation of the hydrophobic effect in an antigen-antibody protein–protein interface. *Biochemistry 39*, 15375–15387.

Svensson, H. G., Wedemeyer, W. J., Ekstrom, J. L., Callender, D. R., Kortemme, T., Kim, D. E., Sjobring, U., and Baker, D. (2004). Contributions of amino acid side chains to the kinetics and thermodynamics of the bivalent binding of protein L to Ig kappa light chain. *Biochemistry 43*, 2445–2457.

Uetz, P., Giot, L., Cagney, G., Mansfield, T. A., Judson, R. S., Knight, J. R., Lockshon, D., Narayan, V., Srinivasan, M., Pochart, P., et al. (2000). A comprehensive analysis of protein–protein interactions in Saccharomyces cerevisiae. *Nature 403*, 623–627.

Warren, A. J. (2002). Eukaryotic transcription factors. *Curr Opin Struct Biol 12*, 107–114.

Yang, J., Swaminathan, C. P., Huang, Y., Guan, R., Cho, S., Kieke, M. C., Kranz, D. M., Mariuzza, R. A., and Sundberg, E. J. (2003). Dissecting cooperative and additive binding energetics in the affinity maturation pathway of a protein–protein interface. *J Biol Chem 278*, 50412–50421.

Zhou, H. X., and Qin, S. (2007). Interaction-site prediction for protein complexes: A critical assessment. *Bioinformatics (Oxford, England) 23*, 2203–2209.

4 Mapping Protein Function by Combinatorial Mutagenesis

Gábor Pál and Sachdev S. Sidhu

CONTENTS

INTRODUCTION

Biological processes are mediated through macromolecular interactions and virtually all of these processes involve proteins. Proteins play a key role in biology through interactions with diverse ligands, including other proteins, nucleic acids, carbohydrates, lipids, and small molecules [1–3]. A genuine understanding of biological processes requires detailed characterization of the interacting molecules. High-resolution structures of molecular complexes provide crucial information about the interacting surfaces, or structural epitopes, but structural data alone cannot explain how affinity and specificity are achieved. To enable in silico design of protein structure and function, there is a crucial need for not only structural data, but also complementary functional data that describe the energetic roles of individual protein residues.

Although the three-dimensional structures of proteins are undeniably complex, the accumulation of a large database of high-resolution structures reveals common themes in folding, and it appears that there are a limited number of basic folds that are used in nature [4,5]. This knowledge has served as an important input for mathematical descriptions that describe protein folding and structure. In an analogous manner, a large-scale functional data set describing the energetics of protein–protein interactions would be an invaluable tool for elucidating common principles of molecular recognition with the accuracy necessary for computational modeling.

To build functional data sets, systematic amino acid replacements through site-directed mutagenesis are used to map the binding energetics of individual side chains that constitute binding interfaces. In particular, alanine scanning studies have been used to define the subsets of side chains that form energetically favorable interactions and constitute the functional epitope within the structural epitope [6]. However, conventional mutagenesis strategies based on biophysical analysis of individual point mutants are slow and ill-suited for acquiring the quantities of data required for developing mathematical descriptions of the kinetics and thermodynamics of protein–protein interactions. Indeed, it has become apparent that high-throughput technologies for protein analysis will be needed to accelerate progress beyond what is possible with traditional biophysical methods [7].

This chapter focuses on phage display strategies for combinatorial mutagenesis, which are designed to enable the high-throughput mapping of binding energetics at protein–protein interfaces. The so-called shotgun scanning approaches assess energetics by DNA sequencing and statistical analysis, rather than by biophysical analysis of purified proteins. These methods harness the power of combinatorial methods for rapid and quantitative analysis or protein function, and unlike conventional biophysical methods, they are compatible with high-throughput strategies.

PHAGE DISPLAY TECHNOLOGY

Developed more than two decades ago, phage display is the first and still dominant molecular display technology [8,9]. The method relies on genetic engineering to produce fusion proteins consisting of polypeptides of interest fused to bacteriophage coat proteins, and this results in the display of heterologous proteins on the surfaces of phage particles that also encapsulate the encoding DNA (Figure 4.1).

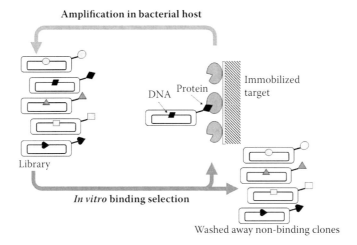

FIGURE 4.1 The phage display selection cycle. Libraries of protein variants (assorted shapes) are displayed on filamentous phage particles as fusions to coat proteins. Each phage particle displays a unique variant and encapsulates the encoding DNA. Highly diverse libraries can be produced and selected for binding to immobilized target. Nonbinding phages are washed away, while binding phages are retained and amplified in host bacteria. Repeated selection and amplification cycles further enrich the population for binding clones. DNA sequencing of individual clones decodes the sequences of displayed polypeptides.

Thus, a physical linkage is established between the phenotype (displayed protein) and genotype (encapsulated DNA), and phage display provides an *in vitro* version of Darwinian evolution. Using combinatorial mutagenesis, a library of billions of protein variants can be produced and represented as phage pools that can be cycled through rounds of binding selection against immobilized ligands to enrich for functional members [10]. The selected phage population can be amplified by passage through bacteria and the amplified pool can be cycled through additional rounds of selection to further enrich the pool for binding clones. Subsequently, individual clones can be amplified and the genomic DNA can be sequenced to decode the sequence of each displayed polypeptide.

SITE-DIRECTED MUTAGENESIS AND ALANINE SCANNING

The dawn of the protein engineering era was brought by recombinant DNA technologies that enable cloning of genes [11–14], production of recombinant proteins in bacterial hosts [15,16], and introduction of site-directed mutations into genes to alter protein sequences in a systematic manner [17–21]. Site-directed mutagenesis was used to alter protein structure, and basic principles of protein chemistry could be deduced from the resulting effects on function [22–55]. The first systematic approach for assessing the energetics of protein interactions was alanine scanning mutagenesis, which removes all side-chain atoms past the β-carbon and can thus be used to assess the roles of individual side chains [56,57].

One of the first alanine scans was performed on the high affinity binding site (Site 1) of human growth hormone (hGH) for its receptor (hGHR) [57]. This was followed by a scan of the complementary binding site on the hGHR [56]. The binding affinities of mutated proteins were compared to that of the wild type (wt) and the effects of alanine (Ala) substitutions on binding energy ($\Delta\Delta G_{\text{Ala-wt}}$) were mapped onto the three-dimensional structures of the molecules. These experiments revealed that only a small subset of residues contribute most of the binding energy on both sides of the interface, and these residues cluster together to form interacting "hot spots" of binding energy [6]. Subsequently, alanine scanning experiments have revealed similar clusters of energetically important residues in many other protein–protein interactions [58], and currently almost 100 such data sets have been compiled [59]. Overall, alanine scanning studies have led to a better understanding of the importance of the spatial organization of functional residues in binding interfaces.

Alanine scanning and other site-directed mutagenesis methods have proven invaluable for probing particular aspects of protein function in individual cases. However, the accumulated database is extremely sparse in comparison with the overwhelming complexity of protein structure and function. Because alanine scanning deals with proteins on an individual basis, the method is by nature slow and laborious. To address these limitations, combinatorial shotgun scanning methods have been developed.

THE SHOTGUN SCANNING METHOD

Shotgun alanine scanning uses phage-displayed libraries in which positions of interest are varied as either the wt or Ala using specially designed degenerate codons (Table 4.1 and Figure 4.2). Using conventional DNA synthesis techniques, the nature of the genetic code necessitates two additional substitutions in the case of some amino acids, but, nevertheless, the additional substitutions do not affect the analysis [60].

The library is subjected to several rounds of two independent selections. One is a "function" selection to assess effects of alanine substitutions on binding to the ligand of interest and the second is a "structure'" selection for binding to an antibody that recognizes an epitope distinct from the mutated region. The structure selection provides a control data set that accounts for biases in the naïve library and for mutational effects that alter levels of protein display on phage. Following each selection, clones exhibiting specific binding to the selection target are subjected to DNA sequencing. By sequencing several hundred clones, the wt/Ala ratios at each position can be determined with statistical accuracy.

To estimate the effect of each alanine mutation on protein function, it is assumed that each wt/Ala ratio is equivalent to the ratio of the corresponding equilibrium binding constants ($K_{\text{a,wt}}/K_{\text{a,Ala}}$). With this assumption, a statistical $\Delta\Delta G$ value can be calculated for the function selection ($\Delta\Delta G_{\text{func}}$) and for the structure selection ($\Delta\Delta G_{\text{struct}}$) by substituting the wt/Ala ratio for the ratio of equilibrium binding constants in the standard equation: $\Delta\Delta G = RT \ln(K_{\text{a,wt}}/K_{\text{a,Ala}})$. Finally, the $\Delta\Delta G$ for the structure selection is used to correct the $\Delta\Delta G$ for the function selection, and the difference between the two is taken as an estimate of $\Delta\Delta G_{\text{Ala-wt}}$, the difference in binding free energy between alanine-substituted and wt protein for binding to the ligand of interest. To

TABLE 4.1
Shotgun Scanning Codons

	Alanin Scan[b]			Serine Scan[b]			Homolog Scan[c]	
wt[a]	Codon[d]	m2	m3	Codon[d]	m2	m3	Codon[d]	Homo
A				KCT			KCT	S
C	KST	G	S	WGT			WGT	S
D	GMT			KMC	A	Y	GAM	E
E	GMA			KMG	A	*	GAM	D
F	KYT	S	V	TYC			TWC	Y
G	GST			RGT			GST	G
H	SMT	D	P	MRC	N	R	MAC	N
I	RYT	T	V	AKC			RTT	V
K	RMA	E	T	ARM	N	R	ARG	R
L	SYT	P	V	TYG	P	V	MTC	I
M	RYG	T	V	AKS	T	V	MTG	L
N	RMC	D	T	ARC	D	T	RAC	D
P	SCA			YCT			SCA	A
Q	SMA	E	P	YMG	E	P	SAA	E
R	SST	G	P	MGT	G	P	ARG	K
S	KCC						KCC	A
T	RCT			WCG			ASC	S
V	GYT			KYT			RTT	I
W	KSG	G	S	TSG	G	S	TKG	L
Y	KMT	D	S	TMC	D	S	TWC	F

Note: For each scan, degenerate shotgun codons were designed to encode the wild-type amino acid and one or more substitutions.

[a] Amino acids are represented by the single-letter amino acid code.

[b] The shotgun codon for each amino acid ideally encodes only the wild type or one type of mutant, but the degeneracy of the genetic code necessitates the occurrence of two other amino acids (m2 and m3) for some substitutions. Asterisks (*) indicate a stop codon.

[c] For the homolog scan, binomial shotgun codons were designed to encode the wild type and a similar amino acid (Homo).

[d] Equimolar DNA degeneracies in shotgun codons are represented by the IUB code (K = G/T, M = A/C, R = A/G, S = G/C, W= A/T, Y = C/T).

visualize the functional epitope, the binding free energy values are mapped onto the three-dimensional protein structure.

EXAMPLES OF SHOTGUN SCANNING

Shotgun scanning has been used to analyze the functions of numerous proteins (Table 4.2), and in many cases, subsequent biophysical analysis has been used to confirm the accuracy of the results. In addition to alanine scanning, the method has been expanded to include other types of scans, including homolog and serine scanning.

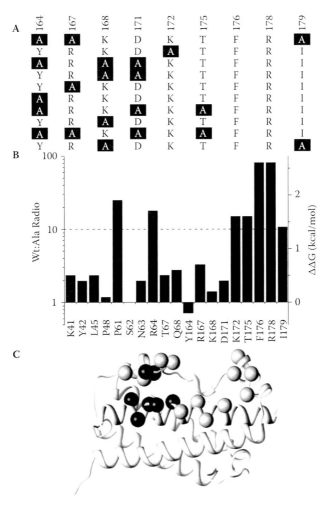

FIGURE 4.2   The shotgun scanning method. The illustration depicts alanine scanning analysis of hGH Site 1 for binding to the hGHR. (A) An hGH library with positions varied as the wt or alanine was selected for binding to the hGHR, and selected clones were sequenced and aligned. Ten representative sequences at nine scanned positions are shown with alanine mutations shaded in gray. (B) For each scanned position, the wt/Ala ratio was determined and used to assess the effect of alanine substitution as a statistical $\Delta\Delta G$ value by substituting the wt/Ala ratio for the $K_{a,wt}/K_{a,Ala}$ ratio in the standard thermodynamic equation: $\Delta\Delta G_{Ala-wt} = RT \ln(K_{a,wt}/K_{a,Ala}) = RT \ln(wt/Ala)$. For greater accuracy, the $\Delta\Delta G$ values can be corrected for effects on protein structure using data from a selection for binding to an antibody that recognizes an epitope distinct from the scanned region. (C) When mapped onto the structure of hGH, the shotgun scan results reveal a hot spot (black spheres) composed of a cluster of side chains that contribute most of the binding energy. Black or white spheres indicate scanned positions with $\Delta\Delta G_{Ala-wt}$ values greater than or less than 0.8 kcal/mol, respectively. The x-ray structure (PDB entry 3HHR) was depicted using the UCSF Chimera package from the Resource for Biocomputing, Visualization, and Informatics at the University of California, San Francisco [141].

TABLE 4.2

Proteins Analyzed by Shotgun Scanning

Scanned Protein	Binding Partner	Number of Scanned Positions	Substitutions	Refs.
hGH	hGHR	35	Alanine	60
			Serine	62
			Homolgo	62
			Comprehensive	67
Affinity-matured hGH variant	hGHR	35	Alanine	64
			Serine	62
			Homolgo	62
Gene-8 major coat protein	Phage coat	50	Alanine	68, 69, 71
Gene-3 minor coat protein	Phage coat	150	Alanine	70
Antibody	ErbB2	61	Alanine	61
			Homolog	61
Antibody	VEGF	30	Homolog	80
Antibody	BR3	40	Alanine	81
BR3	Antibodies (4)	22	Alanine	81
BR3	BAFF	22	Alanine	93
BCMA	BAFF, APRIL	25	Alanine	97
TACI	BAFF, APRIL	31	Alanine	94
Erbin PDZ domain	Peptide	44	Alanine	101
			Homolog	101
Peptide	IGF-1	11	Alanine	103
Peptide	EF-Tu	20	Homolog	104
Caveolin-1	Protein kinase A	20	Homolog	107
	eNOS	20	Homolog	108
Streptavidin	Biotin	30	Alanine	109
Engrailed homeodomain	Specific DNA	30	Alanine	121
			Homolog	121
SGP1-1; SGPI-2	Trypsin	18	Paralog	126
EntB	EntF	18	Alanine	127
PPARγ	SRC-1, TRAP220	14	Alanine	129
SRC-1	PPARγ	7	Alanine	129
EGF	EGF receptor	33	Ortholog	131

Furthermore, combinatorial data has been used to detect intramolecular cooperativity by analysis of double-alanine frequencies. A related quantitative saturation (QS) scanning approach has also been developed to assess the effects of all possible mutations using spatially restricted libraries. The following sections describe the major shotgun scanning studies that have been reported to date.

Shotgun Scanning of hGH

The hGH system was the first to be characterized by shotgun scanning and it provides examples of all of the major variations on the method. In the first shotgun alanine scanning experiment, 19 residues were scanned in Site 1 of hGH for binding to the hGHR, and the results were compared to the previous results of conventional alanine scanning [60]. Importantly, the two data sets were virtually identical, and thus shotgun scanning was shown to accurately map binding energetics without requiring time-consuming biophysical analyses. Subsequently, several additional studies used alternative shotgun scanning approaches to further dissect the function of hGH.

Homolog and Serine Scanning

Since shotgun alanine scanning proved to be very efficient for mapping the hGH functional epitope, other mutagenesis schemes were applied to provide alternative views [61]. The removal of a side chain by alanine substitution is a drastic mutation, and so, homolog scanning with chemically similar substitutions was developed as a more subtle probe of function. In addition, because alanine substitutions introduce apolar groups at the interface, it can be argued that the method overemphasizes the importance of polar side chains. Thus, serine scanning was developed to test the effects of replacing side chains with the smallest polar side chain.

Serine scanning was applied to 35 residues in Site 1 of hGH and, in general, the results were found to track with those of alanine scanning (Figure 4.3) [62]. Thus, it was concluded that serine and alanine scanning are equally effective for assessing side chain contributions to binding. The analysis also showed that the burial of polar serine residues at the interface is no more detrimental than the burial of apolar alanine residues. Thus, it appears that serine is a versatile side chain capable of making a wide variety of packing interactions, and this conclusion agrees with the finding that serine is highly prevalent in the combining sites of antibodies [63–67].

An analogous homolog scan of hGH also provided insights into the nature of the binding site. As expected, most homologous substitutions were much less deleterious than alanine or serine substitutions (Figure 4.3). However, none of the homologous substitutions across the binding site caused a substantial improvement in binding. This suggests that the site is already optimized within the narrow scope of the chemically similar sequence space explored by homolog scanning. Indeed, affinity maturation studies have shown that improved affinity requires substitutions that significantly alter the chemical character of the binding site [68,69]. Thus, it appears that subtle tweaking of the hGH binding site does not alter function appreciably, and, consequently, significant improvements in affinity require nonconservative mutations that cause large changes in the interface.

Shotgun Scanning of a High Affinity hGH Variant

Shotgun scanning has also been used to understand the basis for the improved affinity of an hGH variant (hGH$_v$), derived by *in vitro* evolution [63]. hGH$_v$ contains 15 mutations within Site 1 and binds to the hGHR approximately 400-fold tighter than the wt. The effects of alanine substitutions were determined for 35 residues that constitute

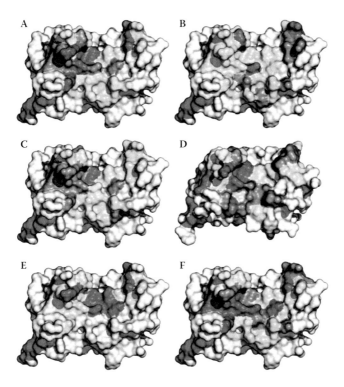

FIGURE 4.3 (SEE COLOR INSERT FOLLOWING PAGE 174.) Alternative views of hGH Site 1 for binding to the hGHR. The first four panels show the energetic effects of (A) alanine, (B) homolog or (C) serine substitutions on hGH, or (D) alanine substitutions on a high affinity hGH variant. All maps were derived by shotgun scanning, except the alanine scanning map for hGH, which was derived by conventional site-directed mutagenesis. The residues are colored according to the $\Delta\Delta G_{mut-wt}$ values as follows: cyan < -0.4 kcal/mol; -0.4 kcal/mol $\leq$ green < 0.4 kcal/mol; 0.4 kcal/mol $\leq$ orange < 1.0 kcal/mol; red ≥ 1.0 kcal/mol; gray untested. Panel E shows the results of a double-alanine frequency analysis of shotgun scanning data to detect cooperativity among 19 side chains. Scanned positions are colored red or green, and red indicates residues predicted to exhibit cooperativity with at least two other residues. Panel F illustrates the results of quantitative saturation scanning, which assesses the tolerance to all possible mutations. The residues are colored according to *SI* values, as follows: cyan < -2; $-2 \leq$ green < 3; $3 \leq$ yellow < 6; red ≥ 6. Larger *SI* values indicate positions that are less tolerant to substitution, and thus are important for binding. The x-ray structures of hGH and the high affinity variant (PDB entries 3HHR and 1kf9, respectively) were rendered in Pymol (DeLano Scientific, San Carlos, CA).

or closely border the binding interface, and the distribution of binding energy was found to differ significantly from that of the wt (Figure 4.3D) [64]. Although the hot spot residues of the wt were still important, their contributions were attenuated and additional binding energy was acquired from residues on the periphery of the hot spot. Side chains that inhibited binding of wt hGH were substituted by side chains that made positive contributions in the new interface. Interestingly, some side chains that were not mutated nevertheless acquired more important functional roles in the high affinity variant. Taken together, these changes produced an expanded and diffused hot spot in which improved affinity resulted from numerous small contributions distributed broadly across the interface. The results were consistent with structural studies, which revealed widespread differences between the wt and variant hormone–receptor interfaces [65]. Clearly, the improved function of hGH_v was achieved through wholesale structural reconfiguration of binding elements rather than through minor adjustments in the wt interface.

Assessing Intramolecular Cooperativity

Alanine scanning provides information on the functional roles of individual residues, but it does not assess whether there is cooperativity between residues. To detect cooperativity, double mutation cycles are required so that the effects of two individual mutations can be compared to the effect of the corresponding double mutant [71–73]. In the absence of cooperativity, the system is additive and the sum of the effects of individual mutations equals the effect of the double mutant. Although double mutation cycles offer a rigorous means for detecting cooperativity, the number of mutants required for a comprehensive analysis makes the approach impractical for large binding sites.

With shotgun alanine scanning, it has been shown that the same data set used for assessing individual residue contributions can be used to detect intramolecular cooperativity by pairwise correlation analysis of double-alanine mutants, as the occurrence of double-alanine mutations is positively or negatively influenced by positive or negative cooperativity, respectively [66]. The validity of this approach was demonstrated for hGH, using the shotgun data set originally acquired for mapping individual contributions of 19 residues in Site 1 [60]. More than 700 unique sequences were analyzed and statistically reliable assessments of additivity were calculated for 144 of 171 residue pairs by comparing predicted and actual counts of double-alanine occurrences. Unfortunately, the remaining pairs could not be evaluated due to a low occurrence of double-alanine mutations at combinations of functionally important residues. Nonetheless, the analysis revealed that the binding site is highly additive, as only 15 of the pairs exhibited evidence of cooperativity and even these pairs deviated less than twofold from additivity. Eight of the nineteen side chains were involved in two or more cooperative interactions, and, notably, five of these eight were charged residues, suggesting that many of the cooperative effects arise from electrostatic interactions (Figure 4.3E). Subsequent biophysical analysis showed that the predictions were accurate for five of six residue pairs tested, thus confirming the accuracy and sensitivity of the method.

Mapping Binding Sites in a Comprehensive Manner

hGH was also scanned with a QS scanning strategy designed to assess the structural and functional consequences of all possible mutations at 35 positions within Site 1 (Figure 4.3F) [67]. The 35 residues were divided into six nonoverlapping libraries containing five or six residues each, and libraries were restricted spatially rather than chemically. Each library contained only one of the six functionally most important residues defined by alanine scanning, and positions were assigned to libraries in a manner that maximized distances between residues in a common library. These design principles were intended to reduce cooperative interactions between proximal positions and to ensure that no single library contained a disproportionate number of hot spot residues. The positions were subjected to saturation mutagenesis using degenerate codons encoding all 20 natural amino acids. Each library was subjected separately to structural and functional selections in a manner analogous to shotgun alanine scanning, and several hundred clones were sequenced from each selection. In total, a database of approximately 2000 hGH variants was compiled and analyzed to reveal a comprehensive view of the binding site.

The information content of the data set was analyzed in terms of randomness of amino acid frequencies at each position using a parameter known as the transformed Shannon entropy (*TH*). The *TH* values vary between 1 (a totally conserved position with only one type of residue) and 20 (a totally random position with equal representation of all 20 residue types). The structure selection yielded high *TH* values across the surface, consistent with the supposition that surface residues do not play a major role in stabilizing the protein fold. However, the sequences were depleted in proline in helical regions, consistent with the helix-breaking properties of proline, and in cysteine at all positions, suggesting that cysteine residues may interfere with the native disulfides of hGH. Surprisingly, most positions were highly abundant in hydrophobic residues, and this observation contradicts the common assumption that protein stability is compromised by solvent-exposed hydrophobes. On average, the *TH* values for the function selection were lower than those for the structure selection, and this finding was consistent with the fact that the sequence requirements for functional hGH molecules are expected to be more constrained than the requirements for structure alone.

As both the structure and function selections require correct folding, the additional constraints imposed by function were quantified by a specificity index (*SI*) defined as the difference between the *TH* values for the structure and function selections. A positive mean *SI* value across the 35 scanned positions indicated that, as expected, receptor binding imposed constraints beyond those imposed by structure, and a large standard deviation indicated that these were position-specific constraints.

The *SI* values were mapped onto the structure of hGH, in a manner analogous to $\Delta\Delta G_{Ala-wt}$ for alanine scanning, to visualize the tolerance of the binding site to mutational pressure (Figure 4.3F). The *SI* is an extremely robust probe of the energy surface, and it is significantly more powerful than alanine scanning for assessing the functional adaptability of the binding site. There is a general correspondence between the functional epitope defined by the two methods, as the high *SI* values superimpose on the alanine scanning hot spot residues, but the epitope defined by *SI* values is somewhat more expansive. This is because, at several high specificity positions, the preferred residue type is not the wt amino acid, and these positions offer potential for affinity maturation. In general, the hot spot residues defined by alanine scanning had the highest *SI* values indicating that these positions require the highest degree of specificity, and furthermore, the wt residue type is usually preferred. Large negative *SI* values were rare and likely indicated mutations that stabilize structure but inhibit function.

By comparing the frequencies of mutations relative to wt, it was possible to predict single-site substitutions that should improve affinity. These predictions were validated by biophysical analysis of mutant proteins, which confirmed that affinity was improved by all six mutations that were tested. Thus, QS scanning provided a full view of the functional adaptability of hGH Site 1, and the data was accurate enough to guide the design of improved variants by rational design.

QS scanning also provided insights into the nature of protein–protein interactions in general, and the results challenged several common assumptions about protein function and evolution. In particular, many apparently conservative substitutions were not tolerated, while many nonconservative changes were accommodated. Thus, the role of amino acids in molecular recognition is highly context dependent and cannot be reliably predicted on the basis of chemical character alone. Furthermore, sequence conservation across species proved to be a poor predictor of mutational tolerance, and thus evolutionary conservation does not necessarily signify that a residue is important for function, but, rather, may reflect other constraints imposed by biology. Taken together, the results indicated that the design of a functional hGH molecule based strictly on biophysical principles would be very different from that of the natural molecule based on evolutionary pressures.

Mapping the Assembly of the Filamentous Phage Coat

In an effort to better understand and improve the phage display platform, shotgun alanine scanning has been applied to the phage particle to study how pVIII and pIII assemble into the coat [68–71]. The filamentous phage particle is a long rod consisting of a single-stranded DNA (ssDNA) genome coated with approximately 2700 copies of the major coat protein (pVIII; Figure 4.4A) [72,73]. One end of the particle is capped with five copies each of the minor coat proteins pVII and pIX and the other end is capped with five copies each of the minor coat proteins pIII and pVI [74]. In a bacterial host, coat proteins insert spontaneously into the inner membrane and ssDNA is recruited to an assembly pore composed of nonstructural viral proteins (pI, pIV, and pXI; Figure 4.4B). At the assembly site, the ssDNA is extruded through the pore and concomitantly surrounded by coat proteins, and, in this way, assembled phage particles are secreted into the extracellular environment without lysis of the host cell.

The length of the phage coat consists of interlocking layers of pVIII molecules arranged around the ssDNA in a symmetrical array (Figure 4.5A) [72,75–77]. Each pVIII molecule makes extensive contacts with other pVIII molecules in the three layers below and in the three layers above, but only minor contacts with other pVIII molecules within the same layer [78]. Shotgun alanine scanning was used to identify the important residues required for the incorporation of pVIII into the phage coat. To enable selection for pVIII incorporation, an hGH–pVIII fusion protein was used in a system that resulted in the display of only a few copies of the fusion protein in a coat composed predominantly of wt pVIII molecules. The entire pVIII sequence was scanned, and selection for hGH display was used as a proxy for incorporation into the phage coat. The analysis predicted that only nine nonalanine side chains were required for efficient incorporation. Indeed, simultaneous alanine substitutions for all side chains except these nine produced a "mini-pVIII" that incorporated into the phage coat almost as efficiently as the wt (Figure 4.5B). When mapped onto the

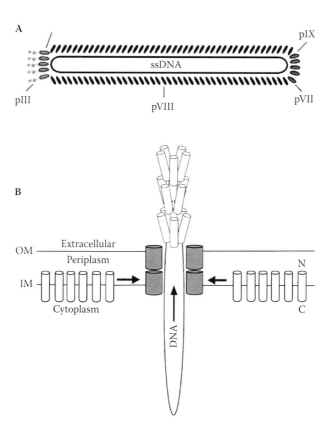

FIGURE 4.4 Filamentous bacteriophage structure and assembly. (A) The filamentous phage particle is a long rod consisting of an ssDNA genome encapsulated in a coat composed of the major coat protein, pVIII. One end of the particle is capped by the minor coat proteins pVII and pIX, and the other end is capped with pIII and pVI. (B) In the bacterial host, coat proteins (white cylinders) insert spontaneously into the inner membrane with the N and C termini located in the periplasm or cytoplasm, respectively. Genomic ssDNA is recruited to an assembly site, where it is extruded through a pore (gray cylinders) and concomitantly surrounded by coat proteins. In this way, the assembled phage particles are secreted into the extracellular environment without lysis of the host cell.

structure of pVIII, the nine side chains form three distinct epitopes (Figure 4.5C). Two of these epitopes are hydrophobic patches at either end of the molecule, which interlock with analogous patches on neighboring pVIII molecules. The third epitope is a basic patch consisting of three lysines near the bottom of the molecule, which interacts with the negatively charged DNA core. The analysis revealed that despite the complex structure of the phage coat, the assembly process is driven by only a few protein–protein and protein–DNA interactions mediated by hot spot clusters in the pVIII helix. Furthermore, the relaxed requirements for incorporation into the phage coat suggested that it may be possible to evolve mutant or even nonnatural coat proteins as improved platforms for phage display, and indeed, both of these suppositions have been verified [79].

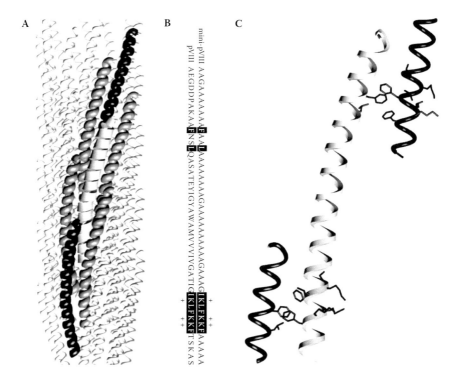

FIGURE 4.5 Shotgun alanine scanning of the phage coat. (A) The length of the phage coat consists of interlocking layers of pVIII molecules arranged around the ssDNA in a symmetrical array. Each pVIII molecule makes extensive contacts with other pVIII molecules in the three layers below and in the three layers above, but only minor contacts with other pVIII molecules within the same layer. (B) Based on shotgun alanine scanning analysis, a "minipVIII" variant with only nine nonalanine side chains was shown to contain all of the structural information necessary for efficient incorporation into the phage coat. (C) A structural model of mini-pVIII reveals that the nine nonalanine side chains form three distinct epitopes. Two of these epitopes are hydrophobic patches at either end of the molecule, which interlock with analogous patches on neighboring pVIII molecules in the phage coat. The third epitope is a negatively charged patch consisting of three lysines near the C terminus, which interacts with the negatively charged DNA core.

An analogous approach was used to decipher the functional epitope responsible for the assembly of the minor coat protein pIII [70]. pIII consists of three domains; the first two are required for host recognition and infection, but only the C-terminal domain is required for incorporation into the phage coat. Shotgun alanine scanning of the 150-residue C-terminal domain revealed that only 24 residues located among the last 70 positions were necessary for the incorporation of pIII into the phage coat. Thus, despite considerable differences in size and structure, both pIII and pVIII rely on only a small set of residues to enable assembly into the phage coat. These findings suggest that heterologous proteins may be readily recruited into virions to enable rapid evolution of new viral functions such as host range expansion.

EXPLORING HOW ANTIBODIES RECOGNIZE ANTIGENS

Antibodies are professional binding proteins produced by the immune system to recognize and neutralize foreign molecules or antigens. Billions of antibodies with unique specificities are present in vertebrate immune repertoires, but unlike most natural proteins that have evolved over millions of years, functional antibodies evolve over a span of weeks. The study of antibody structure and function has contributed significantly to our understanding of molecular recognition. Furthermore, antibodies are ideally suited for phage display, and several shotgun scanning studies have focused on antigen-binding sites to shed light on the mechanisms involved in antigen recognition [61,80,81].

The first antibody studied by shotgun scanning was a monoclonal antibody that binds to human ErbB2, a member of the epidermal growth factor receptor family that has been implicated in the progression of certain cancers [61]. The efficient shotgun scanning method enabled comprehensive analysis of the antigen-binding site by both alanine and homolog scanning. In total, 60 residues were scanned and mapping of the alanine scanning data onto the x-ray crystal structure revealed that the solvent-exposed functional epitope involves only heavy chain residues (Figure 4.6A). Alanine scanning also identified a number of buried residues that act as scaffolding to hold the functional epitope in a binding-competent conformation. The homolog scan further refined the view of the functional epitope afforded by alanine scanning. In particular, the functional epitope defined by homologous substitutions was roughly half the size of that defined by alanine substitutions, and it was concluded that this smaller subset of essential side chains may be involved in precise contacts with the antigen. The validity of the shotgun scanning results was subsequently verified by structure elucidation of the antibody in complex with antigen, demonstrating that the shotgun scanning approach can provide valuable insights into molecular recognition even in advance of structural analysis [82].

Phage display can be used to derive antibodies from "synthetic" libraries with manmade binding sites. In another application of the restricted diversity concept, synthetic antibodies were evolved with binding sites restricted to a tetranomial code (tyrosine, serine, alanine, and aspartate) [83]. Subsequently, the diversity was restricted even further to only a binary combination of tyrosine and serine, which nevertheless proved sufficient for the recognition of diverse antigens [84]. Notably, the binary code was also shown to be effective for generating binding sites supported by a small fibronectin domain scaffold [85]. Structural analysis of several of these minimalist binding sites revealed that tyrosine dominates the interface contacts, while small residues allow for space and conformational flexibility [80,83–86]. One of the antibodies, derived for binding to vascular endothelial growth factor (VEGF), was subjected to shotgun scanning to further explore the mechanisms of antigen recognition [80]. Truncation scanning with short side chains showed that tyrosine residues provide most of the binding energy, but homolog scanning revealed that most of the tyrosines could be replaced with phenylalanine without affecting binding (Figure 4.6B). Furthermore, saturation scanning showed that affinity could be improved by substituting several tyrosines with other amino acid types. Taken together, these results showed that tyrosine is particularly well-suited for naïve

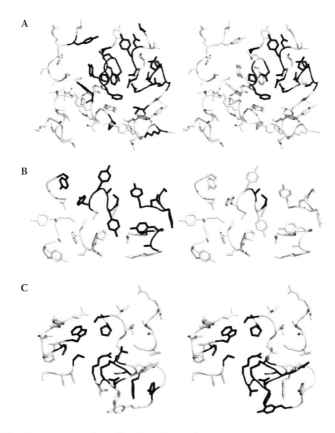

FIGURE 4.6 Shotgun scanning of antibodies. Main chains are shown as tubes and scanned side chains are shown as sticks. Side chains for which substitutions were predicted to reduce binding significantly (>5-fold) are colored gray while others are colored white. (A) Mapping of alanine (left) and homolog (right) scanning data onto the structure of an anti-ErbB2 antigen-binding fragment (Fab; PDB entry 1L7I). (B) Mapping of truncation (left) and homolog (right) scanning data onto the structure of an anti-VEGF Fab (PDB entry 1TZI). (C) Mapping of alanine scanning data for binding to human (left) or mouse (right) antigen onto the structure of an anti-BR3 Fab (PDB entry 2HFG). The x-ray structures were rendered in Pymol (DeLano Scientific, San Carlos, CA).

antigen recognition because the side chain is able to mediate many diverse interactions, but subsequent replacement of tyrosine by residues that improve key contacts may be useful for affinity maturation.

Shotgun scanning has also been used to understand cross-reactivity in antigen recognition by an antibody that recognizes both human and mouse versions of a cell-surface receptor (BR3) involved in B-cell activation [81]. Alanine scanning of the antibody revealed a common hot spot that interacts with a conserved epitope on the human and mouse antigens. Moreover, the conserved epitope on BR3 is also the interaction site for the natural ligand BAFF, and structural analysis showed that the antibody mimics the binding site of BAFF in terms of topology and chemistry. However, aside from the common hot spot, the antibody uses distinct auxiliary regions of its binding site to recognize either human or mouse BR3 (Figure 4.6C). Since the antibody was derived by phage display through a multistep process, the evolution of the binding site could be tracked, and it was found that the common hot spot arose first and additional residues were subsequently recruited during affinity maturation against the different antigens.

Mapping Cross-Reactivity in a Complex Receptor–Ligand Network

The cell-surface receptor BR3 is part of a complex network of cross-reactive receptors and ligands that mediate B-cell maturation and activation [87–92]. In addition to BR3, the system involves two other receptors (BCMA and TACI) and two ligands (BAFF and APRIL). Whereas the receptors BCMA and TACI recognize both ligands, BR3 recognizes only BAFF. Because the signaling pathways regulated by this receptor–ligand system are of considerable therapeutic interest, shotgun scanning was used to map the functional epitopes of all three receptors for binding to their respective ligands.

Shotgun alanine scanning of BR3 was performed using a 26-residue "mini-BR3" fragment that was sufficient for binding to BAFF [93]. The scan revealed a focused functional epitope of seven residues residing mainly on a β-hairpin. These findings enabled further minimization of the BR3 protein by transplantation of the functional epitope into a structured β-hairpin. The resulting "bhp-BR3" peptide was crystallized in complex with trimeric BAFF and the structure revealed a convex epitope on BR3 bound to a cavity formed at the subunit interface of BAFF. Moreover, the BR3 epitope was centered on a "DXL" motif (Figure 4.7A) that was conserved among the three receptors, suggesting that BCMA and TACI likely recognize ligands through a similar structural mechanism. At the same time, sequence differences at other positions could also explain differences in receptor–ligand preferences. These predictions

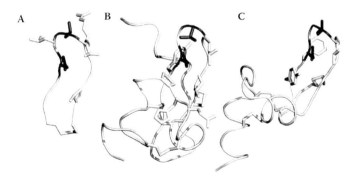

FIGURE 4.7 Shotgun alanine scanning of BR3, BCMA, and TACI. Main chains are shown as tubes and side chains for which substitutions were predicted to reduce binding significantly (>5-fold) are shown as sticks. The Asp and Leu residues of the conserved DXL motif are colored gray. The functional epitopes are shown for (A) the minimized bhp-BR3 binding to BAFF (PDB entry 1OSG), (B) BCMA binding to BAFF and APRIL (PDB entry 1xu2), and (C) TACI binding to BAFF and APRIL (PDB entry 1xut). X-ray structures were rendered in Pymol (DeLano Scientific, San Carlos, CA).

were confirmed by the elucidation of crystal structures for BCMA bound to BAFF and APRIL, and for TACI bound to APRIL [94–96].

Subsequently, shotgun alanine scanning was used to explore the interactions of BCMA and TACI with BAFF and APRIL [94,97]. BCMA binds tightly to APRIL but recognizes BAFF with low affinity. For binding to both ligands, the alanine scan revealed a common hot spot centered on the conserved DXL motif (Figure 4.7B). However, significant differences were also discovered between the two ligand interactions, and these differences were exploited to design an APRIL-selective BCMA mutant with no detectable affinity for BAFF [97]. TACI binds to both BAFF and APRIL with high affinity, and again, alanine scanning revealed that both ligands are recognized by a common functional epitope involving the DXL motif (Figure 4.7C). However, recognition of each of the two ligands also relied on different sets of residues located in a module near the C terminus of TACI. Structural analysis revealed that the two modules together form a concave binding site that mediates high affinity recognition of both ligands [94].

In summary, the comprehensive shotgun alanine scanning of all three receptors revealed both common elements and differences in their functional epitopes, and these findings in turn served to explain the patterns of cross-reactivity and selectivity for ligand recognition.

STUDIES OF PROTEIN–PEPTIDE INTERACTIONS

Many intracellular protein–protein interactions are mediated by modular domains that recognize discrete peptide motifs within proteins. These peptide-binding domains are usually imbedded in large multidomain proteins that are responsible for assembling protein networks [98]. Phage display has proven useful for exploring the specificity of peptide-binding domains, and in addition, phage display has proven useful for deriving peptide ligands against many diverse proteins directly from naïve peptide libraries [10]. In several studies, shotgun scanning techniques have been used to explore both sides of protein–peptide interfaces.

Analysis of the Erbin PDZ Domain

PDZ domains are modules that recognize the extreme C termini of other proteins and act as scaffolding to assemble intracellular complexes [99,100]. Most PDZ domains recognize ligands by a common mechanism, whereby a peptide inserts in an extended manner into a cleft located between a β-strand and an α-helix. To better understand the molecular basis for PDZ domain specificity, the PDZ domain of Erbin (Erbin PDZ) was subjected to detailed structural and functional analysis [101]. First, phage-displayed peptide libraries were used to derive an optimal C-terminal peptide ligand for Erbin PDZ (W^{-4}–E^{-3}–T^{-2}–W^{-1}–V^{0}), and, subsequently, the NMR structure of the complex was solved.

The interaction of the peptide with Erbin PDZ was also investigated by scanning mutagenesis. The ligand was subjected to conventional alanine scanning with synthetic analogs, and Erbin PDZ was subjected to shotgun alanine and homolog scanning (Figure 4.8A). The analysis confirmed that all five ligand side chains contribute favorably to the binding interaction. On the PDZ domain side of the interface, however, the scanning analysis revealed favorable side chain interactions with only three

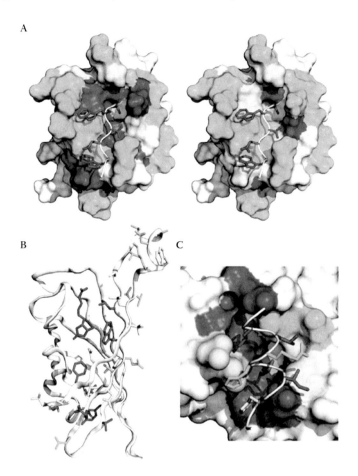

FIGURE 4.8 (SEE COLOR INSERT FOLLOWING PAGE 174.) Shotgun scanning of
proteins binding to peptides and small molecules. (A) Alanine (left) and homolog (right)
scanning data mapped onto the structure of Erbin PDZ bound to a peptide (PDB entry 1N7T).
Erbin PDZ is shown as a surface. The peptide main chain is shown as a tube and side chains
are shown as sticks. Residues are colored according to the predicted fold reduction in binding
due to substitution, as follows: green < 5; 5 ≤ yellow < 25; red ≥ 25. (B) Alanine scanning
data mapped onto the structure of a streptavidin monomer bound to biotin (PDB entry 1STP).
The main chain is shown as a tube and side chains are shown as sticks colored according to
the predicted fold reduction in binding due to substitution, as follows: green < 3; 3 ≤ yellow
< 9; red ≥ 9. Biotin is colored blue. (C) Alanine scanning data mapped onto the structure of
PPARγ bound to SRC1 (PDB entry 1RDT). PPARγ is shown as a surface. A peptide fragment of SRC1 is shown with the main chain depicted as a tube and the side chains depicted
as sticks. The residues are colored according to the predicted fold reduction in binding due to
substitution, as follows: green < 3; 3 ≤ yellow < 10; red ≥ 10. X-ray structures were rendered
using Pymol (DeLano Scientific, San Carlos, CA).

of the ligand side chains (Val^0, Thr^{-2}, and Trp^{-4}). For the other two positions (Trp^{-1} and Glu^{-3}) it appeared that the ligand side chains interacted mainly with the PDZ domain main chain, and, in fact, alanine scanning suggested that several side-chain contacts inhibit binding. Thus, it was suggested that, unlike hormone–receptor and antibody–antigen interfaces that are dominated by side-chain interactions, protein–peptide interactions apparently rely on both side-chain and main-chain interactions.

Analysis of Peptide Ligands

Shotgun scanning can also be applied to the rapid generation and analysis of peptide ligands, which are typically displayed as fusions to pVIII. In one study, naïve phage-displayed libraries were first used to derive a disulfide-constrained peptide ligand for insulin-like growth factor I (IGF-1) [102], and, subsequently, the ligand was analyzed by shotgun alanine scanning [103]. The scan revealed that roughly half of the peptide side chains are required for binding and many of these are located on a helical segment. Complementary NMR analysis suggested that many residues that contribute to function are also required for structural stabilization of the helix, showing that structural and functional effects are often coupled in small peptides.

In another study, naïve peptide-phage libraries yielded ligands for the *Escherichia coli* elongation factor Tu (EF-Tu), a protein that is essential for polypeptide translation and is involved in numerous natural interactions [104]. Perhaps because of its multifunctional nature, EF-Tu gave rise to multiple disulfide-constrained ligands with only limited sequence homology. One of these peptides was subjected to shotgun homolog scanning, and many of the resulting sequences were shown to recognize EF-Tu more effectively than the parent. Thus, homolog scanning was used simultaneously to obtain information about the binding interaction and to derive more effective peptide ligands. EF-Tu is an antibiotic target, but surprisingly, the peptide ligands did not compete with several antibiotics for binding to EF-Tu, suggesting that phage display may have targeted a hitherto uncharacterized binding site on the protein.

Analysis of the Caveolin-1 Scaffolding Domain

Caveolin-1 is a membrane-associated intracellular protein that oligomerizes to form caveolae, flask-shaped invaginations in the plasma membrane [105,106]. Although the structure and function of caveolin-1 is complex, a small "scaffolding" domain has been shown to mediate homooligomerization and several other protein–protein interactions. The 20-residue caveolin scaffolding domain (CSD) was displayed as a peptide on phage and the entire sequence was subjected to shotgun homolog scanning to investigate the interaction between caveolin-1 and the catalytic subunit of protein kinase A (PKAcat) [107]. Mutations at only four positions were predicted to be deleterious for binding and mutations at four other positions were predicted to improve binding, suggesting that the affinity of the interaction between CSD and PKAcat can be readily improved. The mutagenesis data were also used as constraints in computational docking experiments to help define plausible structural solutions for the interaction.

In a subsequent study, the same shotgun homolog scanning approach was used to study the interaction between CSD and another natural ligand, endothelial nitric oxide synthase (eNOS) [108]. The data set from this scan was compared to that from the scan against PKAcat to identify similarities and differences between the binding sites for the two ligands. At six positions, both scans showed similar preferences for the wt CSD sequence, suggesting that these side chains are important for recognition of both ligands. At five other positions, both ligands appeared to prefer the mutant sequence, suggesting that mutations at these positions may improve ligand binding in a general way. Finally, at a third set of seven positions, the two data sets showed marked differences, suggesting that PKAcat and eNOS interact differently with CSD residues at these positions. Taken together, these results suggest that PKAcat and eNOS utilize both common and unique interactions to recognize overlapping binding sites on CSD.

MAPPING THE STREPTAVIDIN–BIOTIN INTERACTION

The interaction between streptavidin and biotin is the tightest noncovalent interaction known in nature [109]. The interaction has proven to be highly useful for affinity labeling applications and has served as a model system for understanding how proteins recognize small molecules. Streptavidin is a homotetramer and the binding contributions of residues in direct contact with biotin have been investigated by conventional site-directed mutagenesis [110–114]. To further explore the mechanisms responsible for high affinity biotin recognition, shotgun alanine scanning was applied to 38 residues, including second-sphere residues that are not directly in the binding site. Because high affinity binding requires tetrameric structure, selections for binding to biotin were sensitive not only to direct effects on the binding pocket, but also to indirect effects on the quaternary structure. The study reiterated results for some previously analyzed residues and revealed a complex network of hydrophobic residues that serve to buttress the biotin binding site and help to align key contact residues (Figure 4.8B). In addition, it was hypothesized that other residues act to strengthen the interactions between subunits to establish a stable tetrameric structure. Overall, the large-scale shotgun scanning analysis facilitated the exploration of areas far from the binding site and established a more comprehensive view of the structure–function requirements for high affinity recognition of small molecules by proteins.

EXPLORING THE BASIS FOR AFFINITY AND SPECIFICITY IN A DNA-BINDING PROTEIN

The engrailed homeodomain recognizes a specific DNA sequence. The interaction has been investigated by structural analysis, site-directed mutagenesis, and *in vitro* selection experiments [115–120]. These studies mapped the subset of residues that contact DNA and are required for binding, but the role of noncontacting residues remained unclear. To provide a more comprehensive view of the interaction, 30 residues were subjected to shotgun alanine scanning and 15 of these were also subjected to homolog scanning [121]. The scans showed that many residues could be readily replaced without affecting function, and some substitutions even resulted in slight improvements to affinity or specificity. However, approximately one-third of the positions were intolerant to substitutions, and these were mainly scaffolding residues involved in maintaining the proper orientation of residues in direct contact with DNA. In particular, residues that were intolerant to alanine substitutions either were buried in the hydrophobic core or were part of a hydrophobic network that supports the formation of a sharp turn between two helices. In addition, the patterns of sequence conservation among selected clones were in good agreement with consensus patterns observed in alignments of human homeodomains, and it was interesting that disease-related natural mutations often occur at positions that were conserved in the scans. These results showed that the sequence constraints for homeodomain function, as defined by shotgun scanning, are also reflected in the evolutionary history of the natural system.

DETECTING COOPERATIVITY IN A PROTEASE INHIBITOR

Two serine protease inhibitor paralogs, *Schistocerca gregaria* serine protease inhibitor-1 (SGPI-1) and -2 (SGPI-2), share high sequence identity but exhibit different inhibition profiles. Both proteins inhibit arthropod trypsins, but only SGPI-2 inhibits mammalian trypsins [122–125]. The 35-residue proteins differ at 18 positions and a shotgun "paralog" scanning strategy was devised to determine the basis for the differing specificities. The two sequences were shuffled to produce all possible chimeras, and this library was selected for binding to either arthropod or mammalian trypsin [126]. A comparison of the results from the two scans revealed different sequence patterns that were likely responsible for the differing specificities. The analysis also revealed significant covariance between certain positions, suggesting that these positions may function in a cooperative manner. In particular, it was found that elements of the hydrophobic core are functionally coupled with a surface loop. This predicted functional coupling, so far unique among reversible protease inhibitors, was verified using point mutations. Thus, the rapid paralog scanning strategy was able to detect complex cooperative relationships that would be impossible to elucidate with conventional site-directed mutagenesis methods.

SHOTGUN SCANNING BEYOND PHAGE DISPLAY

Shotgun scanning was first developed with phage display, and this remains the predominant platform. However, the conceptual basis of the method is compatible with any combinatorial technology that allows for the selection of functional variants in an exhaustive and defined manner. Indeed, shotgun scanning was in part derived from an earlier "binomial" mutagenesis strategy that relied on a survival selection inside cells [66]. In recent years, several studies have used shotgun scanning strategies with other selection techniques, including survival selections, protein complementation assays, and yeast surface display. These alternative approaches extend shotgun scanning to protein systems that are not suitable for phage display and further expand the utility of the method.

SHOTGUN SCANNING BY SURVIVAL SELECTION

The interactions between several components of the *E. coli* enterobactin synthesis pathway were studied using an *in vivo* selection based on the fact that enterobactin is necessary for survival under iron-depleted conditions [127]. Enterobactin is synthesized by a nonribosomal peptide synthetase system through a multistep reaction pathway in which biosynthetic intermediates are covalently linked to carrier proteins. One such carrier protein domain, EntB-ArCP, interacts with two other proteins (EntE and EntF). Shotgun alanine scanning was used to analyze a proposed interaction site on EntB-ArCP, and 5 of 18 scanned residues were found to be important, as judged by conservation of the wt sequence in variants that permitted survival in iron-depleted media. Subsequent *in vitro* analysis of point-mutated proteins revealed that the conserved residues affected the interaction of EntB-ArCP with EntF but not with EntE. This study showed that shotgun scanning can be used *in vivo*, provided that a vital cellular function can be exploited for selection. However, the study also highlighted that, without a parallel selection for structure, individual variants need to be analyzed to assess whether a given trait is due to effects on function or structure.

SHOTGUN SCANNING BY PROTEIN COMPLEMENTATION ASSAYS

In the dihydrofolate reductase (DHFR) protein complementation assay, two complementary fragments of DHFR are fused to two proteins of interest, and functional DHFR only forms if the proteins interact [128]. Recently, the method was combined with shotgun scanning technology to probe the interactions between the transcription factor PPARγ and two coactivators (SRC1 and TRAP220) [129]. All together, 14 or 12 positions in PPARγ were alanine scanned for binding to SRC1 or TRAP220, respectively. The analysis revealed that a common set of six residues was required for binding to both ligands, but other residues appeared to function in a ligand-selective manner. In a complementary experiment, shotgun alanine scanning was applied to a peptide representing the binding region of SRC1, and it was found that binding was mediated predominantly by an "LXXLL" motif, which is common to both SRC1 and TRAP220, but also utilized several residues that are not conserved among the two ligands. Mapping of the alanine scanning data onto the structure of the PPARγ-SRC1 complex revealed that the functionally important residues of the binding partners interact at the interface (Figure 4.8C). An assay for correct protein folding, utilizing fusions of PPARγ with a green fluorescent protein reporter [130], was used to independently assess the effects of mutations on protein expression and stability. Five residues were found to be important for stability, and these were also hot spot residues for coactivator binding. As the protein complementation assay is applicable to many different proteins, this method should be of general use for mapping protein–protein interactions in an *in vivo* environment.

SHOTGUN SCANNING BY YEAST DISPLAY

Shotgun scanning has also been demonstrated with protein libraries displayed on the surfaces of yeast cells. An ortholog scanning strategy for affinity maturation was designed to take advantage of the observation that, in previous studies, affinity was often improved by substitutions resembling variations among natural orthologs [131]. A library of human epidermal growth factor (EGF) variants was designed to incorporate sequence variations from orthologs in other species [132,133]. Yeast-displayed libraries were selected for binding to the EGF receptor and EGF variants with up to 30-fold improvements in affinity were obtained. It was concluded that since natural selection eliminates variations that are deleterious for structure and function, the use of libraries biased toward natural diversity might be an efficient and general means for improving function in protein families. In addition, the study also showed that shotgun scanning analysis in yeast may be a viable alternative for proteins that do not fold correctly in bacteria.

CONCLUSIONS AND FUTURE PERSPECTIVES

Shotgun scanning methods have been utilized in numerous studies to analyze protein function, and the results clearly demonstrate the efficiency and rigor of the approach. Moreover, the general concept has been extended beyond phage display by the use of other combinatorial methods that should enable many additional studies in the future. It is clear that combinatorial analysis with well-defined libraries and selections can be used to explore diverse protein functions in a rapid manner. In parallel, combinatorial libraries with restricted diversities have been used to derive synthetic proteins that are comparable to natural proteins in terms of function, and yet are simplified in terms of structure [134,135]. These synthetic proteins should be ideally suited for analysis by shotgun scanning approaches to reveal basic principles governing molecular recognition in protein–protein interfaces.

Because the main expense for shotgun scanning is DNA sequencing, wider adoption of the method should be enabled by recent advances in sequencing methods that have reduced cost and increased throughput by several orders of magnitude [136,137]. Furthermore, the generation of diverse, well-defined protein repertoires has been made routine by the development of optimized library construction methodologies [10,138] and new DNA synthesis strategies [139]. Finally, robotics and automation technologies are being applied to combinatorial selections, and should enable high-throughput generation and analysis of protein function [140]. In the near future, we envision that these technological advances will be integrated into an automated system that will enable the exponential acceleration of investigations into the principles governing protein structure and function.

ACKNOWLEDGMENTS

The work of Gábor Pál is partially funded by the Hungarian Scientific Research Fund OTKA K68408, and by the János Bolyai Research Fellowship.

REFERENCES

1. Gascoigne, N. R. J., and Zal, T. (2004) *Current Opinion in Immunology* **16**(1), 114–119.
2. Pawson, T., and Nash, P. (2000) *Genes & Development* **14**(9), 1027–1047.
3. Warren, A. J. (2002) *Current Opinion in Structural Biology* **12**(1), 107–114.
4. Andreeva, A., Howorth, D., Chandonia, J. M., Brenner, S. E., Hubbard, T. J. P., Chothia, C., and Murzin, A. G. (2008) *Nucleic Acids Research* **36**, D419–D425.

5. Murzin, A. G., Brenner, S. E., Hubbard, T., and Chothia, C. (1995) *Journal of Molecular Biology* **247**(4), 536–540.

6. Cunningham, B. C., and Wells, J. A. (1993) *Journal of Molecular Biology* **234**(3), 554–563.

7. Sidhu, S. S., Fairbrother, W. J., and Deshayes, K. (2003) *Chembiochem* **4**(1), 14–25.

8. Smith, G. P. (1985) *Science* **228**(4705), 1315–1317.

9. Smith, G. P., and Petrenko, V. A. (1997) *Chemical Reviews* **97**(2), 391–410.

10. Sidhu, S. S., Lowman, H. B., Cunningham, B. C., and Wells, J. A. (2000) *Methods Enzymol* **328**, 333–363.

11. Cohen, S. N., Chang, A. C. Y., Boyer, H. W., and Helling, R. B. (1973) *Proceedings of the National Academy of Sciences of the United States of America* **70**(11), 3240–3244.

12. Jackson, D. A., Symons, R. H., and Berg, P. (1972) *Proceedings of the National Academy of Sciences of the United States of America* **69**(10), 2904–2909.

13. Morrow, J. F., and Berg, P. (1972) *Proceedings of the National Academy of Sciences of the United States of America* **69**(11), 3365–3369.

14. Morrow, J. F., Cohen, S. N., Chang, A. C. Y., Boyer, H. W., Goodman, H. M., and Helling, R. B. (1974) *Proceedings of the National Academy of Sciences of the United States of America* **71**(5), 1743–1747.

15. Goeddel, D. V., Kleid, D. G., Bolivar, F., Heyneker, H. L., Yansura, D. G., Crea, R., Hirose, T., Kraszewski, A., Itakura, K., and Riggs, A. D. (1979) *Proceedings of the National Academy of Sciences of the United States of America* **76**(1), 106–110.

16. Itakura, K., Hirose, T., Crea, R., Riggs, A., Heyneker, H., Bolivar, F., and Boyer, H. (1977) *Science* **198**(4321), 1056–1063.

17. Hutchison, C. A., Phillips, S., Edgell, M. H., Gillam, S., Jahnke, P., and Smith, M. (1978) *Journal of Biological Chemistry* **253**(18), 6551–6560.

18. Kunkel, T. A. (1985) *Proceedings of the National Academy of Sciences of the United States of America* **82**(2), 488–492.

19. Kunkel, T. A., Bebenek, K., and Mcclary, J. (1991) *Methods in Enzymology* **204**, 125–139.

20. Smith, M. (1982) *Trends in Biochemical Sciences* **7**(12), 440–442.

21. Zoller, M. J., and Smith, M. (1983) *Methods in Enzymology* **100**, 468–500.

22. Winter, G., Fersht, A. R., Wilkinson, A. J., Zoller, M., and Smith, M. (1982) *Nature* **299**(5885), 756–758.

23. Wilkinson, A. J., Fersht, A. R., Blow, D. M., and Winter, G. (1983) *Biochemistry* **22**(15), 3581–3586.

24. Fersht, A. R., Shi, J. P., Wilkinson, A. J., Blow, D. M., Carter, P., Waye, M. M. Y., and Winter, G. P. (1984) *Angewandte Chemie-International Edition in English* **23**(7), 467–473.

25. Wilkinson, A. J., Fersht, A. R., Blow, D. M., Carter, P., and Winter, G. (1984) *Nature* **307**(5947), 187–188.

26. Fersht, A. R., Shi, J. P., Knilljones, J., Lowe, D. M., Wilkinson, A. J., Blow, D. M., Brick, P., Carter, P., Waye, M. M. Y., and Winter, G. (1985) *Nature* **314**(6008), 235–238.

27. Wells, T. N. C., and Fersht, A. R. (1985) *Nature* **316**(6029), 656–657.

28. Wells, J. A., Cunningham, B. C., Graycar, T. P., and Estell, D. A. (1986) *Philosophical Transactions of the Royal Society of London Series a-Mathematical Physical and Engineering Sciences* **317**(1540), 415–423.

29. Wells, J. A., and Powers, D. B. (1986) *Journal of Biological Chemistry* **261**(14), 6564–6570.

30. Leatherbarrow, R. J., and Fersht, A. R. (1986) *Protein Engineering* **1**(1), 7–16.

31. Wells, J. A., Bott, R. R., Powers, D. B., Ultsch, M. H., Power, S. D., Adams, R. M., Cunningham, B., Graycar, T. P., and Estell, D. A. (1986) *Journal of Cellular Biochemistry*, 246.

32. Wells, J. A., Powers, D. B., Cunningham, B. C., Bott, R. R., Graycar, T. P., and Estell, D. A. (1987) *Journal of Cellular Biochemistry*, 202.

33. Lau, F. T. K., and Fersht, A. R. (1987) *Nature* **326**(6115), 811–812.

34. Carter, P., and Wells, J. A. (1987) *Science* **237**(4813), 394–399.

35. Carter, P., and Wells, J. A. (1987) *Protein Engineering* **1**(3), 234.

36. Estell, D. A., Graycar, T. P., Power, S. D., Adams, R., Caldwell, R., Bott, R. R., Ultsch, M., Cunningham, B. C., and Wells, J. A. (1987) *Abstracts of Papers of the American Chemical Society* **194**, 191-MBTD.

37. Power, S. D., Adams, R., Caldwell, R., Cunningham, B. C., Wells, J. A., Graycar, T. P., and Estell, D. A. (1988) *Abstracts of Papers of the American Chemical Society* **195**, 4-Btec.

38. Wells, J. A., and Estell, D. A. (1988) *Trends in Biochemical Sciences* **13**(8), 291–297.

39. Carter, P., Nilsson, B., Burnier, J. P., Burdick, D., and Wells, J. A. (1989) *Proteins: Structure, Function, and Genetics* **6**(3), 240–248.

40. Cunningham, B. C., Henner, D. J., and Wells, J. A. (1990) *Science* **247**(4949), 1461–1465.

41. Ward, W. H. J., Timms, D., and Fersht, A. R. (1990) *Trends in Pharmacological Sciences* **11**(7), 280–284.

42. Bass, S. H., Cunningham, B. C., and Wells, J. A. (1990) *Abstracts of Papers of the American Chemical Society* **200**, 69-Biot.

43. Nicholson, H., Anderson, D. E., Daopin, S., and Matthews, B. W. (1991) *Biochemistry* **30**(41), 9816–9828.

44. Daopin, S., Soderlind, E., Baase, W. A., Wozniak, J. A., Sauer, U., and Matthews, B. W. (1991) *Journal of Molecular Biology* **221**(3), 873–887.

45. Abrahmsen, L., Tom, J., Burnier, J., Butcher, K. A., Kossiakoff, A., and Wells, J. A. (1991) *Biochemistry* **30**(17), 4151–4159.

46. Matouschek, A., and Fersht, A. R. (1991) *Methods in Enzymology* **202**, 82–112.

47. Wells, J. A. (1991) *Methods in Enzymology* **202**, 390–411.

48. Zhang, X. J., Baase, W. A., and Matthews, B. W. (1991) *Biochemistry* **30**(8), 2012–2017.

49. Hurley, J. H., Baase, W. A., and Matthews, B. W. (1992) *Journal of Molecular Biology* **224**(4), 1143–1159.

50. Baase, W. A., Eriksson, A. E., Zhang, X. J., Heinz, D. W., Sauer, U., Blaber, M., Baldwin, E. P., Wozniak, J. A., and Matthews, B. W. (1992) *Faraday Discussions* (93), 173–181.

51. Fersht, A., and Winter, G. (1992) *Trends in Biochemical Sciences* **17**(8), 292–294.

52. Schreiber, G., and Fersht, A. R. (1993) *Biochemistry* **32**(19), 5145–5150.

53. Fersht, A. R., and Serrano, L. (1993) *Current Opinion in Structural Biology* **3**(1), 75–83.

54. Ballinger, M. D., Tom, J., and Wells, J. A. (1995) *Biochemistry* **34**(41), 13312–13319.

55. Matthews, B. W. (1996) *FASEB Journal* **10**(1), 35–41.

56. Clackson, T., and Wells, J. A. (1995) *Science* **267**(5196), 383–386.

57. Cunningham, B. C., and Wells, J. A. (1989) *Science* **244**(4908), 1081–1085.

58. Bogan, A. A., and Thorn, K. S. (1998) *Journal of Molecular Biology* **280**(1), 1–9.

59. Thorn, K. S., and Bogan, A. A. (2001) *Bioinformatics* **17**(3), 284–285.

60. Weiss, G. A., Watanabe, C. K., Zhong, A., Goddard, A., and Sidhu, S. S. (2000) *Proceedings of the National Academy of Sciences of the United States of America* **97**(16), 8950–8954.

61. Vajdos, F. F., Adams, C. W., Breece, T. N., Presta, L. G., de Vos, A. M., and Sidhu, S. S. (2002) *Journal of Molecular Biology* **320**(2), 415–428.

62. Pal, G., Fong, S. Y., Kossiakoff, A. A., and Sidhu, S. S. (2005) *Protein Science* **14**(9), 2405–2413.

63. Lowman, H. B., and Wells, J. A. (1993) *Journal of Molecular Biology* **234**(3), 564–578.

64. Pal, G., Kossiakoff, A. A., and Sidhu, S. S. (2003) *Journal of Molecular Biology* **332**(1), 195–204.
65. Schiffer, C., Ultsch, M., Walsh, S., Somers, W., de Vos, A. M., and Kossiakoff, A. (2002) *Journal of Molecular Biology* **316**(2), 277–289.
66. Gregoret, L. M., and Sauer, R. T. (1993) *Proceedings of the National Academy of Sciences of the United States of America* **90**(9), 4246–4250.
67. Pal, G., Kouadio, J. L. K., Artis, D. R., Kossiakoff, A. A., and Sidhu, S. S. (2006) *Journal of Biological Chemistry* **281**(31), 22378–22385.
68. Roth, T. A., Weiss, G. A., Eigenbrot, C., and Sidhu, S. S. (2002) *Journal of Molecular Biology* **322**(2), 357–367.
69. Sidhu, S. S. (2001) *Biomolecular Engineering* **18**(2), 57–63.
70. Weiss, G. A., Roth, T. A., Baldi, P. F., and Sidhu, S. S. (2003) *Journal of Molecular Biology* **332**(4), 777–782.
71. Weiss, G. A., Wells, J. A., and Sidhu, S. S. (2000) *Protein Science* **9**(4), 647–654.
72. Marvin, D. A. (1998) *Current Opinion in Structural Biology* **8**(2), 150–158.
73. Vanwezenbeek, P. M. G. F., Hulsebos, T. J. M., and Schoenmakers, J. G. G. (1980) *Gene* **11**(1-2), 129–148.
74. Barbas, C. F., Burton, D. R., Scott, J. K., and Silverman, G. J. (2001) *Phage Display: A Laboratory Manual*, Cold Spring Harbor Laboratory Press, Cold Spring Harbor, New York.
75. Glucksman, M. J., Bhattacharjee, S., and Makowski, L. (1992) *Journal of Molecular Biology* **226**(2), 455–470.
76. Marvin, D. A., Hale, R. D., Nave, C., and Citterich, M. H. (1994) *Journal of Molecular Biology* **235**(1), 260–286.
77. Marvin, D. A., Welsh, L. C., Symmons, M. F., Scott, W. R. P., and Straus, S. K. (2006) *Journal of Molecular Biology* **355**(2), 294–309.
78. Papavoine, C. H. M., Christiaans, B. E. C., Folmer, R. H. A., Konings, R. N. H., and Hilbers, C. W. (1998) *Journal of Molecular Biology* **282**(2), 401–419.
79. Sidhu, S. S., Weiss, G. A., and Wells, J. A. (2000) *Journal of Molecular Biology* **296**(2), 487–495.
80. Fellouse, F. A., Barthelemy, P. A., Kelley, R. F., and Sidhu, S. S. (2006) *Journal of Molecular Biology* **357**(1), 100–114.
81. Lee, C. V., Hymowitz, S. G., Wallweber, H. J., Gordon, N. C., Billeci, K. L., Tsai, S. P., Compaan, D. M., Yin, J. P., Gong, O., Kelley, R. F., DeForge, L. E., Martin, F., Starovasnik, M. A., and Fuh, G. (2006) *Blood* **108**(9), 3103–3111.
82. Franklin, M. C., Carey, K. D., Vajdos, F. F., Leahy, D. J., de Vos, A. M., and Sliwkowski, M. X. (2004) *Cancer Cell* **5**(4), 317–328.
83. Fellouse, F. A., Wiesmann, C., and Sidhu, S. S. (2004) *Proceedings of the National Academy of Sciences of the United States of America* **101**(34), 12467–12472.
84. Fellouse, F. A., Li, B., Compaan, D. M., Peden, A. A., Hymowitz, S. G., and Sidhu, S. S. (2005) *Journal of Molecular Biology* **348**(5), 1153–1162.
85. Koide, A., Gilbreth, R. N., Esaki, K., Tereshko, V., and Koide, S. (2007) *Proceedings of the National Academy of Sciences of the United States of America* **104**(16), 6632–6637.
86. Fellouse, F. A., Esaki, K., Birtalan, S., Raptis, D., Cancasci, V. J., Koide, A., Jhurani, P., Vasser, M., Wiesmann, C., Kossiakoff, A. A., Koide, S., and Sidhu, S. S. (2007) *Journal of Molecular Biology* **373**(4), 924–940.
87. Haiat, S., Billard, C., Quiney, C., Ajchenbaum-Cymbalista, F., and Kolb, J. P. (2006) *Immunology* **118**(3), 281–292.
88. Mackay, F., and Leung, H. (2006) *Seminars in Immunology* **18**(5), 284–289.
89. Mackay, F., Silveira, P. A., and Brink, R. (2007) *Current Opinion in Immunology* **19**(3), 327–336.

90. Mackay, F., and Tangye, S. G. (2004) *Current Opinion in Pharmacology* **4**(4), 347–354.

91. Schneider, P. (2005) *Current Opinion in Immunology* **17**(3), 282–289.

92. Locksley, R. M., Killeen, N., and Lenardo, M. J. (2001) *Cell* **104**(4), 487–501.

93. Gordon, N. C., Pan, B., Hymowitz, S. G., Yin, J. P., Kelley, R. F., Cochran, A. G., Yan, M. H., Dixit, V. M., Fairbrother, W. J., and Starovasnik, M. A. (2003) *Biochemistry* **42**(20), 5977–5983.

94. Hymowitz, S. G., Patel, D. R., Wallweber, H. J. A., Runyon, S., Yan, M. H., Yin, J. P., Shriver, S. K., Gordon, N. C., Pan, B. L., Skelton, N. J., Kelley, R. F., and Starovasnik, M. A. (2005) *Journal of Biological Chemistry* **280**(8), 7218–7227.

95. Kim, H. M., Yu, K. S., Lee, M. E., Shin, D. R., Kim, Y. S., Paik, S. G., Yoo, O. J., Lee, H., and Lee, J. O. (2003) *Nature Structural Biology* **10**(5), 342–348.

96. Liu, Y. F., Hong, X., Kappler, J., Jiang, L., Zhang, R. G., Xu, L. G., Pan, C. H., Martin, W. E., Murphy, R. C., Shu, H. B., Dai, S. D., and Zhang, G. Y. (2003) *Nature* **423**(6935), 49–56.

97. Patel, D. R., Wallweber, H. J. A., Yin, J. P., Shriver, S. K., Marsters, S. A., Gordon, N. C., Starovasnik, M. A., and Kelley, R. F. (2004) *Journal of Biological Chemistry* **279**(16), 16727–16735.

98. Pawson, T., and Nash, P. (2003) *Science* **300**(5618), 445–452.

99. Hung, A. Y., and Sheng, M. (2002) *Journal of Biological Chemistry* **277**(8), 5699–5702.

100. Saras, J., and Heldin, C. H. (1996) *Trends in Biochemical Science* **21**(12), 455–458.

101. Skelton, N. J., Koehler, M. F. T., Zobel, K., Wong, W. L., Yeh, S., Pisabarro, M. T., Yin, J. P., Lasky, L. A., and Sidhu, S. S. (2003) *Journal of Biological Chemistry* **278**(9), 7645–7654.

102. Deshayes, K., Schaffer, M. L., Skelton, N. J., Nakamura, G. R., Kadkhodayan, S., and Sidhu, S. S. (2002) *Chemistry & Biology* **9**(4), 495–505.

103. Schaffer, M. L., Deshayes, K., Nakamura, G., Sidhu, S., and Skelton, N. J. (2003) *Biochemistry* **42**(31), 9324–9334.

104. Murase, K., Morrison, K. L., Tam, P. Y., Stafford, R. L., Jurnak, F., and Weiss, G. A. (2003) *Chemistry & Biology* **10**(2), 161–168.

105. Liu, P., Rudick, M., and Anderson, R. G. (2002) *Journal of Biological Chemistry* **277**(44), 41295–41298.

106. Williams, T. M., and Lisanti, M. P. (2004) *Genome Biology* **5**(3), 214.

107. Levin, A. M., Coroneus, J. G., Cocco, M. J., and Weiss, G. A. (2006) *Protein Science* **15**(3), 478–486.

108. Levin, A. M., Murase, K., Jackson, P. J., Flinspach, M. L., Poulos, T. L., and Weiss, G. A. (2007) *ACS Chemical Biology* **2**(7), 493–500.

109. Avrantinis, S. K., Stafford, R. L., Tian, X., and Weiss, G. A. (2002) *Chembiochem* **3**(12), 1229–1234.

110. Chilkoti, A., and Stayton, P. S. (1995) *Journal of the American Chemical Society* **117**(43), 10622–10628.

111. Chilkoti, A., Tan, P. H., and Stayton, P. S. (1995) *Proceedings of the National Academy of Sciences of the United States of America* **92**(5), 1754–1758.

112. Freitag, S., Le Trong, I., Chilkoti, A., Klumb, L. A., Stayton, P. S., and Stenkamp, R. E. (1998) *Journal of Molecular Biology* **279**(1), 211–221.

113. Klumb, L. A., Chu, V., and Stayton, P. S. (1998) *Biochemistry* **37**(21), 7657–7663.

114. Sano, T., and Cantor, C. R. (1995) *Proceedings of the National Academy of Sciences of the United States of America* **92**(8), 3180–3184.

115. Ades, S. E., and Sauer, R. T. (1994) *Biochemistry* **33**(31), 9187–9194.

116. Draganescu, A., and Tullius, T. D. (1998) *Journal of Molecular Biology* **276**(3), 529–536.

117. Kissinger, C. R., Liu, B. S., Martin-Blanco, E., Kornberg, T. B., and Pabo, C. O. (1990) *Cell* **63**(3), 579–590.
118. Connolly, J. P., Augustine, J. G., and Francklyn, C. (1999) *Nucleic Acids Research* **27**(4), 1182–1189.
119. Peltenburg, L. T., and Murre, C. (1996) *EMBO Journal* **15**(13), 3385–3393.
120. Peltenburg, L. T., and Murre, C. (1997) *Development* **124**(5), 1089–1098.
121. Sato, K., Simon, M. D., Levin, A. M., Shokat, K. M., and Weiss, G. A. (2004) *Chemistry & Biology* **11**(7), 1017–1023.
122. Fodor, K., Harmat, V., Hetenyi, C., Kardos, J., Antal, J., Perczel, A., Patthy, A., Katona, G., and Graf, L. (2005) *Journal of Molecular Biology* **350**(1), 156–169.
123. Gaspari, Z., Patthy, A., Graf, L., and Perczel, A. (2002) *European Journal of Biochemistry* **269**(2), 527–537.
124. Malik, Z., Amir, S., Pal, G., Buzas, Z., Varallyay, E., Antal, J., Szilagyi, Z., Vekey, K., Asboth, B., Patthy, A., and Graf, L. (1999) *Biochimica Et Biophysica Acta-Protein Structure and Molecular Enzymology* **1434**(1), 143–150.
125. Patthy, A., Amir, S., Malik, Z., Bodi, A., Kardos, J., Asboth, B., and Graf, L. (2002) *Archives of Biochemistry and Biophysics* **398**(2), 179–187.
126. Szenthe, B., Patthy, A., Gaspari, Z., Kekesi, A. K., Graf, L., and Pal, G. (2007) *Journal of Molecular Biology* **370**(1), 63–79.
127. Lai, J. R., Fischbach, M. A., Liu, D. R., and Walsh, C. T. (2006) *Proceedings of the National Academy of Sciences of the United States of America* **103**(14), 5314–5319.
128. Michnick, S. W. (2001) *Current Opinion in Structural Biology* **11**(4), 472–477.
129. Phillips, K. J., Rosenbaum, D. M., and Liu, D. R. (2006) *Journal of the American Chemical Society* **128**(34), 11298–11306.
130. Waldo, G. S., Standish, B. M., Berendzen, J., and Terwilliger, T. C. (1999) *Nature Biotechnology* **17**(7), 691–695.
131. Cochran, J. R., Kim, Y. S., Lippow, S. M., Rao, B., and Wittrup, K. D. (2006) *Protein Engineering Design & Selection* **19**(6), 245–253.
132. Stemmer, W. P. C. (1994) *Proceedings of the National Academy of Sciences of the United States of America* **91**(22), 10747–10751.
133. Stemmer, W. P. C. (1994) *Nature* **370**(6488), 389–391.
134. Sidhu, S. S., and Koide, S. (2007) *Current Opinion in Structural Biology* **17**(4), 481–487.
135. Sidhu, S. S., and Kossiakoff, A. A. (2007) *Current Opinion in Chemical Biology* **11**(3), 347–354.
136. Margulies, M., Egholm, M., Altman, W. E., Attiya, S., Bader, J. S., Bemben, L. A., Berka, J., Braverman, M. S., Chen, Y. J., Chen, Z. T., Dewell, S. B., de Winter, A., Drake, J., Du, L., Fierro, J. M., Forte, R., Gomes, X. V., Godwin, B. C., He, W., Helgesen, S., Ho, C. H., Hutchison, S. K., Irzyk, G. P., Jando, S. C., Alenquer, M. L. I., Jarvie, T. P., Jirage, K. B., Kim, J. B., Knight, J. R., Lanza, J. R., Leamon, J. H., Lee, W. L., Lefkowitz, S. M., Lei, M., Li, J., Lohman, K. L., Lu, H., Makhijani, V. B., McDade, K. E., McKenna, M. P., Myers, E. W., Nickerson, E., Nobile, J. R., Plant, R., Puc, B. P., Reifler, M., Ronan, M. T., Roth, G. T., Sarkis, G. J., Simons, J. F., Simpson, J. W., Srinivasan, M., Tartaro, K. R., Tomasz, A., Vogt, K. A., Volkmer, G. A., Wang, S. H., Wang, Y., Weiner, M. P., Willoughby, D. A., Yu, P. G., Begley, R. F., and Rothberg, J. M. (2005) *Nature* **437**(7057), 376–380.
137. Shendure, J., Porreca, G. J., Reppas, N. B., Lin, X. X., McCutcheon, J. P., Rosenbaum, A. M., Wang, M. D., Zhang, K., Mitra, R. D., and Church, G. M. (2005) *Science* **309**(5741), 1728–1732.
138. Lipovsek, D., and Pluckthun, A. (2004) *Journal of Immunology Methods* **290**(1-2), 51–67.
139. Neylon, C. (2004) *Nucleic Acids Research* **32**(4), 1448–1459.

140. Bradbury, A., Velappan, N., Verzillo, V., Ovecka, M., Chasteen, L., Sblattero, D., Marzari, R., Lou, J., Siegel, R., and Pavlik, P. (2003) *Trends in Biotechnology* **21**(7), 312–317.
141. Pettersen, E. F., Goddard, T. D., Huang, C. C., Couch, G. S., Greenblatt, D. M., Meng, E. C., and Ferrin, T. E. (2004) *Journal of Computational Chemistry* **25**(13), 1605–1612.

5 The Association of Protein–Protein Complexes

Gideon Schreiber

CONTENTS

OVERVIEW

The structure of a protein–protein interaction, its affinity, and thermodynamic characteristics depict a "frozen" state of a complex. This picture ignores the kinetic nature of complex formation and dissociation, which are of major biological and biophysical interest. In this chapter I focus on recent advances in describing the kinetics of protein–protein association, and how a combination of computational tools and experimental data helped us to decipher the pathway.

The rapid formation of specific interactions between proteins is important for many biological processes, including signal transduction and the immune response. For proteins to recognize one another and to interact, their interfaces have to be oriented toward one another at a highly specific conformation. This reaction, which occurs within the milieu of endless competing macromolecules, can be compared to two blind men finding each other in the streets of New York. Yet, it is done rapidly at rates of only one to five orders of magnitude below the Smoluchowski diffusion collision limit of 10^{10} $M^{-1}s^{-1}$. The rate of association of a protein complex is limited by diffusion and geometric constraints of the binding sites (diffusion control). Subsequent chemical processes may further slow the reaction process. Typical association rates are in the order of 10^5–10^6 $M^{-1}s^{-1}$, but rate constants of $>10^9$ $M^{-1}s^{-1}$ have been measured for interactions where the speed of the process is of functional importance. In these cases, strong favorable electrostatic forces enhanced the rate of association.

INTRODUCTION

The association reaction between two proteins can be viewed as a random process, where the rate of association is a function of diffusion limited collisions (which is defined from the Stokes–Einstein relation) divided by the chance of a collision to occur at the exact orientation that will lead to complex formation. For a diffusion-controlled reaction of two similar-sized particles, the rate of collision (k_1) is given by the Smoluchowski relation:

$$k_1 = 4\pi R D \tag{5.1}$$

with R being the sum of the effective radii of the particles, and D is the diffusion coefficient, which is calculated from:

$$D = \frac{k_B T}{6\pi\eta R} \tag{5.2}$$

Here, η is the relative viscosity (to water at 20°C). Taking Equations 5.1 and 5.2 together shows that k_1 is independent of the size of the proteins and inversely linear with η, as was indeed shown to be true experimentally using different types of crowding agents (Kuttner et al., 2005). For two spherical particles in water, Equation 5.1 predicts a rate of collision of ~10^{10} $M^{-1}s^{-1}$. However, the rate of association in the absence of electrostatic forces is only in the order of 10^4–$10^6 M^{-1}s^{-1}$. This means that only 1 out of 10^4–10^6 collisions will transform into a complex. This is not surprising, as binding involves the exact rearrangements of the two interfaces one relatively to the other, and thus involves a component of rotational diffusion. A basic mechanistic question is whether association can be viewed as a simple diffusion-limited reaction or whether it involves a reaction-limited component, accounting for desolvation and structural rearrangement of the interfaces. Moreover, is association a two-state reaction (unbound to bound) or do intermediates (also called encounter complexes

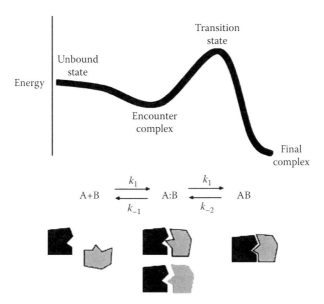

FIGURE 5.1   Free energy diagram describing the pathway of protein–protein binding. Two proteins (A and B) in solution will collide with one another at a rate dictated by diffusion to form an encounter complex, A:B, which following structural rearrangement and desolvation develops into the final complex, AB.

or transient complexes) play a role along the reaction path (Figure 5.1)? For association to be two-state, a random collision has to be sufficiently precise to promote the formation of correct short-range interactions found in the final complex. The random chance for two patches of 1 Å^2 to collide, assuming a total protein surface area of 2000 Å^2 is 1 to 4×10^6. However, complex formation requires many residues to interact simultaneously, reducing the change further. Thus, the assumption that protein–protein association is simply a diffusion process is not realistic. The most basic description for association (according to Figure 5.1) will be to divide the process into two parts, one that is diffusion limited (with a rate of k_1), at which the encounter (transient complex) is formed, and the second (with a rate of k_2) for the formation of the final complex. Under these assumptions the rate of association will be equal to:

$$k_{on} = \frac{k_1 k_2}{k_{-1} + k_2}$$

(5.3)

where k_{-1} is the dissociation rate of the encounter complex, and $K_1 = k_{-1}/k_1$ is the stability of the encounter complex. Under these conditions, k_2 represents the reaction-limited rate. Thus, if $k_2 \gg k_{-1}$, than $k_{on} = k_1$.

An early attempt to satisfy the observed rate of association using computer simulations was done by Northrup and Erickson (1992), who tried to explain the relatively faster rates of association by assuming that binding is speeded up by multiple collisions of large bodies that are proximate to each other (in the order of 10) and that association can be assumed to occur when two to three correct interprotein interactions are formed. These assumptions were used to simulate binding using Brownian dynamics simulations (Gabdoulline and Wade, 1997). The obtained rates were in reasonable agreement with the experimental rates. A somewhat broader view was taken by Zhou (Vijayakumar et al., 1998), who assigns a region where association occurs in a diffusion-controlled limit and defines it as the encounter complex. In this region, he neglects the short-ranged nonelectrostatic effects as the encounter-complex configurations are separated by at least one layer of solvent; therefore, short-ranged forces such as hydrophobic and van der Waals interactions are relatively weak in the diffusion process leading to the encounter complex. However, short-range interactions are essential for determining the location and size of the encounter-complex ensemble in configurational space, which in turn affect the magnitude of k_{on}. An encounter-complex ensemble that is less restricted in translation and rotation will lead to a higher k_{on} (Alsallaq and Zhou, 2007a, 2007b). Variation of the restriction in translation and rotation within the encounter complex with solvent conditions or among different protein complexes can be viewed as a configurational entropy effect.

A similar approach to calculate association rates, adding the assumption that the exact location of a protein complex is located within a binding funnel minimum, was used by Schlosshauer and Baker (2004). Both Zhou and Schlosshauer predict the existence of some sort of encounter/transition complex, which is less restricted than the exact binding complex. The size, energy, and location of the encounter complex were calculated from Brownian dynamics simulation (Spaar et al., 2006). Experimental work using a variety of methods supports the existence of this encounter complex, and assigned its location (Schreiber, 2002; Miyashita et al., 2004; Volkov et al., 2006; Harel et al., 2007; Suh et al., 2007). The picture emerging from the experimental data suggests that the encounter complex is in the region of the final complex, with the two proteins already aligned one toward the other; however, the interface is still mostly desolvated and structural rearrangement of the interface residues to provide an exact match has not yet occurred. Therefore, the transition state would be composed of desolvation and structural rearrangement. Further increase in binding rates can be obtained through columbic forces disseminated through charged residues.

THE CONTRIBUTION OF ELECTROSTATIC EFFECTS ON THE RATE OF ASSOCIATION

The one factor that contributes more than any other to the rate of association is electrostatic attraction between proteins, as seen by analyzing the effects of mutations on k_{on} (Figure 5.2). While mutations of charged residues can affect k_{on} by more than 20-fold (Sheinerman et al., 2000; Selzer and Schreiber, 2001), mutations on noncharged residues have only minor effects on the rate of binding. The picture is fundamentally different when analyzing the effect of mutations on k_{off}, where no significant difference is seen between either group. Moreover, the effect of mutations on k_{off} is much larger than on k_{on}.

Another clear indication on the effect of electrostatic forces on the rate of association is the salt dependence of k_{on}. The relation between ionic strength, protein-charge

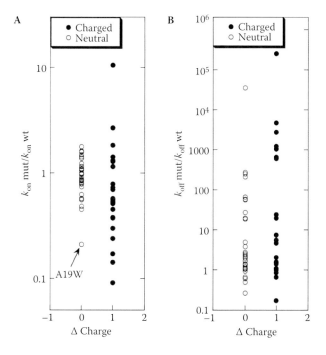

FIGURE 5.2   The change in the rate constants of (A) association and (B) dissociation plotted against the change in charge resulting from a mutation. The rate constants were measured for 55 mutations in barnase–barster, TEM1–BLIP, and IFNα2–IFNAR2 complexes in solution using a stopped-flow instrument.

complementarity, and k_{on} was shown to follow the Debye–Hückel energy of interaction between a pair of proteins according to the following equation (Selzer et al., 2000):

$$\ln k_{on} = \ln k_{on}^0 - \frac{U}{RT}\left(\frac{1}{1+\kappa a}\right)$$ (5.4)

where k_{on} and k_{on}^0 are the rates of association in the presence and absence of electrostatic forces, respectively, U is the electrostatic energy of interaction, κ is the inverse Debye length, and a is the minimal distance of approach. Hence, k_{on} is the sum of two components: (1) the basal rate of association in the absence of electrostatic forces (k_{on}^0) and (2) the contribution of the electrostatic forces between the proteins ($-U/_{RT}$). The later can be attended by mutation (changing U) or changing solution conditions. Equation 5.1 suggests that a plot of $\ln k_{on}$ versus $1 + \kappa a$ (which is proportional to the ionic strength $[I]$) is linear, with the slope being equal to $-U/_{RT}$. The intercept of the line at $1 + \kappa a = 0$ corresponds to the basal rate where electrostatic forces are shielded by salt (Figure 5.3). The intercept at $1 + \kappa a = 1$ corresponds to $\ln k_{on}$ in the absence of salt, with the electrostatic forces being maximized (Selzer and Schreiber, 1999). This linear relation was shown to hold for the association of TEM1–BLIP, interferon–receptor,

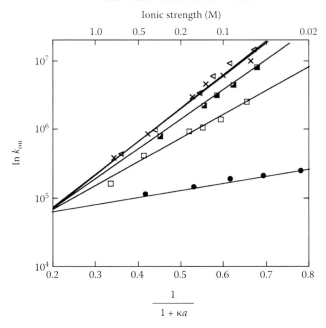

FIGURE 5.3 Association rate constants of wild-type and mutant TEM1–BLIP protein complexes determined at different salt concentrations, with ln k_{on} plotted against $1/(1 + \kappa a)$. The data can be fitted to a line using Equation 5.4.

hirudin–thrombin, barnase–barstar, and a heterodimeric leucine zipper for all salt concentrations tested (Wendt et al., 1997; Piehler and Schreiber, 1999). From Equation 5.4, a computer program (PARE [Protein Association Rate Enhancement]) was developed, which predicts the rate of association of mutant proteins. PARE is able to accurately calculate the rate of association for many mutant proteins (Selzer et al., 2000; Kiel et al., 2004; Stewart and Van Bruggen, 2004; Schreiber et al., 2006). Yet, one should be aware that the basal rate of association is not directly calculated in Equation 5.4, but has to be obtained by other methods such as Brownian dynamic simulations or from experimental data.

The ionic strength in the cell is ~150 mM. At this ionic strength, charge–charge interactions are partially shielded, reducing the negative effect of nonspecific interactions. A good example for this was reported for the complex of barnase–barstar in the presence of the polyion hirudin (Schreiber and Fersht, 1996). Measuring k_{on} at low salt was actually slower than in higher salt, due to nonspecific interactions of barnase with hirudin, which occluded free barnase from the system. The rate peaked at 150–200 mM salt, and slowed down at higher salt (this time due to masking of the charges). Thus, the physiological ionic strength is optimal to obtain fast specific binding, yet reduce nonspecific binding.

ALTERING ASSOCIATION RATES BY PROTEIN DESIGN

We have shown that Equation 5.4 (as implemented in PARE) predicts the rate of association of mutant proteins (Selzer et al., 2000). However, it can also be used for protein design of faster binding complexes. This is of particular interest for mutations placed outside the physical binding site, and thus not affecting the rate of dissociation, as was shown for TEM1–BLIP and Ras–Ral (Selzer et al., 2000; Kiel et al., 2004). For both TEM1–BLIP and Ras–Ral, a strong increase in the rate of association was achieved (250-fold and 17-fold, respectively), with an excellent correlation between the calculated and experimental values. This increased k_{on} was, however, not accompanied by a change in k_{off}, leading to an increased binding affinity of the magnitude described. This observation has far reaching implications on our understanding of the transition state for association, as will be described in the following. It is important to note that Equation 5.4 successfully predicts the rate change also for mutations located within the physical binding site of protein complexes, as was shown for the interactions between barnase–barstar, TEM1–BLIP, Ras–Ral, AChE–fasciculin, hirudin–thrombin, and others (Schreiber and Fersht, 1996). PARE is available at http://www.weizmann.ac.il/home/bcges/PARE.html.

An interesting outcome of the RalGDS-RBD design was that the electrostatic potential map of the designed RalGDS-RBD variant was similar to that observed for Raf (Figure 5.4), which is the native Ras affector. This is despite the very different sequence of the two affectors (<15% homology). The initial aim of this project was to optimize through mutation the electrostatic energy of interaction between RalGDS-RBD and Ras. The similarity of the obtained electrostatic potential maps suggests that the natural complex between Ras and Raf is optimized by natural selection for fast biding.

THE BASAL RATE OF ASSOCIATION

Basal rates of association are, according to Equation 5.4, the rates in the absence of electrostatic forces. These can be calculated either by extrapolating the experimental values of k_{on} to infinite salt or by introducing mutations that reduce the electrostatic energy of interaction to zero. Doing so, showed that the basal rates of association are 4×10^4 $M^{-1}s^{-1}$ for thrombin–hirudin and TEM1–BLIP, 6×10^5 for Ras–Raf, and 2×10^5 $M^{-1}s^{-1}$ for AChE–fass (Shaul and Schreiber, 2005).

It is important to note that Equation 5.4 ignores the contribution of noncharged residues to k_{on} (except for their contribution to the basal rate). Although their contribution is small, it was found to be significant in a number of cases. For example, the mutation A19W in IFNα2 reduced k_{on} by fourfold, a reduction that clearly relates to structural rearrangement during the process of association (as verified using double-mutant cycle analysis with W100A on IFNAR2 [Slutzki et al., 2006]).

THE ABUNDANCE OF HOT SPOTS FOR ASSOCIATION

Analyzing the contribution of electrostatics toward the rate of association of proteins in a database of 68 transient heteroprotein–protein complexes using HyPare (http://bip.weizmann.ac.il/HyPare) has shown a small contribution (<10-fold) for about half

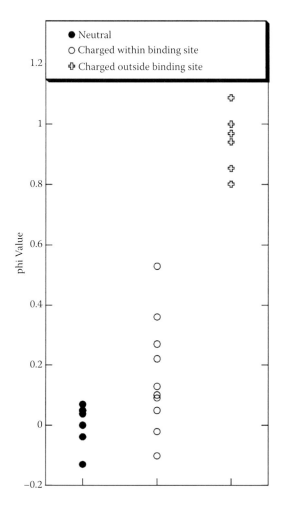

FIGURE 5.4 Φ-value analysis of the transition state for association determined for hot spot mutations (affecting the affinity by >2 kcal/mol) collected from TEM1–BLIP, barnase–barstar, and Ras–Ral binding. Φ values close to one suggest a similar interaction in the transition state as in the native complex, whereas values close to zero indicate the residues do not form any interprotein contacts in the transition state.

of the complexes (Shaul and Schreiber, 2005). In 25% of the complexes electrostatic forces had a major effect on k_{on} by affecting binding by >100-fold. Defining a residue being a hot spot for association as one that changes k_{on} by over 10-fold leaves about half the complexes without any potential hot spot, and a few hot spots per complex in the others. Of those, about 40% are calculated to increase the rate of association upon mutation, and thus increase binding affinity. This is very different from hot spots for dissociation, where experiments show the large majority of mutations to cause weaker binding. Moreover, about 40% of the hot spots for association are located outside the physical boundary of the binding site, making them ideal candidates for protein engineering. These data suggest that a majority of protein–protein complexes are not optimized for fast association. This may not be surprising, as the proteins act in a complex environment, where too many charged residues could actually reduce specificity of binding. Hot spot residues are not evenly distributed between all types of amino acids. About 75% of all hot spots are charged residues. This is understandable, as a charge-reverse mutant changes the total charge by two. More intriguing is the small number of hydrophobic in comparison to polar residues that are hot spots.

BROWNIAN DYNAMICS (BD) SIMULATIONS

Brownian dynamics (BD) is based on the Brownian motion theory, which describes the dynamic behavior of particles immersed in a solution. These particles are subjected to stochastic collisions with the solvent molecules (which are smaller both in their size and their mass) and to the viscous drag effects of the water molecules. This leads to the seemingly random motion of the particles, or in other words Brownian motion. The computation of biomolecular diffusional association rates by BD simulation dates back to the 1980s (the Northrup–Allison–McCammon [NAM] method), which is still widely used to compute bimolecular association rates (Northrup and Erickson, 1992). The association rate is given by the product of an analytically computed rate (k(b)) and a probability (β) that is computed from simulations. k(b) is the rate at which the two molecules approach to within a center-to-center separation distance (b), where b is sufficiently large so that the intermolecular forces are centrosymmetric or negligible. β is the probability that the two molecules, having reached separation b, go on to form a diffusional encounter complex and "react," rather than diffuse away to infinite separation. For the NAM method (a), a large number of simulations are started with the molecules in random orientations at separation b and the fraction of reactive trajectories is recorded. β is computed by correcting this fraction to account for the fact that trajectories are truncated when the molecules reach separation (q).

   When applying equations that describe the motility of the particles in BD simulations, one can describe their movements (Elcock et al., 2001; Gabdoulline and Wade, 2002). The classical use of BD is for k_{on} calculations, which are generally in a good agreement with the experimental rates (Gabdoulline and Wade, 1997, 2001). More recent developments include the use of BD for protein–protein docking, protein adsorption to a solid surface, ion channel permeation studies, and enzyme design (Gabdoulline and Wade, 2002). Spaar and colleagues have used the trajectories generated during the BD simulations to analyze the free energy landscape of encounter complexes (Spaar and Helms, 2005; Spaar et al., 2006) through modeling the trajectories occupancy maps. Another approach, weighted-ensemble Brownian (WEB) dynamics, was proposed by Huber and Kim, and has recently been further developed and applied (Huber and Kim, 1996; Rojnuckarin et al., 2000). Rather than simulate the association of a single pair of molecules, as in the NAM method, one molecule is replaced by an ensemble of pseudoparticles or weighted probability packets. These occupy bins along the intermolecular reaction coordinate that are equally sampled through splitting and combining the weighted pseudoparticles and thus speeding up the time of simulation. Because of their long-range nature, electrostatic forces have an important, if not the most important, influence on biomolecular association. However, solving the Poisson–Boltzmann equation (PBE) along the reaction pathway would take much to long to be feasible. The "test charge" approximation is usually used. In this approximation, the partial charges on one molecule, whose interior low dielectric cavity is neglected, move in the field of the other, as computed from solution of the PBE. Gabdoulline and Wade (1996) introduced "effective charges" that were calculated to reproduce, in a single continuum dielectric, the electrostatic potential of a molecule computed from the PBE for a heterogeneous dielectric. Replacing test charges by effective charges results in a more accurate approximation of the PBE in BD simulations and better agreement with experiment for the ionic strength dependence of protein–protein association rates.

Other forces, in addition to electrostatics, can influence diffusion association. Camacho and colleagues (Camacho et al., 2000; Camacho and Vajda, 2001) studied the influence of hydrophobic desolvation, which has a shorter range than electrostatic interactions, on the free energy landscape for protein–protein association. They found hydrophobic desolvation to be particularly important for electrostatically, weakly attracting proteins. Furthermore, they suggest that the mechanism for molecular recognition requires one of the interacting proteins, usually the smaller of the two, to anchor a specific side chain in a structurally constrained binding groove of the other protein, providing a steric constraint that helps to stabilize a nativelike bound intermediate (Rajamani et al., 2004).

ANALYTICAL MODELS TO CALCULATE PROTEIN ASSOCIATION RATE CONSTANTS

According to the work by Zhou, a protein pair that has reached a reaction region with defined finite volume V_{RR} has a finite rate, g, to form the native complex (Zhou, 1993; Alsallaq and Zhou, 2007b). In this treatment of protein association, the reaction rate g models the conformational rearrangement that brings the protein pair from the encounter complex to the native complex. That is, k_2 (from Equation 5.3) equals g. The equilibrium constant, K_1, is given by $V_{RR}e^{-<U>^*/k_BT}$, where $<U>^*$ is the average interaction energy within the transient complex. By starting Brownian trajectories from within the reaction region one can obtain the surviving fraction, S, of the trajectories. The surviving fraction S depends on the rate g and on how much the absorbing boundary is extended to form the reaction region. In a Brownian dynamics study of protein–protein association under the influence of electrostatic interactions, it was discovered that the survival fraction S is insensitive to the presence of the electrostatic interaction energy (Zhou, 1993). Thus:

$$k_1 = k_0 e^{-<U>^*/k_BT}$$

(5.5)

with k_0 being the basal rate constant in the absence of external forces, which equals to:

$$k_0 = gV_{RR}S_0/(1-S_0)$$

(5.6)

where S_0 is the survival fraction without any biasing force. This suggests that the association is stereo specific and the interaction energy is long ranged. The accuracy of Equation 5.5 has been demonstrated against results from Brownian dynamics simulations and experiments (Zhou et al., 1997; Vijayakumar et al., 1998; Alsallaq and Zhou, 2008). This equation resolves one of the two main obstacles for reliable prediction of protein association rate constants by making it possible to rigorously treat electrostatic interactions. The effect of electrostatic interactions is captured by the Boltzmann factor $e^{-<U^*/k_B T>}$, which can be obtained by averaging over a relatively small number of representative configurations in the encounter complex. The basal rate constant k_0 still needs to be obtained through force-free Brownian dynamics simulations, but these simulations are inexpensive.

MAPPING ENCOUNTER COMPLEXES ALONG THE ASSOCIATION PATHWAY

Structural studies of encounter complexes are routinely done to study the transition state and intermediates of protein folding or enzyme catalysis. A range of experimental tools has been developed for this task. Nuclear magnetic resonance (NMR) is a powerful tool to pin down the residual structures of the unfolded state, as well as to capture transient folding intermediates (Krishna et al., 2004). Phi-value analysis defines whether specific interactions are formed already during the intermediate or transition state of the reaction (Fersht et al., 1992; Petrovich et al., 2006). Time-resolved spectroscopy and single-molecule spectroscopy are powerful tools, which were frequently applied to investigate intermediates and transition states in folding (Nolting et al., 1997). While these experimental tools provide only a partial view, they are extremely valuable for molecular dynamic simulations and other theoretical studies, as they provide experimental reference points to tune the simulation. In comparison, structural studies on the pathway for protein association are much less common. This may be partly attributed to the technical difficulties stemming from the low population of the binding intermediates, and the ill-defined nature of the transition state for binding. Still, the development of protein-engineering tools, NMR, spectroscopy, and single-molecule methods resulted in a number of interesting experimental studies shedding light on the way proteins associate.

Experimental evidence of the structure of the encounter complex of the electron transfer complex of yeast cytochrome c peroxidase (CcP) and iso-1-cytochrome c was recently presented by Volkov et al. (2006) using paramagnetic NMR spectroscopy. The complex is very short lived, with a dominant structure supporting electron transfer and a dynamic encounter complex. The results support the view that the conformational space sampled by the protein molecules during the dynamic part of the interaction is localized around the CcP position in the dominant orientation. This finding is in agreement with the view that an encounter complex facilitates formation of the dominant complex via preorientation of the protein molecules and reduced dimensionality search. For CcP binding electrostatic attraction plays a dominant role in determining the nature of the encounter complex.

Tang et al. (2006) did a similar study for the association between the phospho-carrier protein, Hpr, and three proteins in the bacterial phosphotransferase system (using paramagnetic relaxation enhancement [PRE] NMR). However, they found a much broader definition of the encounter complex, which was spread across its adjacent surface, with electrostatic attraction being the main driving force in its stabilization. However, these experiments were initially being done in the absence of salt, where nonspecific electrostatic attraction of these highly charged proteins is strong over a long range (Schreiber and Fersht, 1996). Indeed, in a follow-up paper (Suh et al., 2007), they showed that the nonspecific part of the encounter complex is reduced to a large extent by adding salt, while more specific encounter complexes (located in the region of the final complex) were less affected. The importance of nonspecific encounter complexes to association is not yet clear and may very well be marginal.

PHI-VALUE ANALYSIS OF BINDING INTERMEDIATES

Phi-value analysis was successfully applied to map the transition state for protein folding and became the golden standard for many theoretical simulations (Fersht and Daggett, 2002). It has been demonstrated that this analysis can be used also for studying the transition state for protein–protein association (Taylor et al., 1998; Mateu et al., 1999; Wu et al., 2002; Kiel et al., 2004; Levy et al., 2005). Equation 5.7 gives the basic formulation for such analysis for binding, with U and C being the unbound and bound states and $\ddagger$ the transition state.

$$\Phi_{ass} = \Delta\Delta G^{\ddagger-U} / \Delta\Delta G^{C-U} \tag{5.7}$$

with $\Delta\Delta G^{\ddagger-U}$ being calculated from:

$$\Delta\Delta G^{\ddagger-U} = -RT \ln (k_{on}^{wt}/k_{on}^{mut}) \tag{5.8}$$

where k_{on}^{mut} is the association rate of the mutated complex and k_{on}^{wt} is the association rate of the wild-type complex. The free energy of binding can be determined directly from the affinity (K_D) using the mass action equation, or from the ratio of $K_D = k_{off}/k_{on}$ (assuming two state-biding), with $\Delta G^{C-U} = -RT\ln(K_D)$. Mutations that induce a similar effect on the transition state and the free energy of binding will have a Φ value of 1, while mutations that have no effect on $\ddagger$, but change the binding affinity, will have a Φ value of zero. Figure 5.4 shows a Φ-value analysis for a large number of hot spot mutations (affecting the affinity by >2 kcal/mol) collected from TEM1–BLIP, barnase–barstar, and Ras–Ral binding (Schreiber and Fersht, 1996; Albeck et al., 2000; Kiel et al., 2004; Reichmann et al., 2005, 2007). The mutations were divided into three groups: one group consists of noncharged residues, the second is for charged residues located within the binding interface, and the third is for charged residues located outside the physical binding interface. The reason we analyze only hot spot residues is to avoid erroneous Φ values, as the experimental error for $\Delta\Delta G$ measurements is in the order of 0.3 kcal/mol. As only very few resides located outside the physical binding site pass this criteria, multiple mutations were used for this group that were designed to increase specifically association (Selzer et al., 2000;

Kiel et al., 2004). The data clearly demonstrate that noncharged mutations always have Φ values close to zero, while charged residues located outside the physical binding site have Φ values close to one. Charged residues located within the binding site have mixed values. These results clearly show that noncharged residues do not form specific contacts during the transition state for association, and hence confirm the mutant data. Conversely, charged residues do affect association; however, their effect is related to long-range columbic forces and not to specific short-range interactions. This explains why charged residues have intermediate Φ values when located within the binding site (they exert both long-range effects on association and short-range effects on dissociation). Thus, the evolving picture from Φ value analysis for binding is much simpler than that found for folding, with the association transition preceding the formation of short-range interactions.

DIRECT EVIDENCE FOR THE EXISTENCE OF AN ENCOUNTER COMPLEX

Direct evidence from kinetic studies for the existence of an encounter complex along the association pathway was presented for RalGDS-RBD binding Ras (Kiel et al., 2004). This interaction shows a nonlinear increase in association with concentration. Using PARE (Selzer et al., 2000), charged mutations were designed that specifically increased k_{on} and showed that the increase in k_{on} was a result of an increased rate of formation of the encounter complex (k_1 in Equation 5.3), while the rate of conversion to final complex (k_2) was unchanged at a rate of ~400 s^{-1}. This demonstrates that increasing electrostatic steering by mutation stabilizes the encounter complex and the transition state to a similar extent as the final complex, and that a rate-limiting transition state exists. A similar conclusion was reached from studying the pKa shift of His102 in barnase in the free and barstar bound form (Schreiber and Fersht, 1993, 1995). The pKa of His102 in unbound barnase was 6.3, while in complex a shift to <5 was measured. The pH dependence of k_{on} showed a similar pKa value as for the unbound protein; thus, the shift in pKa upon binding occurs after the transition state. X-ray crystallography has shown that the shift in pKa can be attributed to the tight interactions of His102 with its surrounding on the barstar protein, suggesting again that these interactions are not yet formed during the transition state. A similar behavior was observed for the association of R67 DHFR, with a pK_a of 6.6 that was attributed to H62, but a dissociation reaction with a pK_a of under 5.5 (Mejean et al., 2001), with the pK_a shift being attributed to specific short-range interactions that are not formed at the transition state for association. These studies provide a clear statement that short range interactions are mostly not formed during the transition state for binding.

DOUBLE-MUTANT CYCLE ANALYSIS AS A TOOL TO DECIPHER THE STRUCTURE OF THE TRANSITION STATE FOR BINDING

Double-mutant cycles measure the coupling energy between a pair of residues from the difference in binding free energy of two single mutations and the double-mutant.

Accordingly, the activation interaction energy, $\Delta\Delta G^{\ddagger}_{int}$, is a measure of the interaction between two residues at the transition state and is equal to:

$$\Delta\Delta G^{\ddagger}_{int} = \Delta\Delta G^{\ddagger}_{X\to A,\, Y\to A} - \Delta\Delta G^{\ddagger}_{X\to A} - \Delta\Delta G^{\ddagger}_{Y\to A} \tag{5.9}$$

where X and Y represent the wild-type residues and A represents a mutant (Carter et al., 1984; Horovitz, 1987; Horovitz and Fersht, 1990; Schreiber and Fersht, 1995). Like Φ-value analysis, this technique was first applied successfully to protein-folding studies (Carter et al., 1984; Horovitz et al., 1990).

From a large number of double-mutant cycles calculated for the activated complex of barnase and barstar, significant coupling energies for association were determined only between charged residues distanced less than 10 Å from one another in the final complex (Schreiber and Fersht, 1996). None of the noncharged residues had a significant $\Delta\Delta G^{\ddagger}_{int}$ value with any other residue. A similar experiment was done for the interaction between cytochrome C_2 and the bacterial reaction center, but only between charged residues. Relating the energy transfer rate (k_2), which for these types of reactions is similar to k_{on} (Miyashita et al., 2004), to the distance between the probed residues showed that residues interact at the activated complex up to a distance of 10 Å (Tetreault et al., 2002). A similar result was obtained for the interaction between TEM1-β-lactamase and its protein inhibitor BLIP (Harel et al., 2007). Repeating the same double-mutant cycles at up to 1 M salt (which masks most of the effects of charges) showed that for barnase–barstar, as for the complex between *P. laminosum* Cyt *f* and plastocyanin, some but not all pairwise charge–charge interactions were maintained, suggesting that structural specificity of the activated complex is preserved even at high salt (Frisch et al., 2001; Miyashita et al., 2003).

MODELING THE TRANSITION STATE FOR BINDING USING SIMULATIONS BASED ON EXPERIMENTAL DOUBLE-MUTANT CYCLE DATA

The experimental mutant and double-mutant cycle data measured for the association process were further used to model the structures of the encounter and association transition-state complexes. In the study of Harel et al. (2007) the transition-state structures were modeled from the experimental $\Delta\Delta G^{\ddagger}_{int}$ values by introducing structural perturbations of one protein relative to the other, and searching for those interprotein orientations that best account for the experimental $\Delta\Delta G^{\ddagger}_{int}$ values (Figure 5.5; Harel et al., 2007). Similarly, Miyashita et al. (2004) related the experimental k_{on} values of mutations to differences in the calculated electrostatic energies for a wide range of cytochrome C_2-reaction center (Cyt-RC) configurations. Both studies gave a very similar description of the transition state for association. In both cases, the transition state was stabilized by electrostatic interactions, with the ensemble of structures spread out around the final complex, but in neither cases short-range interactions were formed during the transition state, suggesting a solvated transition state. The average transition state structure was not necessarily located exactly on the binding site, but may be shifted toward one side of the interface. This was observed for

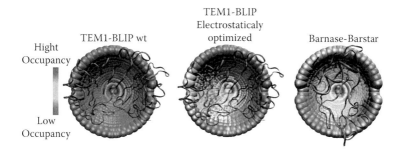

FIGURE 5.5 (SEE COLOR INSERT FOLLOWING PAGE 174.) Mapping the transition state for protein–protein association using double-mutant cycle data as constraints (Harel et al., 2007). Each point represents the center of mass of 1 of 2220 configurations perturbed from the native complex. The point in the middle of each cap represents the x-ray structure of the native complex. The different colors represent configurations selected by different filtering cutoffs; cooler colors designate a configuration that passes a more stringent cutoff (thus has a higher probability of being occupancy in the transition state). TEM1 was the mobile protein in the simulations, while BLIP was fixed. The TEM1–BLIP complex was electrostatically optimized using the program PARE, by introducing mutations located outside the physical binding interface.

both Cyt-RC and for the electrostatically optimized TEM1–BLIP interface, while for barnase–barster the transition state overlaps the final complex (Harel et al., 2007). These data suggest a certain pathway for association (energy funnel), which would help in speeding up association. In contradiction to these results, no indication for specific transition state structures was found for the interaction of wild-type TEM1–BLIP or the complex between IFNα2 and IFNAR2. Therefore, a diffusive transition state was suggested for these interactions. Specific transition states are characterized by defined interprotein orientations, which cannot be modeled for the diffusive transition states. As was clearly shown for the TEM1–BLIP complex, mutations introduced through rational design can change the transition state from diffusive to specific and vice versa (Harel et al., 2007).

FRUITFUL AND FUTILE ENCOUNTERS ALONG THE ASSOCIATION REACTION BETWEEN PROTEINS

Experimental data on kinetic processes can, at best, provide snapshots along the reaction coordinates, with computer simulations using BD filling in the gaps. The association between TEM1 and BLIP and barnase with barstar are perfectly suited for detailed computational simulations that can be compared with the large bulk of experimental data gathered on these systems, including the many mutations that directly affect the rate of association (Selzer et al., 2000; Harel et al., 2007). Spaar and colleagues have used the trajectories generated during the BD simulations in order to analyze the free energy landscape of the encounter complex (Spaar and Helms, 2005; Spaar et al., 2006). This was done through modeling the occupancy map. As the number of the trajectories was very high, the occupancy maps could be interpreted using probability distribution, from which the entropy landscape was calculated. The free energy landscape could be obtained by summing the energy and the entropy contributions, as follows (Spaar and Helms, 2005; Spaar et al., 2006):

$$\Delta G = \Delta E_{el} + \Delta E_{ds} - T\Delta S \qquad (5.10)$$

where ΔG is the free energy, ΔE_{el} is the electrostatic energy, ΔE_{ds} is the desolvation energy, T is the temperature, and ΔS is the entropy. From the free energy landscape one can compute the encounter complex region (the minimum in the free energy landscape) and the optimal association and dissociation pathways. Using these tools, two encounter complex regions were mapped along the association reaction of bar-nase and barstar, one above the interface and the other above the RNA binding loop (Spaar and Helms, 2005; Spaar et al., 2006). Analyzing the effect of mutations on the encounter complex showed that a single mutation could considerably alter the free energy landscape and change the population of the two minima (i.e., the two regions of the encounter complex). As expected for a charged protein pair like barnase–barstar, the free energy landscape was also affected by ionic strength.

The results of the BD simulation for wild-type TEM1–BLIP also shows two encounter regions; however, both are not at the interface. The left region is larger and energetically more favorable, yet more distant from the interface (Figure 5.6). Both

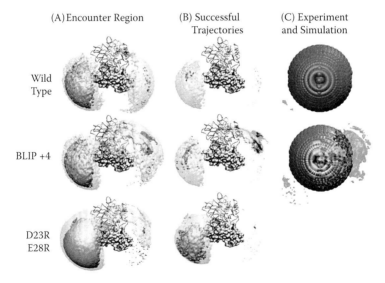

FIGURE 5.6 (SEE COLOR INSERT FOLLOWING PAGE 174.) Brownian dynam-ics simulations of TEM1–BLIP mutants. BLIP is represented as a gray surface, TEM1 wild type is represented as a purple ribbon. All the simulations were done at 150 mM NaCl. (A) Encounter complexes are drawn as yellow isosurfaces representing the center of mass of TEM1 on BLIP at $\Delta G < -2.0$ kcal/mol (transparent yellow) and $\Delta G < -3.0$kcal/mol (dark yellow). (B) Superimposition of the successful configurations at $\Delta G < -3.0$ kcal/mol (dark yellow) and the encounter complex region, defined by $\Delta G < -3.0$ kcal/mol, marked in trans-parent yellow. (C) An overlay of the successful trajectories from the BD simulation and the experimentally mapped transition state (see Figure 5.5).

regions may be valid encounter complexes; however, they imply a diffusive encounter complex, as the two regions are broad and remote from each other. Furthermore, these encounter regions do not guide the interaction toward the final complex, as can be seen from the analysis of successful trajectories, which shows very few of the encounter complex trajectories developing into a complex. This is in line with our inability to identify a specific transition state for wild-type TEM1–BLIP (Harel et al., 2007). To better understand the role of the encounter regions observed in the BD simulations for the association reaction, a number of mutant proteins with perturbed association rates (as experimentally determined) were studied. Most notable were mutants that enlarge the encounter region near or above the physical binding site (BLIP mutant BLIP +4 in Figure 5.6), and mutants that expand the encounter regions (BLIP D23R,E28R, particularly the left one, Figure 5.6). Mutations of group I have a very significant effect on k_{on} (up to 100-fold), while mutations of group II did not change the association rate at all. These mutants clearly show that no simple relation can be found between either the size or the energy of the encounter regions and the rate of association. Moreover, even the analysis of successful encounters does not always correlate with the observed change in k_{on}. For example, the encounter region of the BLIP D23R,E28R double-mutant shows a much higher degree of successful trajectories compared to the wild type, but has the same k_{on}. Conversely, for group I, a correlation between successful encounters and k_{on} was observed. Group I mutations, which are located at the vicinity but outside the physical binding site, were designed to optimize the electrostatic energy of interaction of the complex (Selzer et al., 2000). We have noted that the experimentally determined transition state, which could be assigned only for TEM1–BLIP with optimized electrostatic attraction (group I mutations) is smaller than the BD calculated encounter region. However, the experimentally determined transition state better fits the area of successful trajectories mapped for these mutants. This subgroup within the encounter region can be assumed to be much closer to the transition state, which is defined as the activated form of a molecule that has partly undergone a chemical reaction. As the transition state has to be on the pathway to product formation, only successful trajectories fulfill this requirement.

The mutant data presented here suggest that some of the encounter regions do not contribute to association and thus are futile encounters. In general, the futile regions are distant from the interface, and although the simulations suggest these regions to be energetically favorable, they do not influence the association rate. The reason for this is that in reality, futile encounters do not develop into final complexes and hardly affect the concentration of free protein in solution. Finally, the data presented here could explain why group I mutations do not change the rate of dissociation, despite their large effect on the electrostatic complementarity between the two proteins, that causes an increase in the rate of association. From the comparison of the encounter complex to successful encounter trajectories, it becomes clear that encounters are readily formed, but most of them are futile. The mutations in group I were designed to increase the percent of fruitful encounters and hence k_{on}. However, still most encounters will dissociate (see Figure 5.1). Thus, even for group I mutations, once the final complex dissociates, it will have a small chance of re-forming. This behavior is a result of the relative flat energy landscape leading to association prior to the transition state, which is characterized by desolvation and formation of short-range interactions, versus the steep energy landscape leading to dissociation, which is composed of breaking the short-range interactions between the proteins (Figure 5.1).

SUMMARY

The association of proteins to form a complex is a multistep process, which starts by random collisions of the individual proteins. Multiple collisions and rotational diffusion brings the proteins to an orientation that is close to that of the native complex, leading to the formation of a transient complex. This part of the process is diffusion controlled, and strongly affected by electrostatic interactions. Computer simulations and experimental data suggest that the transient complex develops into the native complex through a transition state. Structurally, the transient complex and the transition state seem to be similar; however, the transition state is smaller, with the encounter complex also occupying futile areas that will not further develop into a complex. The most important parameter that affects the rate of association is the electrostatic force, which can act over a distance. However, local surface complimentarily and desolvation also play an important role, and in some cases mutations can be located that change the rate of association due to these factors.

REFERENCES

Albeck, S., Unger, R., & Schreiber, G. (2000) Evaluation of direct and cooperative contributions towards the strength of buried hydrogen bonds and salt bridges. *J Mol Biol*, **298**, 503–520.

Alsallaq, R. & Zhou, H. X. (2007a) Prediction of protein–protein association rates from a transition-state theory. *Structure*, **15**, 215–224.

Alsallaq, R. & Zhou, H. X. (2007b) Energy landscape and transition state of protein–protein association. *Biophys J*, **92**, 1486–1502.

Alsallaq, R. & Zhou, H. X. (2008) Electrostatic rate enhancement and transient complex of protein–protein association. *Proteins*, **71**, 320–335.

Camacho, C. J., Kimura, S. R., DeLisi, C., & Vajda, S. (2000) Kinetics of desolvation-mediated protein–protein binding. *Biophys J*, **78**, 1094–1105.

Camacho, C. J. & Vajda, S. (2001) Protein docking along smooth association pathways. *Proc Natl Acad Sci USA*, **98**, 10636–10641.

Carter, P. J., Winter, G., Wilkinson, A. J., & Fersht, A. R. (1984) The use of double mutants to detect structural changes in the active site of the tyrosyl-tRNA synthetase (Bacillus stearothermophilus). *Cell*, **38**, 835–840.

Elcock, A. H., Sept, D., & McCammon, J. A. (2001) Computer simulation of protein–protein interactions. *J. Phys. Chem. B*, **105**, 1504–1518.

Fersht, A. R. & Daggett, V. (2002) Protein folding and unfolding at atomic resolution. *Cell*, **108**, 573–582.

Fersht, A. R., Matouschek, A., & Serrano, L. (1992) The folding of an enzyme. I. Theory of protein engineering analysis of stability and pathway of protein folding. *J Mol Biol*, **224**, 771–782.

Frisch, C., Fersht, A. R. & Schreiber, G. (2001) Experimental assignment of the structure of the transition state for the association of barnase and barstar. *J Mol Biol*, **308**, 69–77.

Gabdoulline, R. R. & Wade, R. C. (1996) Analytically defined surfaces to analyze molecular interaction properties. *J Mol Graph*, **14**, 341–353, 374–375.

Gabdoulline, R. R. & Wade, R. C. (1997) Simulation of the diffusional association of barnase and barstar. *Biophys J*, **72**, 1917–1929.

Gabdoulline, R. R. & Wade, R. C. (2001) Protein–protein associaton: Investigation of factors influencing asaociation rates by Brownian dynamics simulations. *J Mol Biol*, **306**, 1139–1155.

Gabdoulline, R. R. & Wade, R. C. (2002) Biomolecular diffusional association. *Curr Opin Struct Biol*, **12**, 204–213.

Harel, M., Cohen, M., & Schreiber, G. (2007) On the dynamic nature of the transition state for protein–protein association as determined by double-mutant cycle analysis and simulation. *J Mol Biol*, **371**, 180–196.

Horovitz, A. (1987) Non-additivity in protein–protein interactions. *J Mol Biol*, **196**, 733–735.

Horovitz, A. & Fersht, A. R. (1990) Strategy for analysing the cooperativity of intramolecular interactions in peptides and proteins. *J Mol Biol*, **214**, 613–617.

Horovitz, A., Serrano, L., Avron, B., Bycroft, M., & Fersht, A. R. (1990) Strength and cooperativity of contributions of surface salt bridges to protein stability. *J Mol Biol*, **216**, 1031–1044.

Huber, G. A. & Kim, S. (1996) Weighted-ensemble Brownian dynamics simulations for protein association reactions. *Biophys J*, **70**, 97–110.

Kiel, C., Selzer, T., Shaul, Y., Schreiber, G., & Herrmann, C. (2004) Electrostatically optimized Ras-binding Ral guanine dissociation stimulator mutants increase the rate of association by stabilizing the encounter complex. *Proc Natl Acad Sci USA*, **101**, 9223–9228.

Krishna, M. M., Hoang, L., Lin, Y., & Englander, S. W. (2004) Hydrogen exchange methods to study protein folding. *Methods*, **34**, 51–64.

Kuttner, Y. Y., Kozer, N., Segal, E., Schreiber, G., & Haran, G. (2005) Separating the contribution of translational and rotational diffusion to protein association. *J Am Chem Soc*, **127**, 15138–15144.

Levy, Y., Cho, S. S., Onuchic, J. N. & Wolynes, P. G. (2005) A survey of flexible protein binding mechanisms and their transition states using native topology based energy landscapes. *J Mol Biol*, **346**, 1121–1145.

Mateu, M. G., Sanchez Del Pino, M. M. & Fersht, A. R. (1999) Mechanism of folding and assembly of a small tetrameric protein domain from tumor suppressor p53. *Nat Struct Biol*, **6**, 191–198.

Mejean, A., Bodenreider, C., Schuerer, K., & Goldberg, M. E. (2001) Kinetic characterization of the pH-dependent oligomerization of R67 dihydrofolate reductase. *Biochemistry*, **40**, 8169–8179.

Miyashita, O., Onuchic, J. N., & Okamura, M. Y. (2003) Continuum electrostatic model for the binding of cytochrome c2 to the photosynthetic reaction center from Rhodobacter sphaeroides. *Biochemistry*, **42**, 11651–11660.

Miyashita, O., Onuchic, J. N., & Okamura, M. Y. (2004) Transition state and encounter complex for fast association of cytochrome c2 with bacterial reaction center. *Proc Natl Acad Sci USA*, **101**, 16174–16179.

Nolting, B. et al. (1997) The folding pathway of a protein at high resolution from microseconds to seconds. *Proc Natl Acad Sci USA*, **94**, 826–830.

Northrup, S. H. & Erickson, H. P. (1992) Kinetics of protein–protein association explained by Brownian dynamic computer simulation. *Proc Natl Acad Sci USA*, **89**, 3338–3342.

Petrovich, M., Jonsson, A. L., Ferguson, N., Daggett, V., & Fersht, A. R. (2006) Phi-analysis at the experimental limits: Mechanism of beta-hairpin formation. *J Mol Biol*, **360**, 865–881.

Piehler, J. & Schreiber, G. (1999) Biophysical analysis of the interaction of human ifnar2 expressed in E. coli with IFNalpha2. *J Mol Biol*, **289**, 57–67.

Rajamani, D., Thiel, S., Vajda, S., & Camacho, C. J. (2004) Anchor residues in protein–protein interactions. *Proc Natl Acad Sci USA*, **101**, 11287–11292.

Reichmann, D. et al. (2005) The modular architecture of protein–protein binding interfaces. *Proc Natl Acad Sci USA*, **102**, 57–62.

Reichmann, D. et al. (2007) Binding hot spots in the TEM1-BLIP interface in light of its modular architecture. *J Mol Biol*, **365**, 663–679.

Rojnuckarin, A., Livesay, D. R., & Subramaniam, S. (2000) Bimolecular reaction simulation using weighted ensemble Brownian dynamics and the University of Houston Brownian dynamics program. *Biophys J*, **79**, 686–693.

Schlosshauer, M. & Baker, D. (2004) Realistic protein–protein association rates from a simple diffusional model neglecting long-range interactions, free energy barriers, and landscape ruggedness. *Protein Sci*, **13**, 1660–1669.

Schreiber, G. (2002) Kinetic studies of protein–protein interactions. *Curr Opin Struct Biol*, **12**, 41–47.

Schreiber, G. & Fersht, A. R. (1993) Interaction of barnase with its polypeptide inhibitor barstar studied by protein engineering. *Biochemistry*, **32**, 5145–5150.

Schreiber, G. & Fersht, A. R. (1995) Energetics of protein–protein interactions: Analysis of the barnase-barstar interface by single mutations and double mutant cycles. *J Mol Biol*, **248**, 478–486.

Schreiber, G. & Fersht, A. R. (1996) Rapid, electrostatically assisted association of proteins. *Nat Struct Biol*, **3**, 427–431.

Schreiber, G., Shaul, Y., & Gottschalk, K. E. (2006) Electrostatic design of protein–protein association rates. *Methods Mol Biol*, **340**, 235–249.

Selzer, T., Albeck, S., & Schreiber, G. (2000) Rational design of faster associating and tighter binding protein complexes. *Nat Struct Biol*, **7**, 537–541.

Selzer, T. & Schreiber, G. (1999) Predicting the rate enhancement of protein complex formation from the electrostatic energy of interaction. *J Mol Biol*, **287**, 409–419.

Selzer, T. & Schreiber, G. (2001) New insights into the mechanism of protein–protein association. *Proteins*, **45**, 190–198.

Shaul, Y. & Schreiber, G. (2005) Exploring the charge space of protein–protein association: A proteomic study. *Proteins*, **60**, 341–352.

Sheinerman, F. B., Norel, R., & Honig, B. (2000) Electrostatic aspects of protein–protein interactions. *Curr Opin Struct Biol*, **10**, 153–159.

Slutzki, M., Jaitin, D. A., Yehezkel, T. B., & Schreiber, G. (2006) Variations in the unstructured C-terminal tail of interferons contribute to differential receptor binding and biological activity. *J Mol Biol*, **360**, 1019–1030.

Spaar, A., Dammer, C., Gabdoulline, R. R., Wade, R. C., & Helms, V. (2006) Diffusional encounter of barnase and barstar. *Biophys J*, **90**, 1913–1924.

Spaar, A. & Helms, V. (2005) Free energy landscape of protein–protein encounter resulting from Brownian dynamics simulations of barnase:barstar. *J Chem Theory Comput*, **1**, 723–736.

Stewart, R. C. & Van Bruggen, R. (2004) Association and dissociation kinetics for CheY interacting with the P2 domain of CheA. *J Mol Biol*, **336**, 287–301.

Suh, J. Y., Tang, C., & Clore, G. M. (2007) Role of electrostatic interactions in transient encounter complexes in protein–protein association investigated by paramagnetic relaxation enhancement. *J Am Chem Soc*, **129**, 12954–12955.

Tang, C., Iwahara, J., & Clore, G. M. (2006) Visualization of transient encounter complexes in protein–protein association. *Nature*, **444**, 383–386.

Taylor, M. G., Rajpal, A., & Kirsch, J. F. (1998) Kinetic epitope mapping of the chicken lysozyme. HyHEL-10 Fab complex: Delineation of docking trajectories. *Protein Sci*, **7**, 1857–1867.

Tetreault, M., Cusanovich, M., Meyer, T., Axelrod, H., & Okamura, M. Y. (2002) Double mutant studies identify electrostatic interactions that are important for docking cytochrome c2 onto the bacterial reaction center. *Biochemistry*, **41**, 5807–5815.

Vijayakumar, M. et al. (1998) Electrostatic enhancement of diffusion-controlled protein–protein association: Comparison of theory and experiment on barnase and barstar. *J Mol Biol*, **278**, 1015–1024.

Volkov, A. N., Worrall, J. A., Holtzmann, E. & Ubbink, M. (2006) Solution structure and dynamics of the complex between cytochrome c and cytochrome c peroxidase determined by paramagnetic NMR. *Proc Natl Acad Sci USA*, **103**, 18945–18950.

Wendt, H. et al. (1997) Very rapid, ionic strength-dependent association and folding of a heterodimeric leucine zipper. *Biochemistry*, **36**, 204–213.

Wu, L. C., Tuot, D. S., Lyons, D. S., Garcia, K. C., & Davis, M. M. (2002) Two-step binding mechanism for T-cell receptor recognition of peptide MHC. *Nature*, **418**, 552–556.

Zhou, H. X. (1993) Brownian dynamics study of the influences of electrostatic interaction and diffusion of protein–protein association kinetics. *Biophys J*, **64**, 1711–1726.

Zhou, H. X., Wong, K. Y., & Vifayakumar, M. (1997) Design of fast enzymes by optimizing interaction potential in active site. *Proc Natl Acad Sci*, **94**, 12372–12377.

6 Computational Simulations of Protein–Protein and Protein–Nucleic Acid Association

Georgi V. Pachov, Razif R. Gabdoulline, and Rebecca C. Wade

CONTENTS

OVERVIEW

The kinetics of the formation of macromolecular complexes contribute to their biological function. We first describe key features and determinants of bimolecular association kinetics. Then we present an overview of theoretical and computational approaches to calculating kinetic properties. Finally, we discuss recent computational advances with selected examples of protein–nucleic acid and protein–protein complexation.

INTRODUCTION

The formation of biological complexes between proteins, proteins and small molecules, and proteins and nucleic acids is critical to many biological processes, including cell signaling, gene transcription, enzyme catalysis, and the immune response. Molecular association is governed by both the kinetic and the thermodynamic properties of the molecules and of the medium in which they are immersed. Inside a cell, the medium is packed with a wide variety of different molecules and is considered to be crowded. Biomacromolecular complexes vary widely in their affinities and lifetimes, ranging from obligate and permanent to transient and short-lived complexes. Here, we will only consider bimolecular association to form a transient complex. Complexation is usually characterized in terms of affinity, as weak (and loose) or strong (and tight). The variation in affinity is often largely determined by the variation in dissociation rate. Association rates can, however, also vary over many orders of magnitude between complexes and can be critical in the biological context. For example, the snake toxin fasciculin must not only strongly inhibit acetylcholinesterase (an enzyme that is critical to neural transmission) but also reach its target quickly (Quinn 1987). Similarly, the intracellular inhibitor barstar protects the bacterium *Bacillus amyloquefaciens* from the enzyme barnase, which it excretes to act as an extracellular ribonuclease (Jucovic and Hartley 1996). The protein interleukin-4 forms a complex with its cellular receptor, and the time of this process is a measure for the regulation of the immune system (Wang, Shen, and Sebald 1997). Furthermore, the speed at which the *lac* repressor binds to its chromosomal *lac* operator regulates gene expression in the cell (Elf, Li, and Xie 2007).

Here, we discuss the use of computational approaches to address the problem of understanding how a biomolecular complex forms and the macromolecular interactions involved. First, important parameters for describing the kinetics of molecular association are introduced. Then, we focus on theoretical and computational approaches for calculating association rates and discuss the current limitations of these approaches. The chapter ends with a review of recent computational advances in studying protein–protein and protein–nucleic acid association.

BIMOLECULAR ASSOCIATION

Molecules diffuse in the cellular environment, and, upon molecular recognition, can form bound complexes. Active transport processes may also contribute to binding but will not be discussed here. Bimolecular association can be considered to entail two steps. In the first step, an intermediate is formed by diffusion; this is called a diffusional encounter complex. In the second step, this intermediate evolves to form a tightly bound complex. Bimolecular association is diffusion controlled when the first step is rate limiting; it is reaction controlled when the second step determines the rate of association.

Diffusional Encounter Complex

Characterization of the diffusional encounter complex is important for protein and nucleic acid design studies aimed at altering the association kinetics. In diffusion-controlled processes, formation of the encounter complex determines the bimolecular association rate constant. The rate of diffusional association has an upper limit, that is, the binding of two molecules cannot be faster than their rate of collision. In aqueous solvent, this limit is around 10^9 $M^{-1}s^{-1}$ for uniformly reactive spheres of the size of a small- to medium-sized protein (Smoluchowski 1917) with no forces between them.

A random collision of two molecules does not usually result in binding. A freely diffusing molecule must come close to its binding patch on a target molecule and form a diffusional encounter complex. Geometrically, the encounter complex can be viewed as an ensemble of configurations able to evolve to the bound state. During a single encounter, the two molecules can undergo rotational reorientation while remaining trapped in the vicinity of each other and undergoing multiple collisions. This effect is known as a diffusive entrapment. A Brownian dynamics (BD) study (Northrup and Erickson 1992) of two noninteracting spheres of the size of small proteins showed that, because of the diffusive entrapment effect, the association rate was about 400 times larger (2×10^6 $M^{-1}s^{-1}$) than the rate calculated by a simple geometric correction of the Smoluchowski rate considering two contacts as the criterion for binding ($1 \times 10^4 M^{-1}s^{-1}$). An association rate constant of about 10^6 $M^{-1}s^{-1}$ is typical of protein–protein pairs that bind without strong electrostatic interactions. Attractive electrostatic forces can lead to higher rates very close to the Smoluchowski rate.

Bound Complex

After formation of the encounter complex, the biomolecules must adjust their positions to form a fully bound complex. As well as translation and reorientation, they may undergo changes in conformation and induced fit to achieve a bound complex. Within the complex, the biomolecules are held together by short-range noncovalent interactions such as salt bridges, hydrogen bonds, and van der Waals interactions. These interactions depend on the chemical properties of the interacting groups on both molecules as well as their spatial arrangement. The interactions may be mediated by individual water molecules. One or several binding sites on a biomolecule may stabilize the complex. A subtle change in the binding sites can change the binding mode significantly. As a result, biological associations are dependent on the structure of both molecules and can be highly specific.

MOLECULAR DIFFUSION

For a particle undergoing normal diffusion, the average value of the squared displacement (r) in n spatial dimensions is proportional to the time (t) elapsed

$$<r^2> = 2nDt \qquad (6.1)$$

where D is the diffusion coefficient. In some studies of molecular diffusion in cells and nuclei, anomalous diffusion has been observed with the displacement showing a smaller or larger dependence on time corresponding, respectively, to subdiffusion or superdiffusion (Dix and Verkman 2008).

The flux of particles (J) across a defined area is related to the concentration (C) gradient by Fick's first law

$$J = -D\nabla C \qquad (6.2)$$

Many transport phenomena are described by the continuity equation

$$\nabla \cdot J + \frac{\partial C}{\partial t} = 0 \qquad (6.3)$$

which describes the conservation of matter. Fick's second law, or the diffusion equation, can be derived from Equations 6.2 and 6.3

$$\frac{\partial C}{\partial t} = -D\Delta C \qquad (6.4)$$

When a particle moves in a fluid, it experiences friction to an extent depending on the properties of the fluid. The macroscopic quantity describing the internal resistance to flow is viscosity (η). For a moving sphere with radius r, it is inversely related to the diffusion coefficient (D) through the Stokes–Einstein formula

$$\eta = \frac{k_B T}{6\pi r D} \qquad (6.5)$$

where k_B is the Boltzmann constant and T is the temperature. The crowded cytoplasmic and nuclear environments have been observed to result in diffusion of small proteins such as GFP (green fluorescent protein) that is slower by a factor of about 4 than that observed in aqueous solution (Dix and Verkman 2008). The cellular environment is heterogeneous and thus it is a simplification to describe it by a macroscopic viscosity. Indeed, the crowded intracellular environment can, depending on solute size, result in subdiffusion (Dix and Verkman 2008).

Electrostatic Interactions

The interaction forces between biomolecules vary in strength, type, and origin, and a wide spectrum of forces contributes to complex formation (Motiejunas and Wade 2007). Here, we will discuss only electrostatic interactions as their contribution to the kinetics of bimolecular association has been shown to be considerable. Electrostatic interactions are important for bimolecular association because they are relatively long range and may therefore guide the association process by means of attractive and repulsive interactions. Their importance is shown by the dependence of association rates on ionic strength and pH and the generally much greater influence on the association rate of mutations of charged than of neutral residues.

Ionic solutions screen the electrostatic interactions between the molecular solutes. One way to treat the ions is to compute the molecular electrostatic potential (ϕ) using the nonlinear Poisson–Boltzmann equation

$$-\nabla \, \varepsilon(r) \, \nabla \, \phi(r) = \rho(r) + \sum_i q_i n_i e^{-\frac{q_i}{k_B T}} \tag{6.6}$$

where $\varepsilon(r)$ is the position dependent dielectric permittivity, $\rho(r)$ is the molecular charge density, and q_i and n_i are the charge and the concentration of the i-th ionic species in the bulk, respectively. Equation 6.6 can be approximated by the linear Poisson–Boltzmann equation if the exponential is expanded as a Taylor series

$$-\nabla \, \varepsilon(r) \, \nabla \, \varphi(r) + \varepsilon \cdot \kappa^2 \varphi = \rho(r) \tag{6.7}$$

where κ is the Debye–Hückel screening length. Equations 6.6 and 6.7 are used in studies of interactions between macromolecules in continuum solvent, that is, when water molecules and ions are not modeled explicitly.

When two approaching molecules come close in an aqueous solvent, an electrostatic desolvation effect arises due to the lower dielectric constant of the solute compared to that of the solvent. Charges located at the bimolecular complex interface become desolvated upon complex formation resulting in an unfavorable electrostatic energy change. This desolvation effect becomes significant at short distances and is mainly dependent on the location and magnitude of the charged groups.

REACTION RATES

If a molecule of type X forms a complex of type Z with a molecule of type Y, then the reaction kinetics are characterized by the association and dissociation rate constants, k_{on} and k_{off}, respectively,

$$X + Y \underset{k_{off}}{\overset{k_{on}}{\rightleftarrows}} Z \tag{6.8}$$

The rate constants can be related to an equilibrium association constant

$$K_a = \frac{k_{on}}{k_{off}}$$

(6.9)

The reciprocal of K_a is the equilibrium dissociation constant K_d. An analytical solution for the diffusion-controlled association constant K_{on} can be obtained for uniform spheres reacting at a center-to-center distance r (Smoluchowski 1917)

$$K_{on} = 4\pi r (D_X + D_Y),$$

(6.10)

where D_X and D_Y are the diffusion constants for species X and Y, respectively. Equation 6.10 is valid when there are no forces between the spheres. For interacting spheres, k_{on} is given by (Berg and von Hippel 1985)

$$k_{on} = \frac{4\pi r (D_X + D_Y)}{\int_r^\infty \frac{e^{U(r)/kT}}{r^2} dr}$$

(6.11)

where $U(r)$ is a centrosymmetric interaction potential between the spheres. For more complicated geometries and interaction forces, numerical approaches are necessary to compute association rates (see the next section).

THEORETICAL AND COMPUTATIONAL APPROACHES

For biological molecules, the bimolecular diffusional association rate constant can be computed using two distinct approaches. In the first approach (see sections "Particle-Based Approaches" and "Density-Distribution Approaches"), absolute rate constants are computed using a model that accounts for the forces between the interacting biomolecules as well as relevant properties of the cellular environment. Diffusional motion is treated with particle-based Brownian dynamics (BD) simulations (Northrup and Erickson 1992) or a density-distribution-based formalism (Schlosshauer and Baker 2004). In the second approach (see "Electrostatic Enhancement of Association Rates" section), the relative rather than the absolute association rate constants are computed from the interaction energy between the molecules.

PARTICLE-BASED APPROACHES

The calculation of bimolecular association rate constants by simulation of the diffusional motion of the interacting particles was first implemented by Northrup, Allison, and McCammon (NAM; Northrup, Allison, and McCammon 1984). In the NAM method, one of the interacting molecules is placed at the center of a sphere, while the other one starts Brownian moves at a distance b (see Figure 6.1A). The b distance is chosen such that there are no forces between the molecules at this separation or that the forces are centrosymmetric. For these cases, the rate constant for the molecules to approach a separation b can be computed from Equation 6.10 or Equation 6.11 with r = b. By generating thousands of trajectories and monitoring those that fulfill criteria for forming an encounter complex, the probability of reaction (β) can be obtained, and thus the association rate constant k_{on} calculated (see Figure 6.1A).

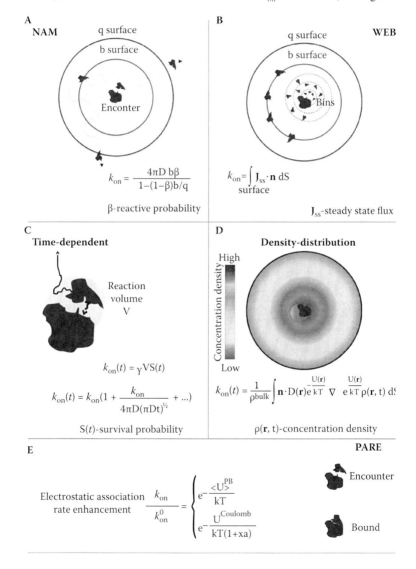

FIGURE 6.1   (SEE COLOR INSERT FOLLOWING PAGE 174.) Schematic figure of the methods for calculating bimolecular association rates (see "Theoretical and Computational Approaches" section for more details).

The trajectories are truncated when the molecules reach a separation q (at the "q sur-face"). Each trajectory is started from a randomly chosen position and orientation on the "b surface." The two diffusing molecules can be modeled in atomic detail.

Superoxide dismutase was the first system to which the NAM method was applied (Allison, Ganti, and McCammon 1985; Antosiewicz, Briggs, and McCammon 1996; Stroppolo et al. 2000). Subsequently, it has been applied to many diffusion-influ-enced enzymes (Wade 1996; Wade et al. 1998).

It has also been applied to compute protein–protein association rates by monitor-ing the formation of native polar contacts in the experimentally determined bound complex (Gabdoulline and Wade 1997, 2001; Elcock et al. 1999) or the formation of electron-transfer complexes (Northrup, Boles, and Reynolds 1988). A wide range of protein–protein interactions has been investigated with this approach, including gly-colytic enzymes interacting with actin filaments and antigen–antibody complexation (Northrup, Boles, and Reynolds 1988; Altobelli and Subramaniam 2000; Fogolari et al. 2000; Ouporov et al. 2001; Rienzo et al. 2001; Sept and McCammon 2001; Haddadian and Gross 2006).

Another formalism for using BD simulations to calculate association rates was developed by Huber and Kim (1996). It is called weighted-ensemble Brownian (WEB) dynamics and, in contrast to the NAM method, the diffusing particle is represented as an ensemble of weighted probability packets or pseudoparticles (see Figure 6.1B). Multiple BD trajectories are simulated in the available configuration space, which is divided into bins along the intermolecular reaction coordinate that are equally sampled (Huber and Kim 1996). The rate constants are obtained from the reactive steady-state flux J_{ss} (Figure 6.1B). The WEB method was found to be very efficient in calculating rates when there are large free energy barriers to asso-ciation (Rojnuckarin, Livesay, and Subramaniam 2000).

An alternative BD method deals with a pair of reactant biomolecules for which trajectories are started in a reaction region and the time-dependent probability of finding them again in this region in the absence of a reaction is calculated (Lee and Karplus 1987). Zhou designated this reaction region as a "reaction volume" (V; see Figure 6.1C) and expressed the time-dependent rate coefficient (k_{on}(t)) via the sur-vival probability (S(t)) of the reactant biomolecules started in V (Zhou and Szabo 1996). The steady-state association rate constant k_{on} is reached at long times with the known asymptotic behavior (Figure 6.1C).

DENSITY-DISTRIBUTION APPROACHES

In the density-distribution approach, a finite difference solution of the partial differ-ential diffusion equation is computed (Song et al. 2004; Cheng et al. 2007). In this continuum model, the concentration density $\rho(\mathbf{r},t)$ under appropriate boundary condi-tions is derived from the time-dependent Smoluchowski equation (see Figure 6.1D). Knowing the interaction potential ($U(\mathbf{r})$) between the reactants and the diffusion coef-ficients, the association rate constant can be computed (Figure 6.1D). This algorithm is computationally less demanding than BD simulations with the NAM approach when applied to enzyme-substrate association. However, the atomic-detail properties of the substrate cannot be treated as it is modeled by a density distribution.

ELECTROSTATIC ENHANCEMENT OF ASSOCIATION RATES

Zhou (1997) proposed that the variation in the rate constant for bimolecular association is dependent on the electrostatic interaction energy between proteins in transient intermediate configurations. Selzer and Schreiber (1999) showed that this approximation is valid for the bound complex of the proteins as well, and their algorithm, called PARE (Predicting Association Rate Enhancement), is implemented as a Web server (http://www.weizmann.ac.il/home/bcges/PARE.html). The difference between these two methods for estimating the rate enhancement is that in the first method (Zhou 1997) the average Poisson–Boltzmann electrostatic interaction energy ($<U>^{PB}$) in the encounter complex is calculated, while in the second method (Selzer and Schreiber 1999), the Coulombic interaction energy ($U^{Coulomb}$) in the bound complex is computed and the ionic environment is accounted for by a Debye–Hückel term (see Figure 6.1e). A disadvantage of these methods is that only the rate enhancement can be predicted, whereas the basal rate (k_{on}^{0}) should be computed by another method or obtained from experiments. On the other hand, an advantage of this approach is that it can be used for rapid, structure-based calculation of the electrostatic steering of the association of two proteins (Schreiber, Shaul, and Gottschalk 2006). Using this approach one can design faster and tighter binding proteins by optimizing the electrostatic interaction between a reactant protein–protein pair (Schreiber, Shaul, and Gottschalk 2006) and faster enzymes can be designed by altering the electrostatic potential in the active site (Zhou, Wong, and Vijayakumar 1997).

RECENT ADVANCES IN COMPUTATIONAL APPROACHES

PROTEIN–PROTEIN INTERACTIONS

Computation of Association Rates

Most of the methods depicted in Figure 6.1 have been applied to kinetic studies of protein–protein association, particularly focusing on electrostatic enhancement of association rates (Gabdoulline and Wade 2002). Most applications to protein–protein association involve solving the Poisson–Boltzmann (PB) equation. It has been shown that the degree of accuracy of prediction of association rates is dependent on both the definition of the solute–solvent dielectric boundary and the use of a linear or nonlinear PB equation (Alsallaq and Zhou 2007, 2008a). As mentioned earlier, electrostatic rate enhancement can be used as a criterion for protein design (Schreiber, Shaul, and Gottschalk 2006), and a study of 68 transient heteroprotein–protein complexes showed electrostatic steering leading to an increase of over 100-fold in k_{on} for about 25% of the complexes studied (Shaul and Schreiber 2005).

Dissection of the Determinants of Binding

The association and dissociation rates together determine the binding affinity. It has been shown that it is possible to design mutants that change the binding affinity by changing only the association rates for a number of different protein complexes (Selzer, Albeck, and Schreiber 2000; Schreiber, Shaul, and Gottschalk 2006). For example, in the case of Cdc25B phosphatase binding to its Cdk2-pTpY/CycA substrate, the regions most important for association and for dissociation are clearly distinguished. In the phosphatase, the former are remote hot spot residues, while the latter are at the active site (Sohn, Buhrman, and Rudolph 2007). Large (>100-fold) changes in association rates were observed upon single point mutation, thus dramatically changing the binding affinity (Sohn, Buhrman, and Rudolph 2007). Another example is a triple mutant of the Ras effector protein Ral, a guanine nucleotide dissociation stimulator (RalGDS) that was designed to bind faster than the wild-type protein to the Ras protein and was found experimentally to bind 14 times faster (Kiel et al. 2004). The mutant had binding properties similar to Raf, another Ras effector.

Quantification of the Encounter Complex

The diffusional encounter complex, an intermediate state marking the endpoint of diffusion of two proteins toward each other, plays an important role in determining the association rates. However, its structure cannot be directly determined experimentally (Gabdoulline and Wade 1999).

The encounter complex is an ensemble of target positions near the bound complex, and achieving this ensemble in BD simulations ensures subsequent binding of the molecules when the association is diffusion controlled. The encounter complex can be expected to be near to, but not coincident with, the transition state for binding. The nature of this transition state for the association of barnase and barstar was investigated in double-mutant cycle experiments by Frisch, Fersht, and Schreiber (2001). They found evidence for contacts between charged groups. The activation entropy of the transition state was found to be small, indicating a small degree of desolvation. The residue–residue contacts maintained in the transition state differed at low and high ionic strength, indicating that the structure of the intermediate state changes with changing solvent conditions. All these findings regarding the transition state are consistent with the models of the encounter complexes generated by BD simulation (Gabdoulline and Wade 1997).

The structure of the transition state has been investigated by introducing mutations that alter association rates and modeling bimolecular configurations that fit experimental data (Miyashita, Onuchic, and Okamura 2004; Harel, Cohen, and Schreiber 2007). Recently, a BD study (Spaar et al. 2006) showed that the structure of the encounter complex is affected by mutations making it difficult to precisely characterize the encounter complex using mutational data. A detailed picture of the association dynamics of hydrogenase HydA2 and ferredoxin PetF1 was revealed by combining BD and molecular dynamics (MD) simulations (Long et al. 2008), and this enabled a transition state ensemble of configurations for electron transfer to be quantified. Very recently, it has become possible to quantify such transient intermediate complexes using long-range distance restraints derived from paramagnetic nuclear magnetic resonance (NMR) methods (Tang, Iwahara, and Clore 2006; Volkov et al. 2006) and this is expected to shed more light on the nature of encounter complexes. Indeed, Kim et al. (2008) have combined these NMR techniques with replica exchange MD simulations with a coarse-grain model to identify specific and nonspecific binding configurations in a transient protein–protein encounter complex.

Induced Fit Phenomena

In general, there is more than one intermediate state in the association process because protein–protein binding consists of multiple steps: diffusion, conformer selection, and refolding or induced fit (Gruenberg, Leckner, and Nilges 2004). It is not simple, however, to quantify all intermediates experimentally, although it can be shown in some cases that a one-step model of association is not sufficient (Kourentzi et al. 2008).

An extreme case of induced fit occurs when the protein folds or refolds upon binding to its partner (Levy et al. 2005). It was shown (Levy, Onuchic, and Wolynes 2007) that this may be followed by a fly-casting effect coupled to electrostatic steering for the Ets domain of SAP-1 protein binding to its specific DNA sequence. A significant induced fit was found in the case of fasciculin 2 (Fas2) binding to acetylcholinesterase (AChE), two proteins that bind with a very high association rate constant. It was found that the conformation of Fas2 able to bind AChE is not stable in the unbound form of Fas2 and that the association process should follow a conformational change of a stable form of Fas2 that is not complementary to AChE (Bui and McCammon 2006; Bui et al. 2006).

Crowding Phenomena

The influence of crowding agents cannot be explained simply as the action of obstacles, volume exclusion, or the change in the solvent viscosity, because there is a complex dependence of the solute molecular dynamics and reactions in crowded solutions on the properties of the molecular interactions in the system. An inverse linear relation was found between translational diffusion of proteins and viscosity in almost all solutions tested, in accordance with the Stokes–Einstein relation. Conversely, no simple relation was found between either rotational diffusion rates or association rates (k_{on}) and viscosity (Kozer et al. 2007). In all crowded solutions, the measured absolute k_{on} values, but not the k_{off} values, were found to be lower than in buffer. In the presence of low mass crowding agents, k_{on} depends inversely on the solution viscosity. In high mass polymer solutions, k_{on} changes only slightly, even at viscosities 12-fold higher than water (Kozer and Schreiber 2004). See also a recent review on this topic (Zhou 2008). Simulations using a model at the one spherical particle/macromolecule level of barnase–barstar association in crowded solutions designed to represent the cytoplasmic environment revealed a biphasic time course, indicating that crowding exerts different effects over different timescales (Ridgway et al. 2008). Crowding influences not only the rates but also the equilibrium parameters of chemical reactions (Chebotareva 2007) making the quantitative description of crowding phenomena for enzymes *in vivo* even more difficult.

PROTEIN–NUCLEIC ACID INTERACTIONS

Computation of Association Rates

The backbone of nucleic acids contains negatively charged phosphate groups. This negative electrostatic potential leads to attraction of nucleic acids to proteins with positive binding sites. Therefore, the formation of a nucleic acid–protein complex is strongly governed by electrostatic interactions, which enhance the association rate.

Such rate enhancement was predicted by applying the PARE method (see Figure 6.1e) to an atomistic model of protein–RNA (U1A-U1SLII) interactions (Qin and Zhou 2008). The results, based on changing the ionic strength and making mutations, have been shown to be consistent with experiments (Qin and Zhou 2008). In another study, BD simulations were carried out for the translation protein eIF4E binding to five analogous mRNA cap molecules. The association rates were computed for varying electrostatic and hydrodynamic interactions in the system and displayed values very close to the rates determined by experiment (Bachut-Okrasinska and Antosiewicz 2007).

Allsallaq and Zhou (2008b) developed a theoretical model showing that nonspecific binding to DNA enhances the protein–nucleic acid association rate and that binding to a linear DNA leads to a slightly higher association rate constant than to a circular DNA. The formation of an open complex of DNA and a bacterial RNA polymerase (RNAP) from the closed one was analyzed by Djordjevic and Bundschuh (2008). The rate of formation of the open complex was derived from a quantitative model for a reversible two-step binding mechanism. The authors found that it depends on the interaction energies of the closed and opened complexes as well as on the DNA duplex melting energy.

Specificity and Nonspecificity

Binding Dynamics

Recently much effort has been put into understanding the dynamics of DNA-binding proteins: how they search for their target molecule, what interactions govern this process, how specifically they locate the binding site, and so on. The way in which the proteins bind to the DNA (specifically or nonspecifically) can explain the observation of association kinetic rates higher than the Smoluchowski rate in some protein–DNA studies (Halford and Marko 2004). These studies suggest three-dimensional (3D) diffusion of the protein to the DNA followed by one-dimensional (1D) diffusion of the protein along the DNA to form a bound complex. This type of diffusion is referred to in the literature as facilitated diffusion.

Facilitated Diffusion

Slutsky and Mirny (2004) proposed that for an optimal search for the target DNA, a protein should spend half of its time in 3D diffusion and the other half in 1D diffusion, sliding along the DNA. Their study aimed at quantitatively investigating the specific and nonspecific binding of proteins to DNA. However, a theoretical lattice Monte Carlo study (Rezania, Tuszynski, and Hendzel 2007) of transcription factors (TFs) binding to DNA molecules showed that even if only 15% of the diffusional search time is spent freely in solution, the timescale of target location is consistent with experimental measurements. In this diffusional search, the TFs might exhibit conformational changes, which could affect the association rate constant. Such conformational changes during the searching and sliding mechanisms have been investigated to detect the shortest binding time to the DNA consistent with thermodynamics (Hu, Grosberg, and Bruinsma 2008). The simultaneous interactions of multiple proteins with a long DNA chain have been investigated using Monte Carlo simulations (van der Heijden and Dekker 2008). Three possible interactions were proposed: noncooperative/cooperative binding, position-dependent dissociation, and linear motion along the DNA. It was found that noncooperative binding leads to gaps on the DNA that are smaller than the size of the protein binding site and therefore to an overestimation of the apparent size of the binding site of the protein by as much as 30%. For cooperatively bound proteins, the protein–DNA dissociation curve showed exponential behavior indicating the importance of the cooperativity in the protein–DNA interactions (van der Heijden and Dekker 2008). Murugan has developed a generalized theory based on the assumption that first the protein binds nonspecifically to DNA by 3D diffusion and second experiences 1D diffusion to locate the specific DNA binding site (Murugan 2007).

Intersegment Transfer

Some nucleic acid binding proteins have multiple binding sites (Brown, Izard, and Misteli 2006) allowing them to bind simultaneously to several nucleic acid binding sites. For example, a protein can jump from one DNA segment to another without dissociating, a process called intersegment transfer. In this way, the protein can bind specifically to the target site and a rate enhancement can be observed (Hu and Shklovskii 2007). Lattice simulations (Wedemeier et al. 2008) showed that increasing the nucleic acid chain density increases the protein diffusion in the case of intersegment transfer. Moreover, the diffusion coefficient appeared to be reciprocal to the chain density in 1D sliding on the DNA (Wedemeier et al. 2008).

In summary, several factors contribute to the high association rates for nucleic acid binding proteins, such as transcription factors, and DNA; these include 1D diffusion, intersegment transfer, and conformational changes upon binding (Alsallaq and Zhou 2008b).

Chromatin Models

Chromatin is a biological structure occurring in the cell nucleus that consists of a highly packed DNA molecule and histone proteins. The positively charged histones attract the negative DNA molecule, which wraps around them, and together they form a single unit called the nucleosome. The conformation and compaction of the chromatin depend on the interactions between the nucleosomes as well as on the presence of other factors influencing chromatin dynamics. Since chromatin has features on different time and length scales, a considerable number of theoretical models exist that aim to elucidate the driving forces for chromatin compaction. Chromatin has been modeled on a coarse-grained level in which several atoms, residues, or the whole nucleosome are represented as a single geometrical object; the interactions involved are included; and the dynamics are simulated either by BD or Monte Carlo (Arya, Zhang, and Schlick 2006; Merlitz et al. 2006; Langowski and Heermann 2007; Kepper et al. 2008; Stehr et al. 2008). Attempts to predict the conformation of chromatin fiber have been made at an atomistic level as well (Wong, Victor, and Mozziconacci 2007). Some studies have focused on the interactions involved at a single nucleosome level, and the binding dynamics of the linker histone to the nucleosome have been investigated experimentally (Brown, Izard, and Misteli 2006) and theoretically (Fan and Roberts 2006). The binding of the linker histone and its stoichiometry as well as the nucleosome repeat lengths influence

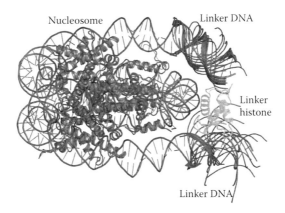

FIGURE 6.2 Position of the linker histone GH5 (light gray) generated by BD simulations. It is located between different conformations of the linker DNAs generated by normal mode analysis (Pachov, Gabdoulline, and Wade 2007, unpublished data).

chromatin compaction leading to topologically differing fibers (Routh, Sandin, and Rhodes 2008). Recently, it was found that the binding mode of the linker histone GH5 is robust to a wide range of linker DNA conformations (see Figure 6.2) (Pachov, Gabdoulline, and Wade 2007, unpublished data). This was revealed by all-atom BD simulations, which also showed two main binding sites on the GH5, in agreement with experimental data (Brown, Izard, and Misteli 2006). In many biological processes, like DNA transcription, replication, and repair, the proteins involved must quickly find their target site. The kinetics of such a process are directly influenced by the level of DNA exposure and histone tail acetylation on the nucleosome (Kampmann 2005), and these are also topics being studied by computational simulation.

OUTLOOK

We have discussed characterization of the kinetics of biomacromolecular complex formation from a theoretical and computational perspective. New experimental techniques and methods are being developed to study the interactions between biomolecules over different time and length scales. However, these techniques are still insufficient to precisely describe and quantify the detailed dynamics of associating biomacromolecules. Here, computational approaches can be of value because they provide a detailed description of the association process. On the other hand, simulations of macromolecular complexation are computationally demanding and require the use of approximations such as the neglect of molecular flexibility. Furthermore, establishing the effects on macromolecular association of the heterogeneous and crowded cellular environment is a challenge for both computational and experimental approaches. Surmounting these hurdles requires the development of multiple-scale and coarse-grained models with more accurate molecular interaction force fields as well as the development of highly parallelized software and new computing hardware to permit detailed simulations over many orders of time and length scales.

ACKNOWLEDGMENTS

We thank the Klaus Tschira Foundation (KTS), the German Research Foundation (DFG), and the Center for Modelling and Simulation in the Biosciences (BIOMS). We are grateful to Professor Dr. Jörg Langowski for helpful discussion, and Anna Feldman-Salit and Paolo Mereghetti for comments on the manuscript.

REFERENCES

Allison, S. A., G. Ganti, and J. A. McCammon. 1985. Simulation of the diffusion-controlled reaction between superoxide and superoxide dismutase. I. Simple models. *Biopolymers* 24 (7):1323–1336.

Alsallaq, R., and H.-X. Zhou. 2007. Prediction of protein–protein association rates from a transition-state theory. *Structure* 15 (2):215–224.

Alsallaq, R., and H.-X. Zhou. 2008a. Electrostatic rate enhancement and transient complex of protein–protein association. *Proteins* 71 (1):320–335.

Alsallaq, R., and H.-X. Zhou. 2008b. Protein association with circular DNA: Rate enhancement by nonspecific binding. *J Chem Phys* 128 (11):115108.

Altobelli, G., and S. Subramaniam. 2000. Kinetics of association of anti-lysozyme monoclonal antibody D44.1 and hen-egg lysozyme. *Biophys J* 79 (6):2954–2965.

Antosiewicz, J., J. M. Briggs, and J. A. McCammon. 1996. Orientational steering in enzyme-substrate association: Ionic strength dependence of hydrodynamic torque effects. *Eur Biophys J* 24 (3):137–141.

Arya, G., Q. Zhang, and T. Schlick. 2006. Flexible histone tails in a new mesoscopic oligonucleosome model. *Biophys J* 91 (1):133–150.

Bachut-Okrasinska, E., and J. M. Antosiewicz. 2007. Brownian dynamics simulations of binding mRNA cap analogues to eIF4E protein. *J Phys Chem B* 111 (45):13107–13115.

Berg, O. G., and P. H. von Hippel. 1985. Diffusion-controlled macromolecular interactions. *Annu Rev Biophys Biophys Chem* 14:131–160.

Brown, D. T., T. Izard, and T. Misteli. 2006. Mapping the interaction surface of linker histone H1(0) with the nucleosome of native chromatin in vivo. *Nat Struct Mol Biol* 13 (3):250–255.

Bui, J. M., and J. A. McCammon. 2006. Protein complex formation by acetylcholinesterase and the neurotoxin fasciculin-2 appears to involve an induced-fit mechanism. *Proc Natl Acad Sci USA* 103 (42):15451–15456.

Bui, J. M., Z. Radic, P. Taylor, and J. A. McCammon. 2006. Conformational transitions in protein–protein association: Binding of fasciculin-2 to acetylcholinesterase. *Biophys J* 90 (9):3280–3287.

Chebotareva, N. A. 2007. Effect of molecular crowding on the enzymes of glycogenolysis. *Biochemistry (Mosc)* 72 (13):1478–1490.

Cheng, Y., J. K. Suen, D. Zhang, S. D. Bond, Y. Zhang, Y. Song, N. A. Baker, C. L. Bajaj, M. J. Holst, and J. A. McCammon. 2007. Finite element analysis of the time-dependent Smoluchowski equation for acetylcholinesterase reaction rate calculations. *Biophys J* 92 (10):3397–3406.

Dix, J. A., and A. S. Verkman. 2008. Crowding effects on diffusion in solutions and cells. *Annu Rev Biophys* 37:247–263.

Djordjevic, M., and R. Bundschuh. 2008. Formation of the open complex by bacterial RNA polymerase—a quantitative model. *Biophys J* 94 (11):4233–4248.

Elcock, A. H., R. R. Gabdoulline, R. C. Wade, and J. A. McCammon. 1999. Computer simulation of protein–protein association kinetics: Acetylcholinesterase-fasciculin. *J Mol Biol* 291 (1):149–162.

Elf, J., G.-W. Li, and X. S. Xie. 2007. Probing transcription factor dynamics at the single-molecule level in a living cell. *Science* 316 (5828):1191–1194.

Fan, L., and V. A. Roberts. 2006. Complex of linker histone H5 with the nucleosome and its implications for chromatin packing. *Proc Natl Acad Sci USA* 103 (22):8384–8389.

Fogolari, F., R. Ugolini, H. Molinari, P. Viglino, and G. Esposito. 2000. Simulation of electrostatic effects in Fab-antigen complex formation. *Eur J Biochem* 267 (15):4861–4869.

Frisch, C., A. R. Fersht, and G. Schreiber. 2001. Experimental assignment of the structure of the transition state for the association of barnase and barstar. *J Mol Biol* 308 (1):69–77.

Gabdoulline, R. R., and R. C. Wade. 1997. Simulation of the diffusional association of barnase and barstar. *Biophys J* 72 (5):1917–1929.

Gabdoulline, R. R., and R. C. Wade. 1999. On the protein–protein diffusional encounter complex. *J Mol Recognit* 12 (4):226–234.

Gabdoulline, R. R., and R. C. Wade. 2001. Protein–protein association: Investigation of factors influencing association rates by Brownian dynamics simulations. *J Mol Biol* 306 (5):1139–1155.

Gabdoulline, R. R., and R. C. Wade. 2002. Biomolecular diffusional association. *Curr Opin Struct Biol* 12 (2):204–213.

Gruenberg, R., J. Leckner, and M. Nilges. 2004. Complementarity of structure ensembles in protein–protein binding. *Structure* 12 (12):2125–2136.

Haddadian, E. J., and E. L. Gross. 2006. A Brownian dynamics study of the interactions of the luminal domains of the cytochrome b6f complex with plastocyanin and cytochrome c6: The effects of the Rieske FeS protein on the interactions. *Biophys J* 91 (7):2589–2600.

Halford, S. E., and J. F. Marko. 2004. How do site-specific DNA-binding proteins find their targets? *Nucleic Acids Res* 32 (10):3040–3052.

Harel, M., M. Cohen, and G. Schreiber. 2007. On the dynamic nature of the transition state for protein–protein association as determined by double-mutant cycle analysis and simulation. *J Mol Biol* 371 (1):180–196.

Hu, L., A. Y. Grosberg, and R. Bruinsma. 2008. Are DNA transcription factor proteins Maxwellian demons? *Biophys J* 95 (3):1151–1156.

Hu, T., and B. I. Shklovskii. 2007. How a protein searches for its specific site on DNA: The role of intersegment transfer. *Phys Rev E Stat Nonlin Soft Matter Phys* 76 (5 Pt 1):051909.

Huber, G. A., and S. Kim. 1996. Weighted-ensemble Brownian dynamics simulations for protein association reactions. *Biophys J* 70 (1):97–110.

Jucovic, M., and R. W. Hartley. 1996. Protein–protein interaction: A genetic selection for compensating mutations at the barnase–barstar interface. *Proc Natl Acad Sci USA* 93 (6):2343–2347.

Kampmann, M. 2005. Facilitated diffusion in chromatin lattices: Mechanistic diversity and regulatory potential. *Mol Microbiol* 57 (4):889–899.

Kepper, N., D. Foethke, R. Stehr, G. Wedemann, and K. Rippe. 2008. Nucleosome geometry and internucleosomal interactions control the chromatin fiber conformation. *Biophys J.* 95 (8):3692–3705.

Kiel, C., T. Selzer, Y. Shaul, G. Schreiber, and C. Herrmann. 2004. Electrostatically optimized Ras-binding Ral guanine dissociation stimulator mutants increase the rate of association by stabilizing the encounter complex. *Proc Natl Acad Sci USA* 101 (25):9223–9228.

Kim, Y. C., C. Tang, G. M. Clore, and G. Hummer. 2008. Replica exchange simulations of transient encounter complexes in protein–protein association. *Proc Natl Acad Sci USA* 105 (35):12855–12860.

Kourentzi, K., M. Srinivasan, S. J. Smith-Gill, and R. C. Willson. 2008. Conformational flexibility and kinetic complexity in antibody-antigen interactions. *J Mol Recognit* 21 (2):114–121.

Kozer, N., Y. Y. Kuttner, G. Haran, and G. Schreiber. 2007. Protein–protein association in polymer solutions: From dilute to semidilute to concentrated. *Biophys J* 92 (6):2139–2149.

Kozer, N., and G. Schreiber. 2004. Effect of crowding on protein–protein association rates: Fundamental differences between low and high mass crowding agents. *J Mol Biol* 336 (3):763–774.

Langowski, J., and D. W. Heermann. 2007. Computational modeling of the chromatin fiber. *Semin Cell Dev Biol* 18 (5):659–667.

Lee, S., and M. Karplus. 1987. Kinetics of diffusion-influenced bimolecular reactions in solution. I. General formalism and relaxation kinetics of fast reversible reactions. *J. Chem. Phys.* 86:1883–1903.

Levy, Y., S. S. Cho, J. N. Onuchic, and P. G. Wolynes. 2005. A survey of flexible protein binding mechanisms and their transition states using native topology based energy landscapes. *J Mol Biol* 346 (4):1121–1145.

Levy, Y., J. N. Onuchic, and P. G. Wolynes. 2007. Fly-casting in protein-DNA binding: Frustration between protein folding and electrostatics facilitates target recognition. *J Am Chem Soc* 129 (4):738–739.

Long, H., C. H. Chang, P. W. King, M. L. Ghirardi, and K. Kim. 2008. Brownian dynamics and molecular dynamics study of the association between hydrogenase and ferredoxin from *Chlamydomonas reinhardtii*. *Biophys* 95 (8):3753–3766.

Merlitz, H., K. V. Klenin, C.-X. Wu, and J. Langowski. 2006. Facilitated diffusion of DNA-binding proteins: Simulation of large systems. *J Chem Phys* 125 (1):014906.

Miyashita, O., J. N. Onuchic, and M. Y. Okamura. 2004. Transition state and encounter complex for fast association of cytochrome c2 with bacterial reaction center. *Proc Natl Acad Sci USA* 101 (46):16174–16179.

Motiejunas, D., and R. C. Wade. 2007. Structural, energetic, and dynamic aspects of ligand receptor interactions. In *Comprehensive Medicinal Chemistry II* (Vol. 4), edited by D. J. Triggle and J. B. Taylor. Elsevier, Oxford.

Murugan, R. 2007. Generalized theory of site-specific DNA-protein interactions. *Phys Rev E Stat Nonlin Soft Matter Phys* 76 (1 Pt 1):011901.

Northrup, S. H., J. O. Boles, and J. C. Reynolds. 1988. Brownian dynamics of cytochrome c and cytochrome c peroxidase association. *Science* 241 (4861):67–70.

Northrup, S. H., and H. P. Erickson. 1992. Kinetics of protein–protein association explained by Brownian dynamics computer simulation. *Proc Natl Acad Sci USA* 89 (8):3338–3342.

Northrup, S. H., S. A. Allison, and J. A. McCammon. 1984. Brownian dynamics simulation of diffusion-influenced bimolecular reactions. *J. Chem. Phys.* 80:1517–1524.

Ouporov, I. V., H. R. Knull, S. L. Lowe, and K. A. Thomasson. 2001. Interactions of glyceraldehyde-3-phosphate dehydrogenase with G- and F-actin predicted by Brownian dynamics. *J Mol Recognit* 14 (1):29–41.

Pachov, G. V., R. R. Gabdoulline, and R. C. Wade. 2007. Simulation of linker histone–chromatin interactions. NIC Series 36: 69–74.

Qin, S., and H.-X. Zhou. 2008. Prediction of salt and mutational effects on the association rate of U1A protein and U1 small nuclear RNA stem/loop II. *J Phys Chem B* 112 (19):5955–5960.

Quinn, D. M. 1987. Acetylcholinesterase: Enzyme structure, reaction dynamics, and virtual transition states. *Chem Rev* 87:955–979.

Rezania, V., J. Tuszynski, and M. Hendzel. 2007. Modeling transcription factor binding events to DNA using a random walker/jumper representation on a 1D/2D lattice with different affinity sites. *Phys Biol* 4 (4):256–267.

Ridgway, D., G. Broderick, A. Lopez-Campistrous, M. Ru'aini, P. Winter, M. Hamilton, P. Boulanger, A. Kovalenko, and M. J. Ellison. 2008. Coarse-grained molecular simulation of diffusion and reaction kinetics in a crowded virtual cytoplasm. *Biophys J* 94 (10):3748–3759.

Rienzo, F. De, R. R. Gabdoulline, M. C. Menziani, P. G. De Benedetti, and R. C. Wade. 2001. Electrostatic analysis and Brownian dynamics simulation of the association of plasto-cyanin and cytochrome f. *Biophys J* 81 (6):3090–3104.

Rojnuckarin, A., D. R. Livesay, and S. Subramaniam. 2000. Bimolecular reaction simulation using weighted ensemble Brownian dynamics and the University of Houston Brownian Dynamics program. *Biophys J* 79 (2):686–693.

Routh, A., S. Sandin, and D. Rhodes. 2008. Nucleosome repeat length and linker histone stoichiometry determine chromatin fiber structure. *Proc Natl Acad Sci USA* 105 (26):8872–8877.

Schlosshauer, M., and D. Baker. 2004. Realistic protein–protein association rates from a simple diffusional model neglecting long-range interactions, free energy barriers, and landscape ruggedness. *Protein Sci* 13 (6):1660–1669.

Schreiber, G., Y. S., and K. E. Gottschalk. 2006. Electrostatic design of protein–protein asso-ciation rates. *Methods Mol Biol* 340:235–249.

Selzer, T., S. Albeck, and G. Schreiber. 2000. Rational design of faster associating and tighter binding protein complexes. *Nat Struct Biol* 7 (7):537–541.

Selzer, T., and G. Schreiber. 1999. Predicting the rate enhancement of protein complex forma-tion from the electrostatic energy of interaction. *J Mol Biol* 287 (2):409–419.

Sept, D., and J. A. McCammon. 2001. Thermodynamics and kinetics of actin filament nucle-ation. *Biophys J* 81 (2):667–674.

Shaul, Y., and G. Schreiber. 2005. Exploring the charge space of protein–protein association: A proteomic study. *Proteins* 60 (3):341–352.

Slutsky, M., and L. A. Mirny. 2004. Kinetics of protein-DNA interaction: Facilitated target location in sequence-dependent potential. *Biophys J* 87 (6):4021–4035.

Smoluchowski, M. V. 1917. Versuch einer mathematischen Theorie der Koagulationskinetik kolloider Losungen. *Z Phys Chem* 92:129–168.

Sohn, J., G. Buhrman, and J. Rudolph. 2007. Kinetic and structural studies of specific pro-tein–protein interactions in substrate catalysis by Cdc25B phosphatase. *Biochemistry* 46 (3):807–818.

Song, Y., Y. Zhang, T. Shen, C. L. Bajaj, J. A. McCammon, and N. A. Baker. 2004. Finite ele-ment solution of the steady-state Smoluchowski equation for rate constant calculations. *Biophys J* 86 (4):2017–2029.

Spaar, A., C. Dammer, R. R. Gabdoulline, R. C. Wade, and V. Helms. 2006. Diffusional encounter of barnase and barstar. *Biophys J* 90 (6):1913–1924.

Stehr, R., N. Kepper, K. Rippe, and G. Wedemann. 2008. The effect of the internucleosomal interaction on the folding of the chromatin fiber. *Biophys J* 95 (8):3677–3691.

Stroppolo, M. E., A. Pesce, M. Falconi, P. O'Neill, M. Bolognesi, and A. Desideri. 2000. Single mutation at the intersubunit interface confers extra efficiency to Cu,Zn superox-ide dismutase. *FEBS Lett* 483 (1):17–20.

Tang, C., J. Iwahara, and G. M. Clore. 2006. Visualization of transient encounter complexes in protein–protein association. *Nature* 444 (7117):383–386.

van der Heijden, T., and C. Dekker. 2008. Monte Carlo simulations of protein assembly, disas-sembly, and linear motion on DNA. *Biophys J* 95 (10):4560–4569.

Volkov, A. N., J. A. R. Worrall, E. Holtzmann, and M. Ubbink. 2006. Solution structure and dynamics of the complex between cytochrome c and cytochrome c peroxidase deter-mined by paramagnetic NMR. *Proc Natl Acad Sci USA* 103 (50):18945–18950.

Wade, R. C. 1996. Brownian dynamics simulations of enzyme-substrate encounter. *Biochem Soc Trans* 24 (1):254–259.

Wade, R. C., R. R. Gabdoulline, S. K. Luedemann, and V. Lounnas. 1998. Electrostatic steer-ing and ionic tethering in enzyme-ligand binding: Insights from simulations. *Proc Natl Acad Sci USA* 95 (11):5942–5949.

Wang, Y., B. J. Shen, and W. Sebald. 1997. A mixed-charge pair in human interleukin 4 dominates high-affinity interaction with the receptor alpha chain. *Proc Natl Acad Sci USA* 94 (5):1657–1662.

Wedemeier, A., T. Zhang, H. Merlitz, C.-X. Wu, and J. Langowski. 2008. The role of chromatin conformations in diffusional transport of chromatin-binding proteins: Cartesian lattice simulations. *J Chem Phys* 128 (15):155101.

Wong, H., J.-M. Victor, and J. Mozziconacci. 2007. An all-atom model of the chromatin fiber containing linker histones reveals a versatile structure tuned by the nucleosomal repeat length. *PLoS ONE* 2 (9):e877.

Zhou, H.-X. 1997. Enhancement of protein–protein association rate by interaction potential: Accuracy of prediction based on local Boltzmann factor. *Biophys J* 73 (5):2441–2445.

Zhou, H.-X. 2008. Effect of mixed macromolecular crowding agents on protein folding. *Proteins* 72 (4):1109–1113.

Zhou, H.-X., and A. Szabo. 1996. Theory and simulation of the time-dependent rate coefficients of diffusion-influenced reactions. *Biophys J* 71 (5):2440–2457.

Zhou, H.-X., K. Y. Wong, and M. Vijayakumar. 1997. Design of fast enzymes by optimizing interaction potential in active site. *Proc Natl Acad Sci USA* 94 (23):12372–12377.

7 Computational Design of Protein–Protein Interactions

Julia M. Shifman

CONTENTS

INTRODUCTION

Protein–protein interactions determine the outcome of all cellular processes including signal transduction, cell division, DNA replication, transcription and translation, biosynthesis, and degradation. Hence, modulating protein–protein interactions is of great interest for both basic science and applied research such as drug design. Directed evolution and combinatorial screening techniques are powerful and well-established means of engineering protein complexes with enhanced affinity and binding specificity. Although very successful in obtaining the end product, these techniques do not address some basic questions such as what makes a particular protein a high-affinity binder or how to obtain a protein with slightly different binding characteristics. Computational approaches to modulating protein–protein interactions are directed toward answering these fundamental questions. These approaches, in principle, provide a fast and efficient way to supply proteins with desired binding properties. However, computational techniques require high-resolution structures for the protein–protein complexes, which are not always available. In addition, they rely on our still incomplete knowledge of the physical basis for protein binding affinity and specificity. Due to these limitations, relatively few successful examples of computationally designed protein–protein interactions have been reported. However, even when not completely successful, investigations of this type greatly advance our understanding of the molecular forces that govern protein binding. With the exponentially growing number of new structures of protein–protein complexes and a constant progress in method development, computational approaches are evolving into a generally accepted strategy for modulating protein–protein interactions. This chapter reviews the methods for design of protein–protein interactions, summarizes keystone studies in this area, and points out future directions of research.

COMPUTATIONAL METHODS FOR MODULATING BINDING INTERACTIONS

Two different structure-based strategies have been applied to the redesign of protein–protein interactions. The first strategy involves optimizing the long-range electrostatic interactions between the two proteins in the complex by introducing charge altering mutations on the protein surface frequently outside of the binding interface. The second strategy, usually referred to as computational protein design, uses side-chain repacking algorithms to predict mutations that lead to better packing, hydrogen bonding, and solvation directly at the binding interface. While the first strategy is rather tolerant to imperfections in structural models of protein–protein complexes, the second strategy relies on the accurate description of interactions across the binding interface, with even small inaccuracies frequently resulting in erroneous predictions. Although more error prone, the second strategy would provide us with a more universal approach to design of protein–protein interactions.

OPTIMIZATION OF LONG-TERM ELECTROSTATIC INTERACTIONS

To enhance association rates of protein complexes, Selzer et al. proposed optimization of long-range electrostatic interactions between the two binding partners.[1] This approach is based on an observation that the association rate, k_{on}, is affected mostly by mutations involving charge alterations, while mutations of uncharged residues have minor effects on k_{on}.[2–4] Optimization of the long-range electrostatic interactions between the two proteins leads to faster formation of an encounter complex that subsequently relaxes to the final stereospecific complex (Figure 7.1A).[5] A program called PARE (Protein Association Rate Enhancement) calculates k_{on} of binding by computing the difference in the Debye–Hückel energy between the two individual proteins and their complex.[1] k_{on} can be increased by introducing mutations that improve charge complementarity of the binding interface. The described method does not allow us to make any predictions about the dissociation rate, k_{off}, and hence the K_d. Nevertheless, by picking the mutation sites wisely, for example, in the vicinity of the binding interface rather than in its center, it has been possible to preserve the k_{off} values similar to the wild type.[1] Hence, in principle, improvement in both k_{on} and K_d could be achieved using the described strategy.

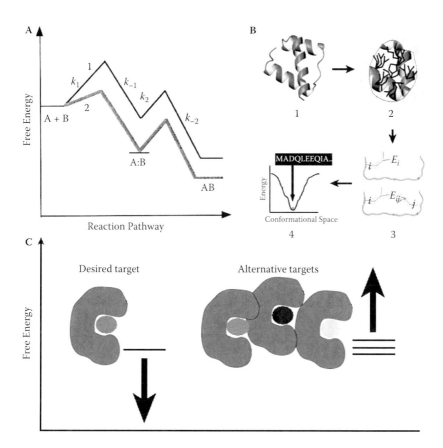

FIGURE 7.1 Methods for binding interface redesign. (A) Enhancement of protein association rate through improvement of long-range electrostatic interactions. Free energy of protein–protein interactions for (1) nonoptimized and (2) electrostatically optimized proteins complexes. A+B represents free proteins, A:B is the encounter complex, and AB is the final complex. The formation of the encounter complex is the rate-limiting state for association. The figure is reproduced from Selzer and Schreiber.[5] (B) The computational protein design approach. 1: Protein coordinates are retrieved from the PDB. 2: Sequence positions to be redesigned are defined. Rotamers (or low-energy side chain conformations) are placed at each designed position. 3: The pairwise energy function, including rotamer–backbone and rotamer–rotamer interactions, is calculated. 4: Fast search algorithms are used to search through the conformational space to find the lowest-energy sequence(s). (C) Illustration of the negative design concept. The wild-type protein binds to the desired target and to several alternative targets. The binding free energies of the desired and the alternative complexes are similar. In procedures not including negative design, the free energy of binding is minimized for the complex with the desired target. In procedures including negative design, free energy of binding is minimized for the complex with the desired target and maximized for the complexes with alternative targets.

To improve electrostatic complementarity at protein interfaces, Tidor and colleagues proposed a more elaborate strategy based on continuum electrostatics models.[6–8] In this strategy, the free energy of binding is evaluated using two opposing energetic terms: favorable electrostatic interactions between the two proteins in the complex and the unfavorable desolvation of the charged groups occurring upon complex formation. The optimal ligand charge distribution is calculated to produce the most favorable balance of these opposing free energy contributions. Mutations are then introduced on the ligand to best fit this calculated charge distribution.

COMPUTATIONAL PROTEIN DESIGN

The second strategy that has been used to reengineer protein binding interfaces is referred to as computational protein design (Figure 7.1B).[9,10] In this approach, proteins are designed by searching through a vast number of amino acid sequences until the optimal protein sequence is selected. During the search, the protein backbone is often kept rigid, while protein side chains are selected from a library of low-energy side chain conformations called rotamers.[11] Rotameric sequences are evaluated with a full-atom energy function that contains terms describing van der Waals interactions, hydrogen bonding, electrostatics, and solvation.[12] Various algorithms have been applied to efficiently search the sequence space and to obtain the lowest-energy protein sequence(s).[13] Several computational protein design packages have been developed by different groups.[14–18] All of them are based on the same approach but differ slightly in the implementation of the energy function and in the search algorithms used for sequence selection. Computational protein design was initially applied to build structural protein units and predict stabilizing mutations.[14,19] More recently, the effort in the field has shifted toward functional design, where the biological function of a protein is being modified or completely altered. Several excellent examples of such work include designing proteins with a novel fold,[20] creating new enzymes[21,22] and sensor molecules.[17,23]

Computational design of protein–protein interactions is one of the directions in the field that presents a particular interest. Ultimately, computational methods should enable the design of receptors and inhibitors for any protein of interest. Nevertheless, at this time, such designs present a major challenge due to several shortcomings in the available computational methods. First, the energy functions used in protein design packages have been initially developed for stabilization of monomeric proteins and are not optimal for design of protein–protein interfaces. Such energy functions, for example, often fail to capture the importance of residues that form salt bridges and hydrogen bonds across the binding interface, resulting in substitution of these residues with hydrophobic amino acids. Second, it remains a challenge to accurately model conformational changes frequently associated with protein binding.[24] The ability to model such changes, however, becomes essential if novel protein–protein complexes are to be designed. Third, water molecules that frequently mediate intermolecular hydrogen bond interactions have been initially ignored by the protein design programs. Although most of the water molecules do not contribute significantly to protein–protein binding,[25] a few highly conserved waters might be very important for binding and should be retained during the binding interface redesign.[26] In spite of these limitations in the protein design methods, several successful examples of redesigning protein–protein interactions have emerged in the past decade.

AT THE BEGINNING

Initial designs of protein–protein interfaces used a combination of computational methods and a manual selection of amino acid substitutions. Several studies focused on design of receptors using coiled coils as a protein scaffold. These small helical domains are an ideal system for testing the basic principles of protein–protein association due to their simplicity and the wealth of biochemical and structural information about them. Among the first engineered receptors was a two-helix hairpin designed to recognize the calmodulin binding domain of calcineurin (CaN).[27] An idealized three-helix bundle geometry was used to generate a model for the backbone conformation of the complex between a helical peptide CaN and a two-helix receptor. Starting from this backbone structure, the buried receptor residues were redesigned using a side chain–repacking algorithm.[28] The solvent-exposed receptor residues were chosen manually by generating favorable electrostatic interactions between the three helixes. When tested experimentally, the designed receptors exhibited binding affinities to CaN ranging from 0.2 to 50 µM. Unfortunately, the receptors were not monomeric in solution as intended, suggesting that modeling of the alternative folds might be necessary to achieve the correct specificity.

Another study reports the design of a two-helix-bundle mimetic of interleukin-4 (IL-4).[29] IL-4 binds to its receptor IL-4Rα with two adjacent antiparallel helixes, exhibiting an affinity of 1.4 nM. The authors mimicked IL-4 by drafting the residues most important for the IL-4Rα recognition on a well-studied coiled coil, GCN4. Molecular dynamics simulations were performed to verify that the designed molecules could fold prior to binding. The IL-4 mimetics showed K_ds ranging from 2 mM to 5 µM, depending on the fraction of the IL-4 binding site incorporated into the molecules. These earlier studies demonstrate that receptors with micromolar affinities could be designed from helical bundles and that computational methods greatly facilitate such designs. Nevertheless, more sophisticated strategies would be required to generate receptors with alternative geometries, tighter association, and better binding specificity.

ENHANCING PROTEIN BINDING AFFINITY

Accurately predicting binding affinities of protein–protein complexes remains an open problem in computational biology. Predicting mutations that enhance binding is a related and more difficult task. Studies where predictions of affinity-enhancing mutations were verified experimentally remain rare but invaluable, especially if they include structural characterization of the redesigned complexes. Both correct and incorrect predictions of such mutations could serve to improve the existing computational methods for design of protein–protein interactions. In addition, such studies facilitate the development of molecules for use in various biotechnological and biomedical applications, including high-affinity antibodies, inhibitors of undesired protein–protein interactions, and sensors for optical imaging.

An approach that enhances association rates and binding affinities through optimization of electrostatic complementarity between the two proteins in the complex was first applied to increase the association rate between the TEM1 β-lactamase and its protein inhibitor BLIP (Figure 7.2A).[1] Electrostatic forces make no contribution to association of these proteins, indicating that the binding interface is not optimized electrostatically. To improve the electrostatic complementarity of the BLIP–TEM1 complex, single mutations were introduced on the surface of BLIP. All the predicted single mutants of BLIP exhibited an increase in association rate for TEM1, with the best mutation showing a 20-fold increase. Combining four beneficial mutations together resulted in a BLIP mutant that showed a 240-fold increase in k_{on} and a 290-fold decrease in K_d.

The same approach was used to optimize electrostatic steering between Ras and its effector, Ral guanine nucleotide dissociation stimulator (RalGDS).[30] Analysis of the charge distribution at the Ras–RalGDS binding interface showed that introduction of positive charge in two regions of RalGDS would be beneficial for binding. Single mutations to Lys in these two regions were predicted to increase the association rate to Ras by up to 10-fold. An excellent agreement between the computational predictions and experimental results was demonstrated. Combining the three most promising mutations in a single RalGDS mutant resulted in a molecule that binds Ras 14 times faster and 25-fold tighter than wild-type RalGDS.

In attempt to develop higher-affinity antibodies, affinity-enhancing mutations were predicted on the antibody surface using either electrostatic optimization of the binding interface or the computational protein design techniques.[31–33] Variable success has been achieved in predicting the effect of single mutations on the antibody–antigen affinity. Nevertheless, some of the affinity-enhancing mutations were always correctly identified. Combining several of such mutations in a single design usually produced an additive effect on binding affinity. Using this strategy, a six-fold improvement in association rate was achieved for the antibody against vascular endothelial growth factor (VEGF),[32] a ten-fold improvement in affinity was engineered into an anti-epidermal growth factor receptor drug cetuximax,[31] and a ten-fold improvement in affinity was demonstrated in the engineered antibody for the I-domain of integrin VLA1.[33] A substantial 140-fold improvement in binding affinity was obtained by introducing six mutations into an antilysozyme antibody.[31] These studies demonstrate that computational methods are becoming a more accepted strategy for improving binding properties of therapeutic molecules.

In an interesting study by Song et al., binding affinity was enhanced between an intercellular adhesion molecule-1 (ICAM-1) and an integrin lymphocyte function-associated antigen (LFA-1).[34] The interaction between ICAM-1 and LFA-1 is critical to many immunological responses, including those evoked in autoimmune diseases and in immune rejection of organ transplantation. Hence, design of competitive antagonists of the ICAM-1/LFA-1 interaction could have an important therapeutical application. The authors used four different computational design programs to introduce mutations into ICAM-1. The majority of single ICAM-1 mutants, when tested in a cell surface binding assay, showed unaltered or increased percent of binding

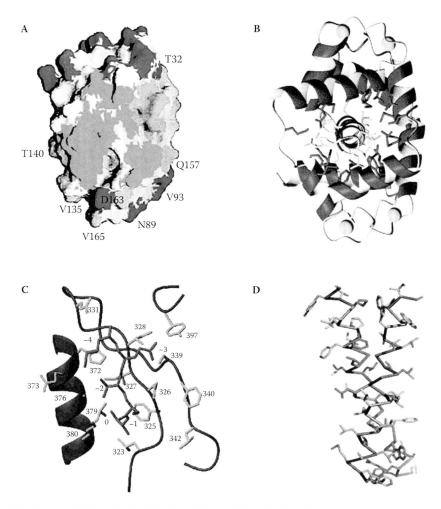

FIGURE 7.2   (SEE COLOR INSERT FOLLOWING PAGE 174.) Examples of protein interface design. (A) Design of BLIP mutants with enhanced association rate for TEM1 β-lactamase.
The figure shows wild-type BLIP with the TEM1 binding interface (green) and mutation sites predicted to increase the association rate. Color coding displays the extent of the predicted association
rate increase: blue, less than 50% increase: yellow, more than 50% increase; red, ten-fold increase.
The figure is reproduced from Selzer et al.[1] (B) Redesign of calmodulin for improved binding
specificity. Calmodulin is embracing the peptide target with its two globular domains. Twenty-
four calmodulin side chains selected for optimization are shown in red. Peptide side chains that
were allowed to change conformation during the calculation are shown in cyan. Calcium atoms
are shown as yellow spheres. The figure is reproduced from Shifman and Mayo.[40] (C) Design of
the PDZ domains with altered binding properties showing wild-type PDZ domain with its natural
ligand, the KQTSV peptide (red). Residues on the PDZ domain selected for the optimization are
shown in green. The figure is reproduced from Reina et al.[47] (D) Design of peptides that recognize
transmembrane domains of integrins. A backbone geometry for the helix–helix interaction was
selected from two helixes in the photosystem I reaction center. The sequence of the integrin α_{IIb}
was threaded into the right helix. Fourteen positions on the second helix were designed (pink). The
figure is reproduced from Yin et al.[51]

to LFA-1 compared to wild-type ICAM-1. All the designed ICAM-1 mutants with multiple amino acid substitutions, however, performed poorly in the cell surface assay. Combining five beneficial single mutations produced an ICAM1 mutant with a 20-fold improvement in binding affinity.

In a recent study, Kuhlman et al. proposed a strategy for improving binding affinity by introducing single mutations that substitute polar amino acids at the binding interface for nonpolar ones and nonpolar amino acids for larger nonpolar amino acids.[35] Such mutations are predicted to enhance affinity if the free energy of binding is calculated to be more favorable than that of the wild-type complex and each monomer is not significantly destabilized by the mutation. This strategy was tested on a protein–protein and a protein–peptide complex. Nine out of 12 predicted mutations selected for experimental testing were correctly identified to be beneficial for binding. Although this method works well in predicting single mutations, it cannot be generalized for redesign of an entire binding interface. First, because combining several of such mutations might lead to problems with protein solubility. Second, the method would automatically eliminate intermolecular hydrogen bonds, which are important for both binding affinity and specificity.

An interesting strategy was proposed for enhancing affinities of proteins that undergo substantial conformational changes upon binding to their ligands. For these proteins, mutations could be computed that stabilize the ligand-bound protein conformation over the ligand-free state without directly affecting interactions with the ligand. This idea was first explored by redesigning the I domain of the hetorodimeric integrin αMβ2.[36] The computational protein design approach was used to introduce hydrophobic mutations that stabilize the I domain in the active conformation. All the mutations were made relatively far from the binding site of the ligand, iC3b. When expressed on the cell surface as part of the intact heterodimeric receptor, the designed I domains bound to iC3b 10 times better compared to wild-type (WT). Furthermore, when expressed in isolation from other integrin domains using an artificial transmembrane domain, designed I domains were active in ligand binding in contrary to wild-type I domains. This study establishes a new viable strategy for increasing protein binding affinity through stabilization of the ligand-bound protein conformation. The potential of such a strategy has yet to be fully explored.

To summarize, two methods have been applied to enhance binding affinities in proteins. Optimizing long-range electrostatic interactions between the two binding partners proved a successful approach for enhancing protein association rate and binding affinity. Nevertheless, the method cannot be applied to protein–protein complexes where electrostatic interactions have already been optimized by nature, such as in the barnase/barstar complex.[37] The computational protein design approach was shown to predict single affinity-enhancing mutations with a relatively high success rate. Nevertheless, attempts to simultaneously redesign the entire binding interface frequently yielded protein complexes with reduced affinity. This is due to possible error accumulation associated with prediction of each single mutation. Until more accurate energy functions for the protein interface design are developed, a winning strategy for binding affinity enhancement is to experimentally test each predicted mutation and only then to combine the beneficial mutations into a single design. Alternatively, a small library of protein sequences could be designed and tested experimentally.[38]

ALTERING PROTEIN BINDING SPECIFICITY

Protein–protein interactions are often determining factors in the outcome of the complicated signaling pathways. In the cellular environment, a large number of proteins coexist in the same cellular compartment and compete for the available binding partners. How do proteins select their cognate binding partners over a multitude of other proteins? Nature solves this problem by supplying each protein with a correct binding specificity. What determines protein binding specificity at the structural and amino acid sequence level remains largely unknown. However, recent studies that report designs of several proteins with altered binding specificity helped to shed light on the origins of this important property. Manipulating protein binding specificity becomes especially important in the area of drug design, since an ideal drug should inhibit the desired pathway without disturbing all other processes in the cell. Computational protein design methods are a potentially invaluable tool for supplying therapeutic molecules with high binding specificity.

To improve protein binding specificity, in principle, we should consider several states: a desired state corresponding to a protein in complex with a target of interest and alternative states corresponding to a protein in complex with all undesired targets (Figure 7.1C). Optimization of binding specificity requires designing sequences with minimum energy in the desired state and maximum energies in the alternative states. The concept of designing against certain protein conformation is referred to as negative design. Although the concept is intuitively clear, its implementation is not so straightforward for several reasons. First, the procedures for using both positive and negative design require substantially more complicated algorithms and more computational power compared to procedures that incorporate positive design only. Second, the alternative states are not always known and the high-resolution structures for the alternative states are often not available. Hence, some of the protein design studies utilize a negative design procedure, while other studies choose to ignore it.

Havranek and Harbury developed an approach that incorporates explicit negative design in the sequence selection procedure.[39] Using genetic algorithms, the procedure selects amino acid sequences with a maximum energy gap between the target protein conformation and a set of alternative undesired structures. The procedure was tested by designing two sequences of coiled coils that preferentially associate into homodimers and do not cross-hybridize with each other. Eight positions (four on each helix) were simultaneously optimized considering four states: the desired homodimeric state, and the alternative states of the unfolded, the aggregated, and the heterodimer. Experimental results on 13 designs show a good correlation between the predicted and the experimental free energies of the homodimeric and the heterodimeric states. Predictions of the protein stability (the energy difference between the folded and the unfolded states) were generally less accurate and showed high dependence on the molecular force field used. To demonstrate the necessity of using the explicit negative design for specificity optimization, the authors performed a similar calculation without considering the alternative states in the optimization procedure. The obtained sequences exhibited worse specificity scores than most of the sequences predicted with the negative design procedure. Unfortunately, no experimental results were obtained to verify the predictions. Even if the negative design procedure might not be necessary or applicable to all design problems, the presented algorithm remains an important keystone in the field of protein design.

Shifman and Mayo were the first to increase binding specificity of calmodulin (CaM), a small α-helical protein that in nature regulates hundreds of targets in response to changes in Ca^{2+} concentration (Figure 7.2B).[40] A number of structures for the CaM-target complexes reveal that CaM binds to ~25 amino acid helical segments by embracing them with two globular domains. Rotation of the CaM globular domains with respect to each other allows CaM to generate slightly different binding surfaces for each target, resulting in low CaM binding specificity. The goal was to increase CaM binding specificity toward a single CaM target, smooth muscle myosin light chain kinase (smMLCK). Starting from the structure of the CaM–smMLCK complex, the 24 fully buried positions on CaM were optimized for better interactions with the selected target. The resulting eightfold CaM mutant retained the wild-type affinity to the desired target and exhibited reduced affinity to six alternative CaM targets, showing up to 120-fold increase in binding specificity.[40] In the second generation design, the number of the CaM residues that were optimized was increased, including the surface accessible interfacial positions.[41] Several slightly different optimization procedures yielded six CaM mutants, of which several lost some binding affinity to smMLCK. The best CaM mutant, with 13 binding interface mutations, retained the wild-type binding affinity to smMLCK and exhibited up to 155-fold increase in binding specificity. In our recent work, we optimized CaM for interactions with another target, Ca^{2+}/CaM-dependent protein kinase II (CaMKII) and tested the binding of the designed CaMs to two targets, the desired target CaMKII and the alternative target calcineurin (CaN).[42] Our CaM variants exhibited a two- to fourfold improvement in binding to CaMKII and substantial decrease in binding to CaN, demonstrating up to 900-fold increase in binding specificity. In all these works, no explicit negative design was incorporated in the sequence selection procedure. However, both the calculated and the experimental free energies of binding for the designed CaM sequences in the alternative states were higher than those of the wild-type CaM sequence. These results show that incorporating the negative design procedure is not necessary to achieve the correct binding specificity in at least some protein–protein complexes.

Extending our work on CaM, two groups introduced complementary mutations to both CaM and its peptide target to create binding interfaces with specificity orthogonal to wild type.[43,44] These CaM–peptide complexes could be used to construct calcium indicators for optical imaging. The native CaM–peptide complexes, previously used for design of such indicators, suffer from susceptibility to unwanted interactions with multiple calmodulin targets in the cell. The indicators made from the redesigned CaM–peptide pairs, with orthogonal to wild-type specificity, do not display this unwanted quality.

To design binding interfaces with novel specificities, Kortemme et al. developed a computational second-site suppressor strategy.[45] In this strategy, mutations disruptive for the interface are first introduced into one partner and are then compensated by mutations in the second interaction partner. The designed molecules associate tightly with each other but show reduced affinity to the original interaction partner.

This method was first tested by redesigning the binding interface of the colicin E7 DNase–Im7 immunity protein complex (E7–Im7).[45] Different colicins interact with their cognate immunity proteins exhibiting high affinity and very high specificity (10^7–10^8-fold affinity differences between the cognate and noncognate pairs). Such binding properties are important since colicins are cytotoxic in the absence of their cognate immunity proteins; the toxicity is inhibited upon colicin–immunity protein interaction. To design a novel colicin–immunity protein complex, three interface perturbing mutations were introduced on colicin E7 and nine positions on the cognate immunity protein Im7 were subsequently redesigned. Four predicted sequences of the designed protein pairs were tested experimentally. The two mutant pairs that contain eight and nine mutations at the binding interface showed subnanomolar to lower nanomolar binding affinity for each other. At the same time, the affinity of the noncognate complexes (one designed and one wild-type molecule) was about 30 times lower. High binding specificity of the designed pairs was confirmed by performing functional essays both *in vitro* and *in vivo*. The crystal structure for one of the redesigned complexes was solved and showed a good agreement between the predicted and the actual side chain conformations of the mutated residues. This study is the first successful attempt to supply a protein–protein complex with an orthogonal binding specificity. Nevertheless, the redesigned E7–Im7 complexes exhibited substantially lower affinities and binding specificities compared to those of the native colicin–immunity protein pairs.

To further improve the specificity of the E7–Im7 complexes, the authors focused on differences in binding orientation exhibited by various native colicin–immunity complexes.[26] In the design procedure, alternative backbone conformations of the two interacting proteins were created by systematically sampling rigid-body rotations of one of the proteins. These backbone conformations were subsequently used to perform the binding interface redesign. Two computational protocols were tested: the second site suppressor protocol described earlier and the affinity protocol that does not incorporate any negative design element. After initial screening of 11 designed pairs, 3 were further pursued. The best design, produced by the affinity protocol, incorporated a new hydrogen bond network across the binding interface. Crystal structure of the designed complex was solved and confirmed most of the predicted intermolecular interactions. The small differences between the predictions and the reality were due to a tightly bound water molecule that was not considered in the design procedure. Using the newly obtained crystal structure and retaining the tightly bound water, the authors reoptimized the binding interface focusing on the residues surrounding the hydrogen bond network. The resulting designs exhibited a 300-fold difference in binding affinity between the cognate and the noncognate pairs. This study demonstrates the power of iterative approach, where an initial interface design is followed by its biochemical and structural characterization and by another round of redesign. This work also supports the hypothesis that negative design is not always required to obtain protein complexes with the desired binding specificity.

Computational approaches could be used to break the symmetry at dimeric interfaces. This was done by Bolon et al., who redesigned the natively homodimeric SspB adaptor protein into a heterodimer.[46] Using a computational protein design approach, the authors optimized eight positions on the dimeric interface (four positions on each monomer) employing two different strategies. One strategy sought to enhance specificity using both positive and negative design, while the second strategy optimized only the stability of the target conformation ignoring the negative design. SspB mutants designed solely with the positive design procedure assembled into heterodimers; however, they also formed equally stable homodimers. In contrast, SspB mutants designed considering explicit negative design assembled most exclusively into heterodimers, but were less stable than the molecules designed for stability only. This work suggests that in the design of protein–protein complexes, a trade-off between affinity and specificity might be always observed. In addition, it presents an example of a protein–protein system where the desired and the alternative states are substantially similar. Hence, the explicit use of the negative design is necessary to prevent the undesired protein assemblies.

Reina et al. redesigned the PDZ domain for binding to new targets (Figure 7.2C). In nature, this small globular domain serves to recognize unstructured C-terminal sequences of many proteins.[47] Starting from the structure of the PDZ domain bound to one of such sequences, the authors redesigned this protein to recognize three targets: the C-terminus of a kinesin-like molecule and two sequences containing either hydrophobic or polar substitutions at two positions of the original peptide. Two of the redesigned PDZ domains were shown to bind to their desired targets with K_d values similar to that of the original PDZ–peptide complex, while the third PDZ domain exhibited two orders of magnitude lower (better) K_d compared to the wild type. The best PDZ domain mutant was demonstrated to specifically recognize its target peptide in a yeast two-hybrid assay, demonstrating that such molecules could be used in various biotechnological applications such as affinity chromatography and western blotting. The ability to not only alter the binding specificity but to also substantially increase the affinity of the redesigned binding partners distinguishes this study from similar efforts in protein design.

Redesign of binding interfaces could be used to generate chimeric proteins with dual functionalities. Chevalier et al. applied the computational protein design approach to engineer a protein that binds to a chimeric DNA target site.[48] To fuse the two domains of distantly related homing endonucleases, the N-terminal domain of an endonuclease I-DmoI was substituted for a single subunit of the homodimeric endonuclease I-CreI producing an initial model for the chimeric endonuclease E-DreI. The helix–helix interface between the two halves of E-DreI was subsequently optimized using the standard protein design procedure. The best E-DreI candidate sequences containing eight to twelve mutations were generated and screened *in vivo* to ensure proper folding and solubility. Biochemical characterization of several soluble E-DreI variants revealed that they were able to bind to and cleave a 23-bp chimeric DNA target site with high specificity and wild-type kinetics. This study shows the promise of the computational protein design approach in creating proteins with novel functions.

In summary, great success has been demonstrated in redesigning protein-binding specificity using computational protein design. Protein complexes with increased or altered binding specificity can now be designed. It remains unclear if incorporation of negative design in the sequence selection procedure is always beneficial. While some studies demonstrate the great utility of negative design, others show that stabilization of the desired state automatically results in destabilization of the alternative states. The utility of negative design highly depends on the protein system, on the similarity of the desired and the alternative states, and the number of protein positions being optimized.

DE NOVO DESIGN OF PROTEIN–PROTEIN COMPLEXES

The redesign of the existing protein complexes has seen a considerable success in the past decade. Nevertheless, it remains a major challenge to computationally design protein complexes that do not exist in nature. The main difficulty comes from our inability to generate a realistic model for the backbone structure of the two proteins in the novel protein–protein complex. Conventional docking algorithms cannot be applied to create such a model since the sequence of the protein–protein interface is not known a priori but is subject to subsequent design. To overcome this problem, Huang et al. developed a strategy to dock the two protein structures with an unknown binding interface sequence.[49] For this purpose, they used a reduced amino acid side-chain representation at the protein interface with side chains artificially restricted to C_β atoms. By systematically translating and rotating one protein with respect to another and evaluating the binding interface complementarity, the best conformation of the two protein backbones is determined. Starting from this conformation, the amino acid sequence for the novel protein–protein interface is then selected. This approach was applied to design a heterodimer of the GB1 domain of the protein G, a protein that is monomeric in nature.[50] Using a helix-to-helix binding arrangement for the model of the dimeric complex, a total of 24 positions on the two monomers were simultaneously redesigned to produce 12-fold and 8-fold mutants of GB1. A binding affinity of 300 μM was experimentally measured for the designed complex. Such weak binding affinity could be partially explained by the low stability of one of the monomers. In spite of the modest experimental success, the described method for design of novel protein complexes is definitely promising and should be tested in other protein systems.

Until recently, all designs of protein–protein interfaces focused on water-soluble proteins. In a recent exciting work, Yin et al. report a method for modulating protein–protein interactions inside membranes.[51] The authors designed helical peptides to bind to transmembrane regions of two closely related integrins ($\alpha_{II}\beta_3$ and $\alpha_v\beta_3$; Figure 7.2D). The starting backbone geometry for this helix–helix complex was generated by searching through the database of membrane–protein structures and selecting helix–helix orientations exhibited by similar sequence motifs. The sequence of the target integrin transmembrane domain was threaded onto one helix, while the second helix was designed using the computational protein design approach. The energy function for this design included a van der Waals term and a membrane depth-dependent, knowledge-based potential.[52] The resulting peptides were demonstrated to bind specifically to the desired integrin in micelles and in bacterial membranes. In addition, the peptides interacted with the transmembrane integrin domains in mammalian cells, where they were shown to inhibit integrin heterodimer formation, stimulating integrin activation. The reported methodology provides a general way to design binding partners to membrane proteins and to probe the functional consequences of blocking protein–protein interactions in membranes.

CONCLUSIONS AND FUTURE DIRECTIONS

In the last decade, great progress has been achieved in computational design of protein–protein interactions. Many studies have demonstrated that binding affinity and specificity could be improved or altered in a predictive way. In a few studies, the desired binding properties of the computationally designed proteins have been also verified in the cellular environment. Crystal structures of some of the complexes have been solved and proved extremely valuable in pointing out the strengths and the weaknesses of the computational methods for protein interface design.

The future challenge lies in designing high-affinity complexes from proteins that show no considerable binding in nature. These include novel receptors for a protein of interest or novel binders and inhibitors of the existing protein–protein interactions. To enable such designs, several shortcomings in the existing computational procedures should be addressed. The computational methods still frequently fail to correctly predict the effect of various mutations on the free energy of binding. To overcome this problem, the energy functions for protein design should be fine-tuned to capture the delicate balance between the favorable electrostatic interactions and the unfavorable desolvation of the polar groups at protein binding interfaces. In addition, a more accurate yet easily computable description of electrostatic and hydrogen bond interactions is needed to model energetics and specificity at protein–protein interfaces. Progress in this direction has been recently reported.[53–55] Explicit modeling of water molecules at binding interfaces might also help to achieve better designs. A solvated rotamer approach, developed for this purpose, allows introduction of water-mediated contacts across the binding interface.[56] Experimental validation of such newly designed contacts is still to come.

Finally, the progress in design of new protein binders is highly dependent on our ability to generate backbone structures for the novel protein complexes. Here, we are faced with two major questions: what protein scaffold to pick for design of a novel binding molecule and how to dock this molecule into the selected binding site. The answer to the first question largely depends on the protein complex to be designed. Frequently, the scaffold for the new binding partner could be inferred from a homologous protein that already binds to the protein of interest. Alternatively, it could be taken from an unrelated protein with some structural similarity in the region of the binding interface. Once the scaffold is selected, it remains a major challenge to create a good model for the backbone structure of a novel protein–protein complex since the sequence of the binding interface is unknown prior to design. Each backbone conformation would result in a distinct set of the low-energy solutions for the binding interface. It is not known a priori, which backbone would produce the lowest-energy sequence. Introducing backbone flexibility, including flexibility of each single molecule as well as flexibility in relative orientation between the two molecules in the complex, becomes extremely important for design of novel binding interfaces. It has been recently shown that exploring several alternative backbone structures during a helical ligand design leads to a larger and a more diverse set of low-energy solutions than can be achieved using the native backbone as a template.[57] Hence, a winning strategy for design of novel protein–protein complexes might be to select a number of backbone structures and to perform the design on each of the structures, selecting the lowest-energy sequences only at the end. Alternatively, a single amino acid sequence could be designed to be compatible with an ensemble of different backbone conformations. These and other ideas should be explored and experimentally tested in the next few years. In spite of the described challenges, with a growing number of research groups working in the field and with constant improvement in methodology, I envision the universal use of computational methods for modulating protein–protein interactions in the near future.

REFERENCES

1. Selzer, T., Albeck, S., & Schreiber, G. (2000). Rational design of faster associating and tighter binding protein complexes. *Nature Structural Biology 7*, 537–541.
2. Schreiber, G. & Fersht, A. R. (1995). Energetics of protein–protein interactions: Analysis of the barnase-barstar interface by single mutations and double mutant cycles. *Journal of Molecular Biology 248*, 478–486.
3. Schreiber, G. & Fersht, A. R. (1996). Rapid, electrostatically assisted association of proteins. *Nature Structural Biology 3*, 427–431.
4. Selzer, T. & Schreiber, G. (1999). Predicting the rate enhancement of protein complex formation from the electrostatic energy of interaction. *Journal of Molecular Biology 287*, 409–419.
5. Selzer, T. & Schreiber, G. (2001). New insights into the mechanism of protein–protein association. *Proteins: Structure, Function, and Genetics 45*, 190–198.
6. Kangas, E. & Tidor, B. (1998). Optimizing electrostatic affinity in ligand-receptor binding: Theory, computation, and ligand properties. *Journal of Chemical Physics 109*, 7522–7545.
7. Lee, L. P. & Tidor, B. (1997). Optimization of electrostatic binding free energy. *Journal of Chemical Physics 106*, 8681–8690.
8. Gilson, M. K., Sharp, K. A., & Honig, B. H. (1988). Calculating the electrostatic potential of molecules in solution: Method and error assessment. *Journal of Computational Chemistry 9*, 327–335.
9. Street, A. G. & Mayo, S. L. (1999). Computational protein design. *Structure 5*, R105–R109.
10. Kraemer-Pecore, C. M., Wollacott, A. M., & Desjarlais, J. R. (2001). Computational protein design. *Current Opinion in Chemical Biology 5*, 690–695.
11. Dunbrack, R. L. & Karplus, M. (1993). Backbone-dependent rotamer library for proteins. Applications to side-chain predictions. *Journal of Molecular Biology 230*, 543–574.
12. Gordon, D. B., Marshall, S. A., & Mayo, S. L. (1999). Energy functions for protein design. *Current Opinion in Structural Biology 9*, 509–513.
13. Voigt, C. A., Gordon, D. B., & Mayo, S. L. (2000). Trading accuracy for speed: A quantitative comparison of search algorithms in protein sequence design. *Journal of Molecular Biology 299*, 789–803.
14. Dahiyat, B. I. & Mayo, S. L. (1997). De novo protein design: Fully automated sequence selection. *Science 278*, 82–87.
15. Rohl, C. A., Strauss, C. E. M., Misura, K. M. S., & Baker, D. (2004). Protein structure prediction using Rosetta. In *Numerical Computer Methods* (Pt D, Vol. 383), edited by L. Brand & M. L Johnson, 66–93. New York: Academic Press.

16. Chowdry, A. B., Reynolds, K. A., Hanes, M. S., Voorhies, M., Pokala, N., & Handel, T. M. (2007). Software news and update. An object-oriented library for computational protein design. *Journal of Computational Chemistry 28*, 2378–2388.

17. Looger, L. L., Dwyer, M. A., Smith, J. J., & Hellinga, H. W. (2003). Computational design of receptor and sensor proteins with novel functions. *Nature 423*, 185–190.

18. de la Paz, M. L., Lacroix, E., Ramirez-Alvarado, M., & Serrano, L. (2001). Computer-aided design of beta-sheet peptides. *Journal of Molecular Biology 312*, 229–246.

19. Malakauskas, S. M. & Mayo, S. L. (1998). Design, structure and stability of a hyperthermophilic protein variant. *Nature Structural Biology 5*, 470–475.

20. Kuhlman, B., Dantas, G., Ireton, G. C., Varani, G., Stoddard, B. L., & Baker, D. (2003). Design of a novel globular protein fold with atomic-level accuracy. *Science 302*, 1364–1368.

21. Bolon, D. N. & Mayo, S. L. (2001). Enzyme-like proteins by computational design. *Proceedings of the National Academy of Sciences USA 98*, 14274–14279.

22. Jiang, L., Althoff, E. A., Clemente, F. R., Doyle, L., Rothlisberger, D., Zanghellini, A., Gallaher, J. L., Betker, J. L., Tanaka, F., Barbas, C. F., Hilvert, D., Houk, K. N., Stoddard, B. L., & Baker, D. (2008). De novo computational design of retro-aldol enzymes. *Science 319*, 1387–1391.

23. Dwyer, M. A., Looger, L. L., & Hellinga, H. W. (2003). Computational design of a Zn2+ receptor that controls bacterial gene expression. *Proceedings of the National Academy of Sciences USA 100*, 11255–11260.

24. Schueler-Furman, O., Wang, C., & Baker, D. (2005). Progress in protein–protein docking: Atomic resolution predictions in the CAPRI experiment using RosettaDock with an improved treatment of side-chain flexibility. *Proteins: Structure, Function, and Bioinformatics 60*, 187–194.

25. Reichmann, D., Phillip, Y., Carmi, A., & Schreiber, G. (2008). On the contribution of water-mediated interactions to protein-complex stability. *Biochemistry 47*, 1051–1060.

26. Joachimiak, L. A., Kortemme, T., Stoddard, B. L., & Baker, D. (2006). Computational design of a new hydrogen bond network and at least a 300-fold specificity switch at a protein–protein interface. *Journal of Molecular Biology 361*, 195–208.

27. Ghirlanda, G., Lear, J. D., Lombardi, A., & DeGrado, W. F. (1998). From synthetic coiled coils to functional proteins: Automated design of a receptor for the calmodulin-binding domain of calcineurin. *Journal of Molecular Biology 281*, 379–391.

28. Desjarlais, J. R. & Handel, T. M. (1995). De novo design of the hydrophobic cores of proteins. *Protein Science 4*, 2006–2018.

29. Domingues, H., Cregut, D., Sebald, W., Oschkinat, H., & Serrano, L. (1999). Rational design of a GCN4-derived mimetic of interleukin-4. *Nature Structural Biology 6*, 652–656.

30. Kiel, C., Selzer, T., Shaul, Y., Schreiber, G., & Herrmann, C. (2004). Electrostatically optimized Ras-binding Ral guanine dissociation stimulator mutants increase the rate of association by stabilizing the encounter complex. *Proceedings of the National Academy of Sciences USA 101*, 9223–9228.

31. Lippow, S. M., Wittrup, K. D., & Tidor, B. (2007). Computational design of antibody-affinity improvement beyond in vivo maturation. *Nature Biotechnology 25*, 1171–1176.

32. Marvin, J. S. & Lowman, H. B. (2003). Redesigning an antibody fragment for faster association with its antigen. *Biochemistry 42*, 7077–7083.

33. Clark, L. A., Boriack-Sjodin, P. A., Eldredge, J., Fitch, C., Friedman, B., Hanf, K. J. M., Jarpe, M., Liparoto, S. F., Li, Y., Lugovskoy, A., Miller, S., Rushe, M., Sherman, W., Simon, K., & Van Vlijmen, H. (2006). Affinity enhancement of an in vivo matured therapeutic antibody using structure-based computational design. *Protein Science 15*, 949–960.

34. Song, G., Lazar, G. A., Kortemme, T., Shimaoka, M., Desjarlais, J. R., Baker, D., & Springer, T. A. (2006). Rational design of intercellular adhesion molecule-1 (ICAM-1) variants for antagonizing integrin lymphocyte function-associated antigen-1-dependent adhesion. *Journal of Biological Chemistry 281*, 5042–5049.

35. Sammond, D. W., Eletr, Z. M., Purbeck, C., Kimple, R. J., Siderovski, D. P., & Kuhlman, B. (2007). Structure-based protocol for identifying mutations that enhance protein–protein binding affinities. *Journal of Molecular Biology 371*, 1392–1404.

36. Shimaoka, M., Shifman, J. M., Jing, H., Takagi, J., Mayo, S. L., & Springer, T. A. (2000). Computational design of an integrin I domain stabilized in the open high affinity conformation. *Nature Structural Biology 7*, 674–678.

37. Lee, L. P. & Tidor, B. (2001). Barstar is electrostatically optimized for tight binding to barnase. *Nature Structural Biology 8*, 73–76.

38. Treynor, T. P., Vizcarra, C. L., Nedelcu, D., & Mayo, S. L. (2007). Computationally designed libraries of fluorescent proteins evaluated by preservation and diversity of function. *Proceedings of the National Academy of Sciences USA 104*, 48–53.

39. Havranek, J. J. & Harbury, P. B. (2003). Automated design of specificity in molecular recognition. *Nature Structural Biology 10*, 45–52.

40. Shifman, J. M. & Mayo, S. L. (2002). Modulating calmodulin specificity through computational protein design. *Journal of Molecular Biology 323*, 417–423.

41. Shifman, J. M. & Mayo, S. L. (2003). Exploring the origins of binding specificity through the computational redesign of calmodulin. *Proceedings of the National Academy of Sciences USA 100*, 13274–13279.

42. Yosef, E., Politi, R., Choi, M. H., & Shifman, J. M. (2009). Computational design of calmodulin mutants with up to 900-fold increase in binding specificity. *Journal of Molecular Biology, 385*, 1470–1480.

43. Palmer, A. E., Giacomello, M., Kortemme, T., Hires, S. A., Lev-Ram, V., Baker, D., & Tsien, R. Y. (2006). Ca2+ indicators based on computationally redesigned calmodulin-peptide pairs. *Chemistry & Biology 13*, 521–530.

44. Green, D. F., Dennis, A. T., Fam, P. S., Tidor, B., & Jasanoff, A. (2006). Rational design of new binding specificity by simultaneous mutagenesis of calmodulin and a target peptide. *Biochemistry 45*, 12547–12559.

45. Kortemme, T., Joachimiak, L. A., Bullock, A. N., Schuler, A. D., Stoddard, B. L., & Baker, D. (2004). Computational redesign of protein–protein interaction specificity. *Nature Structural & Molecular Biology 11*, 371–379.

46. Bolon, D. N., Grant, R. A., Baker, T. A. & Sauer, R. T. (2005). Specificity versus stability in computational protein design. *Proceedings of the National Academy of Sciences USA 102*, 12724–12729.

47. Reina, J., Lacroix, E., Hobson, S. D., Fernandez-Ballester, G., Rybin, V., Schwab, M. S., Serrano, L., & Gonzalez, C. (2002). Computer-aided design of a PDZ domain to recognize new target sequences. *Nature Structural Biology 9*, 621–627.

48. Chevalier, B. S., Kortemme, T., Chadsey, M. S., Baker, D., Monnat, R. J., & Stoddard, B. L. (2002). Design, activity, and structure of a highly specific artificial endonuclease. *Molecular Cell 10*, 895–905.

49. Huang, P. S., Love, J. J. & Mayo, S. L. (2005). Adaptation of a fast Fourier transform-based docking algorithm for protein design. *Journal of Computational Chemistry 26*, 1222–1232.

50. Huang, P. S., Love, J. J., & Mayo, S. L. (2007). A de novo designed protein–protein interface. *Protein Science 16*, 2770–2774.

51. Yin, H., Slusky, J. S., Berger, B. W., Walters, R. S., Vilaire, G., Litvinov, R. I., Lear, J. D., Caputo, G. A., Bennett, J. S., & DeGrado, W. F. (2007). Computational design of peptides that target transmembrane helices. *Science 315*, 1817–1822.

52. Senes, A., Chadi, D. C., Law, P. B., Walters, R. F. S., Nanda, V., & DeGrado, W. F. (2007). E-z, a depth-dependent potential for assessing the energies of insertion of amino acid side-chains into membranes: Derivation and applications to determining the orientation of transmembrane and interfacial helices. *Journal of Molecular Biology 366*, 436–448.

53. Vizcarra, C. L., Zhang, N. G., Marshall, S. A., Wingreen, N. S., Zeng, C., & Mayo, S. L. (2008). An improved pairwise decomposable finite-difference Poisson-Boltzmann method for computational protein design. *Journal of Computational Chemistry 29*, 1153–1162.

54. Kortemme, T., Morozov, A. V., & Baker, D. (2003). An orientation-dependent hydrogen bonding potential improves prediction of specificity and structure for proteins and protein–protein complexes. *Journal of Molecular Biology 326*, 1239–1259.

55. Morozov, A. V., Kortemme, T., Tsemekhman, K., & Baker, D. (2004). Close agreement between the orientation dependence of hydrogen bonds observed in protein structures and quantum mechanical calculations. *Proceedings of the National Academy of Sciences USA 101*, 6946–6951.

56. Jiang, L., Kuhlman, B., Kortemme, T. A., & Baker, D. (2005). A "solvated rotamer" approach to modeling water-mediated hydrogen bonds at protein–protein interfaces. *Proteins-Structure Function and Bioinformatics 58*, 893–904.

57. Fu, X. R., Apgar, J. R., & Keating, A. E. (2007). Modeling backbone flexibility to achieve sequence diversity: The design of novel a-helical ligands for Bcl-XL. *Journal of Molecular Biology 371*, 1099–1117.

8 Protein–Protein Docking

Howook Hwang, Brian Pierce, and Zhiping Weng

CONTENTS

INTRODUCTION

Biological and biochemical processes rely on networks of molecular interactions. A crucial component of these networks includes proteins, which recognize and associate with one another to perform roles such as cell cycle regulation, signal transduction, and antigen recognition in living organisms. Several systematic experimental techniques, namely, yeast two-hybrid (Fields and Song 1989; Bartel and Fields 1995), mass spectrometry (Gavin, Bosche, et al. 2002; Ho, Gruhler, et al. 2002), protein chips (Zhu, Bilgin, et al. 2001), and phage display (Mullaney and Pallavicini 2001), investigate protein–protein interactions on a genomic scale. These high throughput techniques are rapidly accumulating information on protein–protein interaction.

Complementary to the information gained by these techniques, it is of interest to know how proteins are interacting at the atomic level, by determining or predicting the three-dimensional (3D) structures of the protein complexes of interest. Such structures help further our understanding of important residues guiding the protein interaction and provide deeper insight into the protein–protein interaction network. Additionally, the structure of a protein complex can be used in structure-based drug design of inhibitor molecules or design of proteins for improved binding affinities or altered specificities.

Without the high-resolution structure of a protein complex, one can perform protein–protein docking, which is an in silico method to predict a protein complex structure, given individual protein structures as input. The input structures are generally obtained from x-ray diffraction, nuclear magnetic resonance (NMR), or homology modeling. One may also perform DNA–protein docking (Liu, Guo, et al. 2008; van Dijk and Bonvin 2008), RNA–protein docking (Jonker, Ilin, et al. 2007), RNA–small ligand docking (Guilbert and James 2008), and protein–small ligand docking (Chen and Zhi 2001). This chapter will address fundamental components of computational protein–protein docking methods, with specific examples based on the ZDOCK suite of algorithms developed in our lab.

PROTEIN–PROTEIN DOCKING

Docking programs often take input files in the Protein Data Bank (PDB) macromolecular structure format (Berman, Westbrook, et al. 2000), which contain Cartesian (x, y, z) coordinates of atom positions in angstroms. X-ray crystallography usually provides a single protein structure, whereas the NMR method provides multiple copies of a protein in different conformations. Thus, to use protein structures from NMR for docking, users need to select a representative structure among the multiple conformations. Although it is recommended to use x-ray crystal structures with resolution better than (i.e., less than) 2.5 Å for docking, it has been shown that protein docking simulation results are not severely affected by the protein structures with resolutions as low as 3.25 Å (Chen, Mintseris, et al. 2003).

Protein–protein docking can be subdivided into two categories, bound docking and unbound docking, based on the source of the target proteins to be docked. Bound docking takes the experimentally determined structure of the complex, separates the component proteins, and attempts to reproduce the complex structure. Unbound docking takes as input individually determined protein structures. As the complex structure is already known (by definition) prior to bound docking, bound docking is primarily a way to assess a docking algorithm and is of little predictive use. In terms of difficulty level, unbound docking is much more difficult, since unbound docking must model protein side chain and backbone movements that occur upon binding.

EVALUATION OF DOCKING PERFORMANCE AND ACCURACY OF PREDICTED PROTEIN COMPLEX

Quantitative measurement is required to evaluate the accuracy of predicted protein complex models by a docking algorithm. The most common measure is the root mean square deviation (RMSD):

$$RMSD = \sqrt{\frac{1}{N} \sum_{i=1}^{N} \left\{ [x_p(i) - x_n(i)]^2 + [y_p(i) - y_n(i)]^2 + [z_p(i) - z_n(i)]^2 \right\}}$$

where N is the total number of atoms, and (x, y, z) are the Cartesian coordinates of atoms in predicted (p) and native (n) protein complexes. The predicted complex and the native complex have to be structurally aligned to minimize the RMSD. Depending upon the focus, RMSD can be calculated using all atoms in the complex or only atoms in the binding interface from both receptor and ligand, or atoms only from the ligand. One can choose to use all atoms, only backbone atoms, or only Cα atoms, with minor impact on the resulting RMSD values.

The Critical Assessment of PRedicted Interactions (CAPRI; Janin, Henrick, et al. 2003) is a community-wide blind test of docking algorithms, where an unreleased structure of a protein complex is predicted by various participant groups. In CAPRI, the predictions are evaluated by the fraction of native contacts (fnat) and the fraction of nonnative contacts (fnon-nat) along with interface RMSD (iRMSD) and ligand RMSD (lRMSD). All submitted predictions are grouped into four classes: incorrect, acceptable, medium, and high, based on a Boolean expression that contains the four metrics (Mendez, Leplae, et al. 2003).

CURRENT APPROACHES FOR UNBOUND RIGID-BODY DOCKING

Proteins often undergo conformational changes upon interaction with other molecules, including other proteins, DNA, RNA, or small ligands. The conformational changes mostly occur on surface atoms, yet this conformational variability between unbound and bound forms must be accounted for in a successful protein–protein docking algorithm. Explicitly searching the backbone and side chain degrees of freedom, even if restricted to surface residues, is too computationally intensive and can yield false positives if not performed correctly.

Alternatively, rigid-body docking approaches keep the protein conformation fixed during the docking process and allow small clashes between the two proteins. This allowance of small clashes provides implicit modeling of the generally small side chain and backbone movements that take place to accommodate the proteins as they form a complex. Rigid-body docking (or "initial stage" docking) is typically followed by a refinement stage, which takes the top models from the initial stage and optimizes side-chain and backbone conformation, without large-scale movements of the predicted complex (Chen, Li, et al. 2003; Li, Chen, et al. 2003; Figure 8.1).

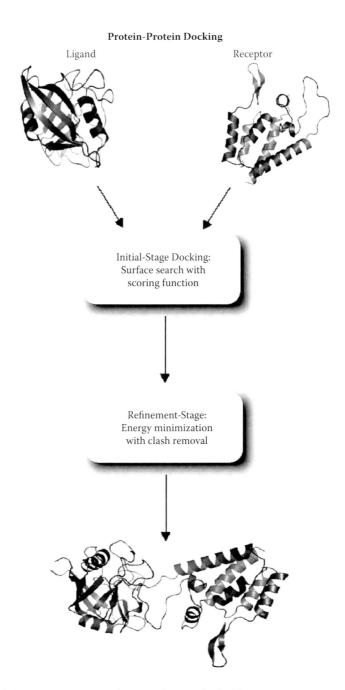

FIGURE 8.1 Two-stage approach to protein–protein docking.

There are two components to any rigid-body docking approach: searching the possible protein orientations in the 6D space (3 translational and 3 rotational degrees of freedom) and scoring the search results so that nativelike complex predictions can be discriminated from nonnative complex predictions.

SEARCHING THE SPACE OF POSSIBLE COMPLEX STRUCTURES

A variety of different methods have been used to successfully search the possible binding interfaces between two proteins in docking algorithms. Monte Carlo methods (Gray, Moughon, et al. 2003; Zacharias 2003) utilize random sampling in the 6D space to create candidate orientations. Geometric hashing has been used for protein docking (Fischer, Lin, et al. 1995), in addition to alignment and functional motif detection of biological molecules (Nussinov and Wolfson 1991; Rosen, Lin, et al. 1998). The algorithm indexes protein surfaces with their unique characteristics in various reference states in a hash table that is used to calculate a correlation between two proteins.

Another search strategy, the fast Fourier transform (FFT), has been one of the more popular approaches in rigid-body protein–protein docking (Vakser 1995; Gabb, Jackson, et al. 1997; Chen, Li, et al. 2003; Kozakov, Brenke, et al. 2006) since its first application for this purpose (Katchalski-Katzir, Shariv, et al. 1992). In FFT docking, the two input proteins (referred to as receptor and ligand) are discretized onto individual 3D grids, resulting in 3D functions of R(x, y, z) for receptor and L(x, y, z) for ligand. Because the FFT algorithm does not speed up the rotational space, it must be sampled explicitly and the ligand is rediscretized after each rotation. In the case of ZDOCK (Chen, Li, et al. 2003), a Euler angle set with sampling spacing of 15° or 6° is used to perform rotation search.

For each rotational conformation of the ligand, the best translation is found using correlation between the receptor and ligand grids. The score of a particular ligand translation of (i, j, k) is obtained by adding the product of overlapping grid points from receptor and ligand:

$$S(i,j,k) = \sum_{x,y,z} L(x+i, y+j, z+k) \times R(x,y,z)$$

The score is calculated for all possible combinations of (i, j, k) for the global translational search and this step has a computational cost that scales with N6 (O(N6)), where N is the number of grid points in each dimension. Alternatively, this can be performed in one step using the discrete Fourier transform (DFT) and inverse Fourier transform (IFT) of the discretized proteins:

$$S(i,j,k) = \frac{1}{N^3} IFT[IFT(L) \times DFT(R)]$$

FFT is an efficient way to compute DFT and IFT with computational cost of O(log(N3)). As a result, FFT reduces the total computational cost from O(N6) to O(N3log(N3)). It should also be noted that the FFT has also been used to dock spherical harmonic representations of the receptor and ligand (Ritchie and Kemp 2000), as opposed to 3D Cartesian grids.

SCORING FUNCTION

With an effective search strategy in place, it is also crucial to have an accurate and fast method to compute scoring function for ranking docking predictions. The most frequently used term in docking scoring functions is shape complementarity. As proteins undergo relatively small conformational changes during complexation, the binding sites can be identified by the complementarity of the protein surfaces at the interface. Two other terms that are often used in docking scoring functions are electrostatics and desolvation.

Shape Complementarity

From the earliest computational protein docking efforts in the late 1970s (Greer and Bush 1978; Wodak and Janin 1978), shape complementarity (SC) has successfully been used to score docking predictions. This optimization of fit between surfaces is based, at the atomic level, on the physical van der Waals (vdW) potential. It was mentioned earlier in this chapter that softening protein surface is essential for unbound rigid-body docking to allow marginal clashes between target proteins and it is the SC term that can contribute to the softening of the protein surface. The vdW is distance dependent and is composed of two terms: repulsive force for short distances and attractive force for long distances. vdW potentials are often approximated with the Lennard-Jones 6-12 potential:

$$V_{L-J} = \frac{A}{r^{12}} - \frac{B}{r^6}$$

where the r^{12} term represents short-distance repulsive energy and the r^6 term represents attractive energy. The parameters A and B are dependent on the particular atoms being considered, and the minimum of this potential occurs at the sum of the vdW radii of the two atoms.

In FFT docking algorithms, SC is often implemented in two ways: grid-based shape complementarity (GSC) and pairwise shape complementarity (PSC). GSC was the first to be developed and was used in some docking algorithms (Gabb, Jackson, et al. 1997), but later it was shown that PSC performs better than GSC (Chen and Weng 2003) and PSC was implemented in the ZDOCK docking algorithm (Chen, Li, et al. 2003). The difference between GSC and PSC is that PSC computes the total number of receptor–ligand atom pairs within a distance cutoff, minus a clash penalty, whereas GSC computes the number of grid points on the molecular surfaces that overlap, minus clash penalty for core points that overlap (Chen and Weng 2003).

Electrostatics

In addition to shape complementarity, electrostatics is a crucial component of the energetics of many protein–protein interactions. Electrostatic energy can be computed by solving the Poisson–Boltzmann equation (Honig and Nicholls 1995); however, for computing efficiency it is often approximated using the Coulombic equation.

Programs that use electrostatics include DOT (Mandell, Roberts, et al. 2001), MolFit (Heifetz, Katchalski-Katzir, et al. 2002), and ZDOCK (Chen, Li, et al. 2003). One study used electrostatics to generate molecular trajectories to perform docking simulations with some success (Fitzjohn and Bates 2003).

Desolvation Energy

In solution, proteins are surrounded by water molecules, and in order for a protein to interact with other proteins, the water molecules around the interaction interface must rearrange. Desolvation is an energetic (largely entropic) term that considers such changes (breaking water–protein bonds to form protein–protein and water–water bonds) and is often referred to as the hydrophobic effect (Chothia and Janin 1975).

Atomic level representations of contact propensities are valuable sources of information in protein–protein docking. Statistical potentials, which are also known as knowledge-based potentials, have been used in computational protein structure prediction for many years. These potentials are a measure of significance of occurrence for an observed contact between a pair of atoms (or residues) versus a reference state. One example is atom contact energy (ACE; Zhang, Vasmatzis, et al. 1997), which was derived from propensities of atoms within 6 Å, as seen in monomeric crystal structures. Other statistical potentials have more advanced functions, such as distance-scaled, finite, ideal gas reference (DFIRE; Zhou and Zhou 2002); this function has proven effective in protein–protein docking and folding.

We recently developed a pair potential IFACE (Interface Atomic Contact Energies), specifically geared toward detecting protein–protein interfaces (Mintseris, Pierce, et al. 2007). IFACE distinguishes itself from other pair potentials in that it optimizes atom types in a principled way. For a total of M atom types, a Monte Carlo simulation is performed to divide all 167 nonhydrogen amino acid atoms into M types. A simulated annealing procedure was used to ensure that the mutual information was maximized for the set of atom types (Mintseris and Weng 2004). The resultant atom types were then assigned pairwise IFACE energies based on the observed atom contacts within a nonredundant data set of 150 transient protein–protein complexes. The energy terms are computed as a log ratio of the actual numbers of contacts and numbers of contacts in a reference state. This reference state is a novel aspect of our approach; it captures all possible contacts between the surface atoms of the interacting proteins.

PROTEIN–PROTEIN DOCKING BENCHMARK

A protein–protein docking benchmark is a curated set of protein structures for evaluating the performance of docking algorithms (Chen, Mintseris, et al. 2003). Each test case includes a pair of unbound proteins, whose structures are available both in complex and in unbound forms. The benchmark is a nonredundant data set in terms of complexes, as classified by SCOP (Structural Classification of Proteins; Murzin, Brenner, et al. 1995). Since the benchmark was first released in 2003 (Chen, Mintseris, et al. 2003), two updates were made, in 2005 (Mintseris, Wiehe, et al. 2005) and 2008 (Hwang, Pierce, et al. 2008).

Benchmark version 3.0, the most recent version, is composed of 124 test cases (Hwang, Pierce, et al. 2008). They are classified based on two different schemes: docking difficulty and biological function. For docking difficulty, three levels are considered: 88 rigid-body cases, 19 medium cases, and 17 difficult cases, largely determined by the extent of interface conformational changes upon complex formation. For the biological classification, three categories include 34 enzyme–inhibitor cases, 25 antigen–antibody cases, and 65 other cases.

ZDOCK, RDOCK, AND ZRANK

ZDOCK (Chen, Li, et al. 2003) is an FFT-based initial stage rigid-body docking algorithm with an optimized scoring function. To overcome the limitation of treating unbound proteins as rigid bodies, ZDOCK softens the protein surfaces with a scoring function that allows light overlaps between two protein interfaces to take into account possible conformational changes. Throughout its different versions, the scoring function in ZDOCK has contained various combinations of three major terms—shape complementarity, electrostatics, and desolvation:

ZDOCK 1.3 (Chen and Weng 2002): Grid-Based Shape Complementarity (GSC) + Electrostatics + Desolvation
ZDOCK 2.1 (Chen and Weng 2003): Pairwise Shape Complementarity (PSC)
ZDOCK 2.3 (Chen, Li, et al. 2003): Pairwise Shape Complementarity (PSC) + Electrostatics + Desolvation
ZDOCK 3.0 (Mintseris, Pierce, et al. 2007) : Pairwise Shape Complementarity (PSC) + Electrostatics + IFACE

One notable advance in the most recent ZDOCK version 3.0 scoring function is the incorporation of the pairwise statistical potential IFACE (Mintseris, Pierce, et al. 2007), which contains 12 atom types. This potential was developed using the atomic propensities across transient protein–protein interfaces and replaces the ACE (Zhang, Vasmatzis, et al. 1997) term in the previous version of ZDOCK, which has 18 atom types. Inclusion of IFACE in ZDOCK 3.0 results in significant improvement in docking success across a docking benchmark (Mintseris, Pierce, et al. 2007).

Once ZDOCK has produced docking predictions, which are either 3,600 or 54,000 depending on the angular sampling density selected (15° or 6°), the predictions are then processed in the refinement stage, which resolves clashes by optimizing side-chain conformations and backbone conformations with RDOCK (Li, Chen, et al. 2003). After the refinement, the docking results are rescored to improve the ranking of the near-native predictions.

RDOCK is composed of two parts: energy minimization of docking predictions from ZDOCK with CHARMM (Brooks, Bruccoleri, et al. 1983) and reranking the optimized predictions with its own scoring function. The scoring function for RDOCK reranking is composed of desolvation energy and electrostatics within interface atom pairs:

$$\Delta G_{binding} = \Delta G_{ACE} + \beta \times \Delta E_{elec}$$

where β is a scaling factor (set to 0.9 as default in RDOCK).

While RDOCK is being used for energy minimization and reranking the minimized predictions, it takes approximately one minute to minimize one test case and this leads to an inevitable limitation for RDOCK. It was recommended to use RDOCK for 1000–2000 predictions out of 54,000 from ZDOCK (Li, Chen, et al. 2003), which could result in losing near-native predictions that were ranked above 2000 by ZDOCK. To overcome this limitation, ZRANK (Pierce and Weng 2007) was developed.

ZRANK is a reranking method that can process 54,000 ZDOCK predictions efficiently and accurately, scoring roughly 180 predictions per minute on a single Intel Pentium III 2.0 GHz machine. The scoring function of ZRANK is a linear weighted sum of van der Waals attractive and repulsive energies, electrostatics short-range and long-range attractive and repulsive energies, and desolvation:

$$\text{Score} = W_{vdW\_a}E_{vdW\_a} + W_{vdW\_r}E_{vdW\_r} + W_{elec\_sra}E_{elec\_sra}$$
$$+ W_{elec\_srr}E_{elec\_srr} + W_{elec\_lra}E_{elec\_lra} + W_{elec\_lrr}E_{elec\_lrr} + W_{ds}E_{ds}$$

The weights in the ZRANK scoring function (denoted by W in the equation) were determined using a downhill simplex minimization algorithm (Press 2002) to optimally rank ZDOCK 2.3 predictions for a set of Protein Docking Benchmark 1.0 benchmark cases. The weights that were obtained are (Pierce and Weng 2007):

van der Waals attractive: 1.0
van der Waals repulsive: 0.009
Electrostatics short-range attractive: 0.31
Electrostatics short-range repulsive: 0.34
Electrostatics long-range attractive: 0.44
Electrostatics long-range repulsive: 0.50
Desolvation: 1.02

Utilizing this weighted scoring function to rescore ZDOCK predictions led to significant improvements in docking success rates when tested using Benchmark 2.0 cases (the cases in the training set were excluded in this evaluation). Because no structural minimization of the predictions was necessary prior to scoring with ZRANK, this indicates that near-native predictions from rigid-body docking possess adequate structural and energetic information to be discriminated from incorrect predictions in many cases.

Following the development of ZRANK, the question of whether this function could be used for refined cases was addressed. After using the docking program RosettaDock (Gray, Moughon, et al. 2003) to refine the structural models from

ZDOCK, using the ZRANK program to rescore these refined models resulted in improved success for Benchmark 2.0 cases (Pierce and Weng 2008). The success rate improved further when reoptimizing the ZRANK weights specifically to rescore refined models and incorporating the IFACE potential as another term in the ZRANK function. In addition to the docking benchmark, ZRANK has been successfully used to score docking models in the CAPRI docking experiment (Wiehe, Pierce, et al. 2007).

In addition to rescoring and refinement, filtering false positive predictions from a set of docking predictions is often an important step in protein complex prediction. For instance, mutagenesis data on a complex may be available to help remove predictions without specified residues in the interface. There are also general properties of proteins that can be used to filter predictions, such as removing predictions of antigens bound to non-CDR portions of antibodies (Chen, Li, et al. 2003).

Clustering is another method to guide analysis of protein docking predictions. This is based on the concept of a low-energy funnel in the vicinity of the binding site between two proteins, thus the abundance of structurally similar low-energy docking models may indicate a binding site (Zhang, Chen, et al. 1999). This has been implemented, for instance, in the clustering server ClusPro (Comeau, Gatchell, et al. 2004). Usually, iRMSD or lRMSD between docking predictions is used as a metric for clustering predictions within a given radius cutoff, for example, 8 Å. Two possible uses for clustering docking results are: (1) clustering docking predictions with a metric and cutoff so that the average binding energy can be calculated for cluster comparison and (2) selecting cluster representative structures efficiently to assess redundancy within a cluster (Tong and Weng 2004). While not strictly related to locating energy funnels, eliminating structure redundancy can be useful when comparing many different protein–protein docking predictions at once.

RMSD AND PERFORMANCE EVALUATION

We typically use interface Cα atoms for iRMSD calculation between predicted complexes and native complexes to evaluate predicted structures from ZDOCK. Interface Cα atoms are defined as the Cα atoms of residues that have any atom within 10 Å of the binding partner protein in the complex. iRMSD of 2.5 Å is used as a cutoff to determine a near-native prediction (hit).

We use two ways to measure performance of docking algorithms: success rate and average hit count. Success rate measures the percent of test cases that have a hit in the given top N predictions, and average hit count divides the total number of hits for all test cases in the given top N predictions by the number of test cases.

ZDOCK/ZRANK PERFORMANCE ON BENCHMARK 3.0

We tested the performance of ZDOCK 3.0 and ZRANK along with other ZDOCK versions on a Benchmark 3.0 data set. During the construction of the benchmark, the unbound structures were superposed onto bound complexes. Hence, prior to docking, unbound ligands and receptors were randomly rotated to avoid biased docking results due to specifically sampling a near-native configuration (the starting configuration).

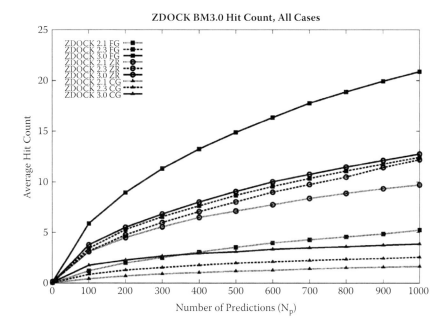

FIGURE 8.2 Average hit count comparison between different ZDOCK versions and ZRANK with two rotational sampling spacings: fine grain (FG; 6°) and coarse grain (C⁻G; 15°).

The average hit count for Benchmark 3.0 in Figure 8.2 shows the performance of ZDOCK 3.0 at 6° (FG or fine grain) rotational sampling density to produce the largest number of near-native predictions within the top 1000 predictions. The other versions of ZDOCK at FG sampling and ZRANK have substantially more hits compared with the 15° (CG or coarse grain) rotational search. Since the FG rotational search samples more densely than the CG search, it has more opportunities to produce near-native structural predictions. In addition, the new scoring function of ZDOCK 3.0 provides better discrimination of near-native predictions from false-positive predictions, causing this scoring scheme to produce the highest hit count.

The success rate results tell a different story. In the success rate comparison (Figure 8.3), ZDOCK 3.0 with the CG rotational search performs the best, followed by ZRANK and ZDOCK 3.0 with FG rotational search. This implies the FG rotational search increases the number of highly ranked false-positive predictions in addition to the highly ranked near-native predictions, with more false-positive predictions ranked in the top few.

Figure 8.4 shows success rate comparison between different versions of ZDOCK with the FG rotational search and ZRANK on the 88 rigid cases in Benchmark 3.0. Reranking ZDOCK predictions with ZRANK outperforms ZDOCK with FG rotational search and the reranking elevates success rate by ~12% at Np = 1000 in the case of ZDOCK 3.0. This indicates that ZRANK is most effective on structures with small conformational changes.

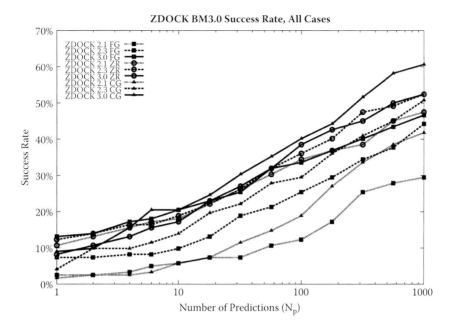

FIGURE 8.3   Success rate comparison between different ZDOCK versions with different rotational search spacings and reranking results with ZRANK.

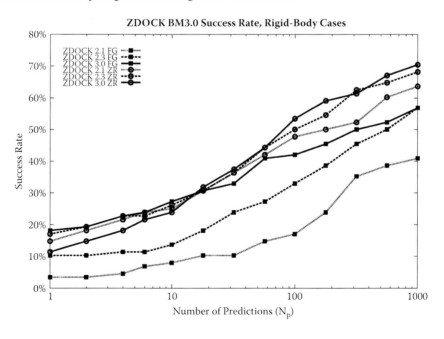

FIGURE 8.4   Success rate comparison between different ZDOCK versions with the fine grain (FG, 6°) rotational search spacing, with and without ZRANK, considering rigid-body cases only.

CASE STUDIES WITH ZDOCK AND ZRANK

DOCKING WITH REBUILT MISSING ATOMS/RESIDUES ON BENCHMARK 2.0

For unbound–unbound docking cases, independently crystallized structures that are used for docking can sometimes have missing residues and/or atoms that may be located in the vicinity of the protein–protein interface of interest. To verify whether this has a noticeable impact on docking performance, we tested if rebuilding these missing residues and atoms improves docking results.

Among the cases from Benchmark 2.0, there are two rigid cases (PDB codes for the complex structures: 1F51 and 1K4C) with interface residues that have missing atoms for which ZDOCK 2.3 does not produce a hit within its top 2000 predictions. We rebuilt all residues with missing atoms on both unbound receptor and ligand using the Accelrys Insight II software package.

1F51 is a complex of sporulation response factor B (PDB code for the unbound protein: 1IXM) and sporulation response factor F (PDB code for the unbound protein: 1SRR), which is involved in phosphoryl group transfer (Zapf, Sen, et al. 2000). In all, 114 residues from the receptor (1IXM) and 10 residues from the ligand (1SRR) had missing atoms to be rebuilt, of which 16 residues of the receptor and 3 residues of the ligand were in the interface. After rebuilding, these protein structures were docked using ZDOCK 2.3 with 6° sampling, resulting in six hits within the top 2000 predictions. Figure 8.5 shows the highest-ranked hit, which is ranked 315, which has an iRMSD of 1.56 Å compared with the complex structure.

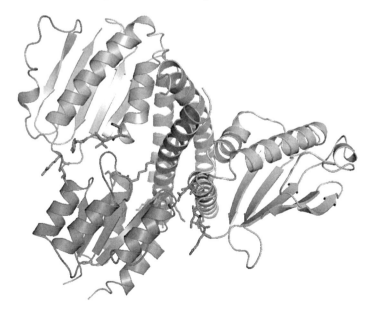

FIGURE 8.5 (SEE COLOR INSERT FOLLOWING PAGE 174.) ZDOCK prediction for test case 1F51 using rebuilt residues prior to docking. 1SRR is colored magenta, 1IXM chain A is colored green, and 1IXM chain B is colored cyan. The rebuilt interface residues with missing atoms are displayed as red sticks.

1K4C is a bound–unbound antibody–antigen case with Fab taken from the complex structure (PDB code for the complex structure: 1K4C) and potassium channel Kcsa (PDB code for the unbound protein: 1JVM; Zhou, Morais-Cabral, et al. 2001). We followed the same procedure we performed with the 1F51 case and one interface residue with missing atoms from 1JVM was located. The identified residue was rebuilt and docking was performed using ZDOCK 2.3. As a result, we obtained nine hits within the top 2000 predictions. Figure 8.6 shows the highest-ranked hit, which is ranked 601, with iRMSD 0.88 Å compared with the complex.

It is of interest to see the docking prediction improvement by rebuilding missing atoms in unbound structures from docking test cases, which were unable to produce hits. Strikingly, the results for the 1K4C test case improved by building only one interface residue, tyrosine. This shows the high sensitivity of ZDOCK and protein–protein recognition itself. Similar studies may be implemented in a systematic way to locate hot spot residues in protein–protein interfaces using docking algorithms.

DOCKING WITH FLEXIBLE INTERFACE LOOPS

Protein–protein docking with highly mobile proteins is one of the major challenges in the docking field (Ehrlich, Nilges, et al. 2005; Bonvin 2006). This includes proteins with flexible loops in interfaces, which was addressed in a study where a conformational

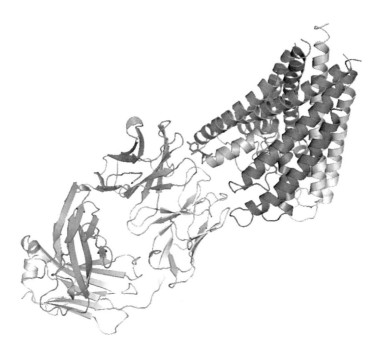

FIGURE 8.6   (SEE COLOR INSERT FOLLOWING PAGE 174.) Structure prediction for test case 1K4C. 1JVM is colored in blue, cyan, magenta, and green and 1K4C is colored in salmon and gray. The rebuilt interface residue with missing atoms is displayed as sticks in red.

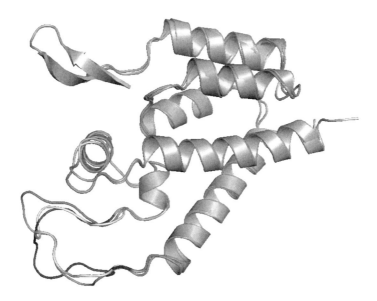

FIGURE 8.7 Loop modeling of the 1AK4 ligand (PDB code: 1E6J), which is the HIV capsid protein. Residue numbers V86-E98 in the unbound structure were modeled (modeled loop: light gray; unbound loop: black; bound loop: dark gray). The modeled loop with the best docking result is shown.

search of flexible interface loops in unbound protein structures was performed in low resolution and incorporated into docking (Bastard, Prevost, et al. 2006).

In Benchmark 2.0, there are 13 medium and 8 difficult cases with moderate or large conformational changes upon protein–protein association (Mintseris, Wiehe, et al. 2005). For rigid-body docking algorithms, it is inherently difficult to solve the docking problem with highly mobile proteins. To address this problem, we selected Benchmark 2.0 test cases (PDB codes: 1AK4, 1K5D, and 1ATN) for which ZDOCK 2.3 is unable to produce hits within the top 2000 predictions due to the conformational change of a single mobile loop in the interface. We used the Accelrys Insight II software package to remodel the mobile interface loops for these three test cases. For the selected cases, each of the mobile interface loops from the unbound structure was targeted to be modeled, while the remainder of the protein was kept fixed. We generated nine unbound structures with remodeled loops, plus one unbound structure with the bound loop inserted. The docking results are summarized in Table 8.1.

1AK4 is the complex of cyclophilin (PDB code: 2CPL) and an HIV capsid protein (PDB code: 1E6J). A loop from the HIV capsid protein (residues V86-E98 of chain P; shown in Figure 8.7) was targeted for remodeling in the unbound structure. ZDOCK 2.3 obtained at least one hit within the top 2000 predictions with eight out of nine unbound structures with modeled loop, and with the unbound structure with the bound loop inserted as well.

1K5D is the complex of Ran GTPage (PDB code: 1RRP) and Ran GAP protein (PDB code: 1YRG). A loop from Ran GTPase (residues Q69-I81 of chain A) was targeted for remodeling in the unbound structure. ZDOCK 2.3 obtained at least one

TABLE 8.1
Summary of Docking Results with Flexible Interface Loops

Complex[a]	Modeled Loop[b]	Number of Loops[c]	Number of Hits[d]	Best Ranked Hit (iRMSD)[e]
1AK4	1E6J_P: V86-E98	8	20	32 (2.32 Å)
1K5D	1RRP_A: Q69-I81	5	8	163 (1.88 Å)
1ATN	1IJJ_B: R439-Y453	1	3	1479 (2.19 Å)

[a] PDB codes of the complexes. Detailed information about complexes is available at http://zlab.bu.edu/benchmark/.

[b] Targeted loops for remodeling in the unbound structures. (PDBcode_Chain: starting residue-ending residue.)

[c] Number of unbound structures with a remodeled loop that produced at least one hit in the top 2000 ZDOCK predictions.

[d] Number of hits with the unbound structure with remodeled loop that produced the largest number of hits.

[e] The best ranked hit with the unbound structures with remodeled loop, along with its iRMSD.

hit within the top 2000 predictions with five out of nine unbound structures with modeled loop, and with the unbound structure with the bound loop inserted as well.

1ATN is the complex of Actin (PDB code: 1IJJ) and Dnase I (PDB code: 3DNI). A loop from actin (residues R439-Y453 of chain B) was targeted for remodeling in the unbound structure. ZDOCK 2.3 obtained at least one hit within the top 2000 predictions with one out of nine unbound structures with modeled loops, and with the unbound structure with the bound loop inserted as well.

Overall, the docking results seen in Table 8.1 reflect the docking difficulties of the respective cases. Specifically, 1AK4 is a rigid-body case with mild conformational change, 1K5D is a medium case with moderate conformational change, and 1ATN is a difficult case with severe backbone conformational change in the interface.

CONCLUSION

Protein–protein interactions play critical roles in biological and biochemical systems. Understanding these interactions on a molecular level can aid redesign or modulation of the interaction network, possibly providing therapeutic value. Protein–protein docking is an active research area and is advancing rapidly. The recent progress and success in the CAPRI blind test (Janin and Wodak 2007) indicates that many protein–protein interactions can be successfully solved by protein–protein docking. In addition, docking experiments such as CAPRI stimulate the application of protein–protein docking knowledge to other fields, such as protein–RNA docking or protein–DNA docking. The incorporation of accurate docking methods with homology modeling and genomic scale macromolecular interaction maps will allow biologists to have greater insight into molecular functions and, in the broader scope, biological systems.

At this stage, much of the success in protein–protein docking is limited to rigid-body protein docking, which allows a small degree of conformational changes on protein surfaces. Protein surface structural plasticity upon association is one of the major challenges that have to be addressed in the next step of docking. The ZDOCK suite of algorithms is capable of producing hits with unbound structures with conformational changes, as long as the conformational search results are provided. The next step for improvement will be incorporating an ensemble of multiple conformations of an unbound protein in the ZDOCK framework.

ACKNOWLEDGMENTS

We thank Kevin Wiehe for helpful discussions and Mary Ellen Fitzpatrick for computing support. We are grateful to the Scientific Computing Facilities at Boston University and the Advanced Biomedical Computing Center at National Cancer Institute, National Institutes of Health for computing support. This work was funded by National Science Foundation grants DBI-0078194, DBI-0133834, and DBI-0116574. All protein structure figures were generated with Pymol (www.pymol.org).

REFERENCES

Bartel, P. L. and S. Fields (1995). "Analyzing protein–protein interactions using two-hybrid system." *Methods Enzymol* 254: 241–63.

Bastard, K., C. Prevost, et al. (2006). "Accounting for loop flexibility during protein–protein docking." *Proteins* 62(4): 956–69.

Berman, H. M., J. Westbrook, et al. (2000). "The Protein Data Bank." *Nucleic Acids Res* 28(1): 235–42.

Bonvin, A. M. (2006). "Flexible protein–protein docking." *Curr Opin Struct Biol* 16(2): 194–200.

Brooks, B. R., R. E. Bruccoleri, et al. (1983). "CHARMM: A program for macromolecular energy, minimization, and dynamics calculations." *Journal of Computational Chemistry* 4: 187–217.

Chen, R., L. Li, et al. (2003). "ZDOCK: An initial-stage protein-docking algorithm." *Proteins* 52(1): 80–7.

Chen, R., J. Mintseris, et al. (2003). "A protein–protein docking benchmark." *Proteins* 52(1): 88–91.

Chen, R. and Z. Weng (2002). "Docking unbound proteins using shape complementarity, desolvation, and electrostatics." *Proteins* 47(3): 281–94.

Chen, R. and Z. Weng (2003). "A novel shape complementarity scoring function for protein–protein docking." *Proteins* 51(3): 397–408.

Chen, Y. Z. and D. G. Zhi (2001). "Ligand-protein inverse docking and its potential use in the computer search of protein targets of a small molecule." *Proteins* 43(2): 217–26.

Chothia, C. and J. Janin (1975). "Principles of protein–protein recognition." *Nature* 256(5520): 705–8.

Comeau, S. R., D. W. Gatchell, et al. (2004). "ClusPro: A fully automated algorithm for protein–protein docking." *Nucleic Acids Res* 32(Web Server issue): W96–99.

Ehrlich, L. P., M. Nilges, et al. (2005). "The impact of protein flexibility on protein–protein docking." *Proteins* 58(1): 126–33.

Fields, S. and O. Song (1989). "A novel genetic system to detect protein–protein interactions." *Nature* 340(6230): 245–6.

Fischer, D., S. L. Lin, et al. (1995). "A geometry-based suite of molecular docking processes." *J Mol Biol 248*(2): 459–77.

Fitzjohn, P. W. and P. A. Bates (2003). "Guided docking: First step to locate potential binding sites." *Proteins 52*(1): 28–32.

Gabb, H. A., R. M. Jackson, et al. (1997). "Modelling protein docking using shape complementarity, electrostatics and biochemical information." *J Mol Biol 272*(1): 106–20.

Gavin, A. C., M. Bosche, et al. (2002). "Functional organization of the yeast proteome by systematic analysis of protein complexes." *Nature 415*(6868): 141–7.

Gray, J. J., S. Moughon, et al. (2003). "Protein–protein docking with simultaneous optimization of rigid-body displacement and side-chain conformations." *J Mol Biol 331*(1): 281–99.

Greer, J. and B. L. Bush (1978). "Macromolecular shape and surface maps by solvent exclusion." *Proc Natl Acad Sci USA 75*(1): 303–7.

Guilbert, C. and T. L. James (2008). "Docking to RNA via root-mean-square-deviation-driven energy minimization with flexible ligands and flexible targets." *J Chem Inf Model 48*(6): 1257–68.

Heifetz, A., E. Katchalski-Katzir, et al. (2002). "Electrostatics in protein–protein docking." *Protein Sci 11*(3): 571–87.

Ho, Y., A. Gruhler, et al. (2002). "Systematic identification of protein complexes in Saccharomyces cerevisiae by mass spectrometry." *Nature 415*(6868): 180–3.

Honig, B. and A. Nicholls (1995). "Classical electrostatics in biology and chemistry." *Science 268*(5214): 1144–9.

Hwang, H., B. Pierce, et al. (2008). "Protein–protein docking benchmark version 3.0." *Proteins 73*(3): 705–9.

Janin, J., K. Henrick, et al. (2003). "CAPRI: A Critical Assessment of PRedicted Interactions." *Proteins 52*(1): 2–9.

Janin, J. and S. Wodak (2007). "The third CAPRI assessment meeting Toronto, Canada, April 20–21, 2007." *Structure 15*(7): 755–9.

Jonker, H. R., S. Ilin, et al. (2007). "L11 domain rearrangement upon binding to RNA and thiostrepton studied by NMR spectroscopy." *Nucleic Acids Res 35*(2): 441–54.

Katchalski-Katzir, E., I. Shariv, et al. (1992). "Molecular surface recognition: Determination of geometric fit between proteins and their ligands by correlation techniques." *Proc Natl Acad Sci USA 89*(6): 2195–9.

Kozakov, D., R. Brenke, et al. (2006). "PIPER: An FFT-based protein docking program with pairwise potentials." *Proteins 65*(2): 392–406.

Li, L., R. Chen, et al. (2003). "RDOCK: Refinement of rigid-body protein docking predictions." *Proteins 53*(3): 693–707.

Liu, Z., J. T. Guo, et al. (2008). "Structure-based prediction of transcription factor binding sites using a protein-DNA docking approach." *Proteins 72*(4): 1114–24.

Mandell, J. G., V. A. Roberts, et al. (2001). "Protein docking using continuum electrostatics and geometric fit." *Protein Eng 14*(2): 105–13.

Mendez, R., R. Leplae, et al. (2003). "Assessment of blind predictions of protein–protein interactions: Current status of docking methods." *Proteins 52*(1): 51–67.

Mintseris, J., B. Pierce, et al. (2007). "Integrating statistical pair potentials into protein complex prediction." *Proteins 69*(3): 511–20.

Mintseris, J. and Z. Weng (2004). "Optimizing protein representations with information theory." *Genome Inform 15*(1): 160–9.

Mintseris, J., K. Wiehe, et al. (2005). "Protein–Protein Docking Benchmark 2.0: An update." *Proteins 60*(2): 214–6.

Mullaney, B. P. and M. G. Pallavicini (2001). "Protein–protein interactions in hematology and phage display." *Exp Hematol 29*(10): 1136–46.

Murzin, A. G., S. E. Brenner, et al. (1995). "SCOP: A structural classification of proteins database for the investigation of sequences and structures." *J Mol Biol 247*(4): 536–40.

Nussinov, R. and H. J. Wolfson (1991). "Efficient detection of three-dimensional structural motifs in biological macromolecules by computer vision techniques." *Proc Natl Acad Sci USA 88*(23): 10495–9.

Pierce, B. and Z. Weng (2007). "ZRANK: Reranking protein docking predictions with an optimized energy function." *Proteins 67*(4): 1078–86.

Pierce, B. and Z. Weng (2008). "A combination of rescoring and refinement significantly improves protein docking performance." *Proteins 72*(1): 270–9.

Press, W. H. (2002). *Numerical recipes in C: The art of scientific computing.* Cambridge: Cambridge University Press.

Ritchie, D. W. and G. J. Kemp (2000). "Protein docking using spherical polar Fourier correlations." *Proteins 39*(2): 178–94.

Rosen, M., S. L. Lin, et al. (1998). "Molecular shape comparisons in searches for active sites and functional similarity." *Protein Eng 11*(4): 263–77.

Tong, W. and Z. Weng (2004). "Clustering protein–protein docking predictions." *Conf Proc IEEE Eng Med Biol Soc 4*: 2999–3002.

Vakser, I. A. (1995). "Protein docking for low-resolution structures." *Protein Eng 8*(4): 371–7.

van Dijk, M. and A. M. Bonvin (2008). "A protein–DNA docking benchmark." *Nucleic Acids Res* in press.

Wiehe, K., B. Pierce, et al. (2007). "The performance of ZDOCK and ZRANK in rounds 6-11 of CAPRI." *Proteins 69*(4): 719–25.

Wodak, S. J. and J. Janin (1978). "Computer analysis of protein–protein interaction." *J Mol Biol 124*(2): 323–42.

Zacharias, M. (2003). "Protein–protein docking with a reduced protein model accounting for side-chain flexibility." *Protein Sci 12*(6): 1271–82.

Zapf, J., U. Sen, et al. (2000). "A transient interaction between two phosphorelay proteins trapped in a crystal lattice reveals the mechanism of molecular recognition and phosphotransfer in signal transduction." *Structure 8*(8): 851–62.

Zhang, C., J. Chen, et al. (1999). "Protein–protein recognition: Exploring the energy funnels near the binding sites." *Proteins 34*(2): 255–67.

Zhang, C., G. Vasmatzis, et al. (1997). "Determination of atomic desolvation energies from the structures of crystallized proteins." *J Mol Biol 267*(3): 707–26.

Zhou, H. and Y. Zhou (2002). "Distance-scaled, finite ideal-gas reference state improves structure-derived potentials of mean force for structure selection and stability prediction." *Protein Sci 11*(11): 2714–26.

Zhou, Y., J. H. Morais-Cabral, et al. (2001). "Chemistry of ion coordination and hydration revealed by a K+ channel-Fab complex at 2.0 A resolution." *Nature 414*(6859): 43–48.

Zhu, H., M. Bilgin, et al. (2001). "Global analysis of protein activities using proteome chips." *Science 293*(5537): 2101–5.

9 Prediction of Protein Interaction Sites

Yanay Ofran

CONTENTS

OVERVIEW

Proteins recognize and bind to each other through interaction sites. Hence, understanding the mechanisms that underlie protein–protein interaction requires the elucidation of the characteristics of interaction sites. Analysis of interaction sites have revealed some of their commonalities and suggested that it may be possible to identify these sites a priori. Prediction methods that identify protein interaction sites from the structure or even the sequence of a protein will enhance the study of protein–protein interaction, and may break new grounds in protein design and even in the development of drugs.

TARGETING BINDING SITES

Benzodiazepinedione and Nutlin are new anticancer drugs that are currently in advanced clinical trials. They represent a new promising approach in drug development: both of them are small molecules designed to thwart protein–protein interaction by binding specifically and selectively to a protein interaction site, thus preventing the interaction. Although Benzodiazepinedione and Nutlin are two very different molecules, they both target the same binding site: the one where a protein called HDM2 binds the tumor suppressor protein p53. The interaction between these two proteins is believed to prevent p53 from suppressing the tumor. When Benzodiazepinedione and Nutlin bind specifically to the binding site of p53 on HMD2, they prevent the interaction and allow the suppression of the tumor by p53 (Vassilev 2004; Vassilev, Vu, et al. 2004; Koblish, Zhao, et al. 2006). Attacking protein–protein interaction sites is becoming increasingly popular in drug development (Archakov, Govorun, et al. 2003; Arkin and Wells 2004; Rudolph 2007; Wells and McClendon 2007). However, to target interaction sites, one needs first to identify the residues that compose them. In this chapter I will review the computational attempts to identify protein–protein interaction sites using various approaches, tools, and sources of data.

A KEY TO UNDERSTANDING BIOLOGICAL PROCESSES

Hopes for new types of drugs, however, are not the original raison d'tere of the field of interaction site prediction. Since biological processes are realized by the interaction of proteins, to fully understand or to manipulate biological processes one needs to unravel the mechanisms that underlie protein interactions. The first step in this direction is the identification of interaction sites. Prediction of binding sites would improve the understanding of molecular recognition and interactions. It may enhance the computational prediction of protein–protein interactions and lay the foundation for a rational design of interaction sites.

IDENTIFYING INTERFACES FROM 3-D STRUCTURE OF COMPLEXES

Protein–protein interaction sites are rather different from sites that bind small ligands, nucleic acids, metal ions, and even small peptides. Interfaces between proteins and smaller substrates are typically cavities and concave clefts (Laskowski, Luscombe, et al. 1996; Peters, Fauck, et al. 1996; Pettit and Bowie 1999). Proteins, however, tend to bind to each other through much larger and more structurally intricate surfaces (Janin 1995; Jones and Thornton 1996; Bahadur, Chakrabarti, et al. 2004; Keskin, Tsai, et al. 2004). The most straightforward way to identify protein interaction sites is by analyzing the three-dimensional (3-D) structure of the complex of two or more chains. For example, the infection of humans by the HIV virus is mediated by the interaction of two proteins: gp120 from the HIV and the human protein CD4, which is a receptor expressed on the surface of the immune system's T cells. When the 3-D structure of this complex was solved (Kwong, Wyatt, et al. 1998), it was greeted with excitement as a crucial step in highlighting the interaction sites and thus revealing the mechanism of HIV infection (Balter 1998). While 3-D is indeed a powerful means to identify binding sites, it is hardly a silver bullet: 3-D structures are available for less than 1% of all known pairs of interacting proteins. This gap is growing by the day as the throughput of experimental methods for the detection of interacting proteins grows rapidly, leaving the technologies for structure determination of complexes far behind. Between 2006 and 2008 intAct (Kerrien, Alam-Faruque, et al. 2007), a major database of protein–protein interactions, grew by more than 100,000 pairwise interactions. At the same time less than 1000 new 3-D structures of heterocomplexes were added to the Protein Data Bank (PDB), the major database of protein structure.

WHEN THERE IS NO STRUCTURE AVAILABLE

In the absence of experimentally solved 3-D structures of the complex, it has been suggested to rely on methods for computational prediction of the 3-D structure of the complex of two proteins (Fernandez-Recio, Totrov, et al. 2004). The field of modeling the structure of complexes, or docking as it is called by its practitioners, is far from being able to provide accurate 3-D models on a large scale. Currently, its ability to provide reliable models is limited to those cases in which there is a good experimentally solved 3-D structure of the respective unbound proteins. Even in those cases, most docking algorithms provide numerous possible models for the complex, each of which may suggest different surface patches as interaction sites. Indeed, a common practice in docking is to use predictions of interaction sites to choose the right model. Thus, while docking could sometimes help identify interaction sites, it is more common for docking to use interaction site prediction than vice versa. Computational approaches for the prediction of interaction sites are based on an attempt to identify general features that are shared by many interaction sites and then use these features to identify new putative interaction sites. Searching for such features requires the analysis of known interaction sites. To curate a large data set of interaction sites that will allow for such analysis, one first needs to define interaction sites.

HOW TO DEFINE AN INTERACTION SITE

Most definitions rely to some extent on structural considerations, attempting to come up with a formulation that will capture all and only the residues that are localized in the protein–protein interface. Numerous definitions have been suggested, but each of them suffers from some shortcomings that may bias the set of residues that it identifies as interaction sites. A common definition is: All residues that are accessible to solvent in the unbound state but are buried in the interface in the bound state. Obviously, the number of residues that will be captured by this definition depends to a large extent on the definition of "solvent-accessible residue." Deeming a residue exposed depends on the choice of a cutoff—either in terms of its absolute accessible area or in terms of percentage of its theoretically calculated surface area that is accessible to solvent. Typically, only a few residues have zero accessibility to the solvent and virtually none is 100% exposed. The vast majority of residues can, at least theoretically, interact with a water molecule in the solvent.

Yet only in some of these residues is the exposed area large enough to allow for an effective interaction of the residue with another protein (even if the interaction is mediated by a water molecule). Different choices of the minimal exposed area that renders a residue exposed will result in identification of different interaction sites. Therefore, a modification of the earlier definition was suggested: All residues whose accessible area was reduced upon binding. Although this definition overcomes the problem of choosing a cutoff for exposure, it may be too permissive by introducing into the interaction sites some residues at the rim of the interface that do not form any physical contact with the other protein, yet upon interaction go through marginal reduction in their accessible area. Another shortcoming of both of these definitions is their inability to account for interaction-dependent conformational changes. Many proteins undergo conformational changes upon interaction, which result in substantial changes in the exposure of residues even if they are not located in the interface. It may, therefore, happen that a buried residue in the unbound state will move to the interface in the bound state. Alternatively, conformational changes may affect the accessibility of some residues that are remote from the binding site. These definitions will miss such cases.

Looking only at the bound state may offer a solution to this problem. Such is the approach of the following definition: All residues in a protein chain that are in contact with a residue in another protein chain. The problem with this definition is that it relies on the somewhat fuzzy notion of contact. We assume that if two residues are very close to each other, they are in physical interaction. Thus, to use this definition one has to define a distance cutoff for rendering two residues contacting. The choice of a cutoff, again, depends on many physical, structural, and statistical considerations. It is common to require a certain minimal distance between the C_α or C_β atoms. A similar approach is to require a certain minimal distance between the center of mass of the residues. This minimal required distance is typically set between 5 and 8 angstroms. However, one has to bear in mind that if this distance cutoff is applied uniformly to any pair of amino acids, the result would be a data set that is biased toward smaller residues, as their C_α atoms, for example, are more likely to be spatially closer to each other than those of bulky residues. To account for this possible bias some studies have required a minimal distance, typically 4–6 angstroms, between any heavy atom from a residue on one chain and any heavy atom from a residue on the other chain. This definition would lead to identifying interaction sites without any size bias, but it may allow some pairs of residues that are too far apart to be in actual physical interaction to be identified as part of the interaction sites.

More intricate definitions—for example, setting different distance cutoffs for each possible combination of amino acids—are possible. However, comparative analyses of the residues that will be identified by such different definitions suggest that the differences between the interaction sites they identify are fairly small. Any of these definitions solves some biases but introduces others. Therefore, when choosing a definition of interaction sites for a specific study or when using data sets of interaction sites curated by others, it is important to consider the biases that the definition entails and try to account for them.

DIFFERENT TYPES OF INTERFACES

Protein–protein interaction comes in different flavors: There are interactions that are permanent, namely, interactions between chains that could not function without each other, and interactions that are transient, namely, interactions between chains that have a molecular function also in their unbound state or when they are bound to another partner. Some interactions, while transient, modify one or both of the proteins involved. Such is the case in phosphorylation, cleavage, or unibiquitination, to name just a few. Interactions may occur in different environments, under different chemical and physical conditions; proteins interact in different cellular compartments, even inside the lipid bilayer of the membrane. Each of these types of interactions may be stabilized by different mechanisms, and hence the interaction sites involved may be different in their characteristics. Many studies have attempted to characterize the differences between the mechanisms that stabilize different types of interactions. In particular, they looked at residue–residue contacts, that is, noncovalent interactions between residues that stabilize structures. A basic distinction was drawn between the contacts that stabilize the structure of a single chain and those that stabilize a complex of chains. Several studies have shown that there are different types of contacts in play in these two types of interactions (Jones and Thornton 1996, 1997; McCoy, Chandana Epa, et al. 1997; Keskin, Bahar et al. 1998; Lo Conte, Chothia, et al. 1999; Sheinerman, Norel, et al. 2000; Glaser, Steinberg, et al. 2001; Ofran and Rost 2003a).

Another distinction was drawn between permanent and transient interactions, or between obligatory and nonobligatory ones. Early studies that have looked at small data sets of only a few complexes have shown that obligatory or permanent interactions tend to be mediated by larger interfaces (Jones and Thornton 1996, 1997). When larger data sets were used, more differences were found: The transient complexes not only have smaller contact areas, but the interfaces themselves are different. In particular, they tend to be more polar on average. It was also shown that obligatory interactions tend to require more conformational changes upon association/dissociation than transient ones (Jones and Thornton 1996, 1997). More comprehensive studies that analyzed hundreds of complexes have reaffirmed the difference in the characteristics of the interfaces between complexes that are obligatory (sometimes referred to as obligomers) and complexes that are transient. Differences were also found between interfaces in homo-oligomeric interactions and those in hetero-oligomeric interactions.

INTERACTION SITES VERSUS OTHER SURFACE RESIDUES

When looking at the sequences of the interaction sites, studies have suggested that interface residues tend to be more conserved than other surface residues (Jones and Thornton 1996, 1997). It seems, based on later and more comprehensive analyses, that the level of conservation, while significant, is fairly small (Caffrey, Somaroo, et al. 2004; Ofran and Rost 2007b). Protein–protein interactions have been shown to be one of the functional descriptors that are least conserved among homologues proteins: Typically, sequence identity of less than 40% is sufficient to determine that two proteins share the same 3-D fold (Rost 1999). Sixty percent sequence identity is sufficient in most cases to determine that two homologous proteins are located in the same subcellular localization (Rost, Liu, et al. 2003). Eighty percent sequence identity would usually suffice to infer molecular function (Rost, Liu, et al. 2003). However, to infer that two proteins interact with the same partners, the level of sequence identity must be higher than 80% and by some accounts even greater than 90% (Mika and Rost 2006). These observations point in the same direction: Interaction sites may be conserved but typically not highly conserved. Whether this level of conservation is instrumental in interaction site prediction is still debated (Armon, Graur, et al. 2001; Caffrey, Somaroo, et al. 2004; Res, Mihalek, et al. 2005; de Vries and Bonvin 2008).

Interaction sites have structural features that distinguish them from other surface residues. By and large, they are more planar and tend to have different secondary structure compositions. As opposed to sequence conservation, analysis of structural conservation—that is, identification of residues that are structurally aligned across a protein family—found that interface residues are highly conserved structurally to the extent that their structural conservation alone may be a good way to distinguish between interaction sites and other surface residues (Ma, Elkayam, et al. 2003). Again, different types of interfaces tend to have different structural features. For example, antigen–antibody interactions are mediated by interfaces that are very different than other types of interfaces in many of their traits, including their secondary structures (Ofran, Schlessinger, et al. 2008). The same is true for protease–inhibitor complexes (Jackson 1999). Antibodies and antigenic proteins are two opposite cases in terms of prediction. The interaction sites on the antibody are fairly easy to identify, even when there is no 3-D available; they typically fall within a few well-defined loops on the antibody, known as complementarity determining regions (CDRs). The interaction sites on antigenic proteins (aka B-cell epitopes, the regions on the protein surface that bind specifically to the antibody), on the other hand, are extremely hard to predict (Greenbaum, Andersen, et al. 2007). In fact, an assessment of existing methods for the prediction of B-cell epitopes has concluded that most of them are at most marginally better than random (Blythe and Flower 2005). This poor performance has to do with peculiar molecular characteristics of epitopes (Burgoyne and Jackson 2006; Ofran, Schlessinger, et al. 2008) and their interdependence in complex immunological cellular and molecular processes. Until the early 2000s, most of the known structural data about protein–protein interactions came from complexes of either antigen–antibody or protease–inhibitor (Smith and Sternberg 2002). However, over the last years the available data has grown to include many other types of protein–protein interfaces.

As the peculiarities of these two types of interactions became clearer, it became a common practice to exclude protease–inhibitor and antigen–antibody complexes from large-scale analyses and from training sets for new predictors. The justifiability of this practice is debated (de Vries and Bonvin 2008).

PREDICTIVE FEATURES

Computational prediction of interface requires the identification of common denominators between interfaces. However, given the variations between interaction sites of different types, lumping all of them together may only allow for the identification of very general common denominators. Consequently, methods attempting to predict all types of protein interaction sites may pay in accuracy for the general applicability. Methods focusing on one type of interaction could theoretically produce more accurate predictions. In practice, however, for most types of interactions there are not enough known examples. Only a few types of interactions are covered by enough known examples to allow for the training of a specific prediction method. Table 9.1 lists types of interactions that are targeted by specialized methods.

EXPERIMENTAL DATA IMPROVES PREDICTION BUT LIMITS APPLICABILITY

Clearly, the more experimentally collected features that are taken into account, the better the predictions will be. If a method relies on an experimentally determined 3-D structure of the unbound chain, it can take into account the overall physicochemical characteristics of each residue, and not just its sequence neighbors. If a method also utilizes functional knowledge of the protein, such as its subcellular localization or molecular function, it can take into account considerations such the conditions and the environment under which the interactions occur. The downside, obviously, is that a method that requires a wide range of experimental data would be applicable only to proteins for which all these data are available. The vast majority of known proteins are only known by their amino acid sequence. As of summer 2008, for every protein deposited in SWISS-PROT, a database of functionally annotated proteins, there are more than 10 protein sequences in databases that have no functional annotation. For every protein in the PDB—the database of structurally annotated proteins—there are more than 100 unannotated proteins in other databases with no such annotation. By and large, methods could be delimited according to the extent of experimental data they require, and their consequential applicability and performance.

TABLE 9.1

Specialized Method for Prediction of Specific Types of Interaction Sites

Type of Interaction	Example Methods
Heterooligomer	Chung, Wang, et al. 2006; Res, Mihalek, et al. 2005
All types of interactions	Chen and Zhou 2005; Kufareva, Budagyan, et al. 2007
Transient heterooligomer	Fariselli, Pazos, et al. 2002; Neuvirth, Raz, et al. 2004; Ofran and Rost 2007a
Homodimers	Pettit, Bare, et al. 2007
Functionally important surface elements	Armon, Graur, et al. 2001; Pettit, Bare, et al. 2007
Obligomers (obligatory interactions)	Dong, Wang, et al. 2007
Protease–inhibitor	Yan, Honavar, et al. 2004
Antigen–antibody	Yan, Honavar, et al. 2004
B-cell epitopes	Haste Andersen, Nielsen, et al. 2006
CDRs	Kabat 1985; Ofran, Schlessinger, et al. 2008

EARLY METHODS

Early methods have been developed when available data was fairly meager. Therefore, these methods relied mostly on general, and often theoretical, parameters. Such was a method introduced by Kini and Evans (1996), who relied on their observation that proline is abundant in sequence segments that are flanking the interaction sites. Their suggestion was to simply search for sequence elements enclosed by proline rich segments and to identify them as putative interaction sites. This method is applicable to any sequence and does not require any additional data, but it was soon outperformed by more elaborate methods. Theoretical considerations were behind another method (Gallet, Charloteaux, et al. 2000) that suggested computing the hydrophobic moment of sequence stretches to determine whether there are likely to be interaction sites. This method is based on a simple computation that could be performed on any protein sequence and was aimed to identify any type of interaction site. While utilizing an elegant idea, it was proven too simple for this robust task.

Later methods that incorporated a similar idea and combined it with other features achieved better performance. Jones and Thornton (1997), relying on their earlier structural analysis of protein–protein interfaces (Jones and Thornton 1996), introduced a method that uses topology, solvent accessible surface area (ASA), and hydrophobicity to predict whether a given surface patch is likely to be an interaction site. To perform this analysis, the method requires an experimentally determined 3-D structure of the unbound chain. These pioneering and rudimentary methods were based on small data sets, simple computational procedures, and theoretical physicochemical considerations. The next generation of prediction methods is different in all three aspects: they are based on increasingly larger datasets, they employ sophisticated algorithms (predominantly machine learning ones), and they rely more on knowledge-based parameters than on theoretically derived ones.

THE NEXT GENERATION

The structure-based method of Jones and Thornton (1997) analyzed interaction sites as patches of residues on the surface. Some subsequent methods used a similar patch-based definition, but improved performance by using much larger data sets to train sophisticated algorithms such as Bayesian networks or support vector machines (SVM) (Bradford and Westhead 2005; Bradford, Needham, et al. 2006). The patch-based approach is also employed by some general methods that attempt to identify all functionally important residues, including interaction sites of any ilk, such as Ben-Tal's group ConSurf method (Armon, Graur, et al. 2001) and the hotPatch server (Pettit, Bare, et al. 2007) that search for conserved and functionally important surface patches. Most recent methods, however, replaced the notion of patch by the analyses of individual residues. Some of these methods use only sequence and sequence-derived features to predict interaction sites from sequence (Ofran and Rost 2003b; Koike and Takagi 2004; Res, Mihalek, et al. 2005; Ofran and Rost 2007a), but most of them require a full 3-D model of the protein (Fariselli, Pazos, et al. 2002; Bordner and Abagyan 2005; Bradford and Westhead 2005; Chen and Zhou 2005; Chung, Wang, et al. 2006; de Vries, van Dijk, et al. 2006; Li, Huang, et al. 2006; Liang, Zhang, et al. 2006;

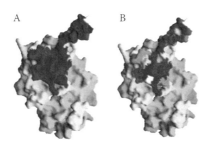

FIGURE 1.3 The core/rim model and the conservation of interface residues. The surface of the Gα subunit of transducin (PDB entry 1got; Lambright et al., 1996) is rendered in gray except for the region in contact with Gβγ. The feature protruding on the top right is the N-terminal helix. (A) The interface core, made of residues containing atoms buried at the interface, is in red; the rim, made of residues in which all interface atoms remain solvent accessible, is in blue. (B) The interface is colored according to the Shannon entropy that measures the divergence of each position in aligned sequences, ranging from 0 (red) to 0.4 (pink) to 1.4 (dark blue). Figure made by M. Guharoy (Bose Institute, Calcutta) with GRASP (Nicholls et al., 1991).

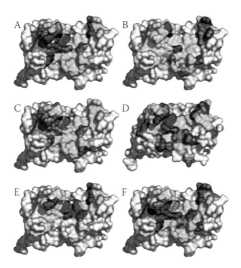

FIGURE 4.3 Alternative views of hGH Site 1 for binding to the hGHR. The first four panels show the energetic effects of (A) alanine, (B) homolog or (C) serine substitutions on hGH, or (D) alanine substitutions on a high affinity hGH variant. All maps were derived by shotgun scanning, except the alanine scanning map for hGH, which was derived by conventional site-directed mutagenesis. The residues are colored according to the $\Delta\Delta G_{mut-wt}$ values as follows: cyan < −0.4 kcal/mol; −0.4 kcal/mol ≤ green < 0.4 kcal/mol; 0.4 kcal/mol ≤ orange < 1.0 kcal/mol; red ≥ 1.0 kcal/mol; gray untested. Panel E shows the results of a double-alanine frequency analysis of shotgun scanning data to detect cooperativity among 19 side chains. Scanned positions are colored red or green, and red indicates residues predicted to exhibit cooperativity with at least two other residues. Panel F illustrates the results of quantitative saturation scanning, which assesses the tolerance to all possible mutations. The residues are colored according to SI values, as follows: cyan < −2; −2 ≤ green < 3; 3 ≤ yellow < 6; red ≥ 6. Larger SI values indicate positions that are less tolerant to substitution, and thus are important for binding. The x-ray structures of hGH and the high affinity variant (PDB entries 3HHR and 1kf9, respectively) were rendered in Pymol (DeLano Scientific, San Carlos, CA).

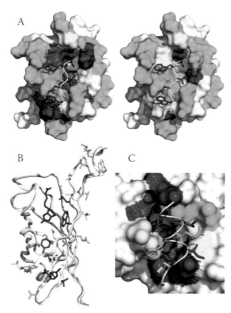

FIGURE 4.8 Shotgun scanning of proteins binding to peptides and small molecules. (A) Alanine (left) and homolog (right) scanning data mapped onto the structure of Erbin PDZ bound to a peptide (PDB entry 1N7T). Erbin PDZ is shown as a surface. The peptide main chain is shown as a tube and side chains are shown as sticks. Residues are colored according to the predicted fold reduction in binding due to substitution, as follows: green < 5; 5 ≤ yellow < 25; red ≥ 25. (B) Alanine scanning data mapped onto the structure of a streptavidin monomer bound to biotin (PDB entry 1STP). The main chain is shown as a tube and side chains are shown as sticks colored according to the predicted fold reduction in binding due to substitution, as follows: green < 3; 3 ≤ yellow < 9; red ≥ 9. Biotin is colored blue. (C) Alanine scanning data mapped onto the structure of PPARγ bound to SRC1 (PDB entry 1RDT). PPARγ is shown as a surface. A peptide fragment of SRC1 is shown with the main chain depicted as a tube and the side chains depicted as sticks. The residues are colored according to the predicted fold reduction in binding due to substitution, as follows: green < 3; 3 ≤ yellow < 10; red ≥ 10. X-ray structures were rendered using Pymol (DeLano Scientific, San Carlos, CA).

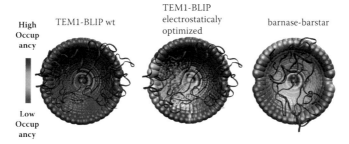

FIGURE 5.5 Mapping the transition state for protein–protein association using double-mutant cycle data as constraints (Harel et al., 2007). Each point represents the center of mass of 1 of 2220 configurations perturbed from the native complex. The point in the middle of each cap represents the x-ray structure of the native complex. The different colors represent configurations selected by different filtering cutoffs; cooler colors designate a configuration that passes a more stringent cutoff (thus has a higher probability of being occupancy in the transition state). TEM1 was the mobile protein in the simulations, while BLIP was fixed. The TEM1–BLIP complex was electrostatically optimized using the program PARE, by introducing mutations located outside the physical binding interface.

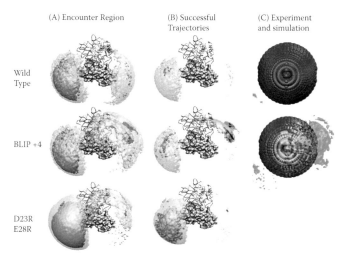

FIGURE 5.6 Brownian dynamics simulations of TEM1–BLIP mutants. BLIP is represented as a gray surface, TEM1 wild type is represented as a purple ribbon. All the simulations were done at 150 mM NaCl. (A) Encounter complexes are drawn as yellow isosurfaces representing the center of mass of TEM1 on BLIP at $\Delta G < -2.0$ kcal/mol (transparent yellow) and $\Delta G < -3.0$ kcal/mol (dark yellow). (B) Superimposition of the successful configurations at $\Delta G < -3.0$ kcal/mol (dark yellow) and the encounter complex region, defined by $\Delta G < -3.0$ kcal/mol, marked in transparent yellow. (C) An overlay of the successful trajectories from the BD simulation and the experimentally mapped transition state (see Figure 5.5).

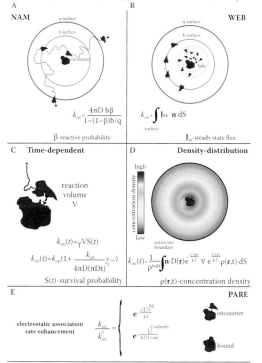

FIGURE 6.1 Schematic figure of the methods for calculating bimolecular association rates (see "Theoretical and Computational Approaches" section for more details).

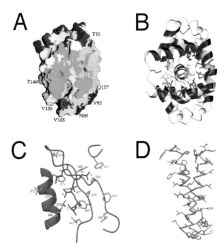

FIGURE 7.2 Examples of protein interface design. (A) Design of BLIP mutants with enhanced association rate for TEM1 β-lactamase. The figure shows wild-type BLIP with the TEM1 binding interface (green) and mutation sites predicted to increase the association rate. Color coding displays the extent of the predicted association rate increase: blue, less than 50% increase: yellow, more than 50% increase; red, tenfold increase. The figure is reproduced from Selzer et al.[1] (B) Redesign of calmodulin for improved binding specificity. Calmodulin is embracing the peptide target with its two globular domains. Twenty-four calmodulin side chains selected for optimization are shown in red. Peptide side chains that were allowed to change conformation during the calculation are shown in cyan. Calcium atoms are shown as yellow spheres. The figure is reproduced from Shifman and Mayo.[40] (C) Design of the PDZ domains with altered binding properties showing wild-type PDZ domain with its natural ligand, the KQTSV peptide (red). Residues on the PDZ domain selected for the optimization are shown in green. The figure is reproduced from Reina et al.[47] (D) Design of peptides that recognize transmembrane domains of integrins. A backbone geometry for the helix–helix interaction was selected from two helixes in the photosystem I reaction center. The sequence of the integrin α_{IIb} was threaded into the right helix. Fourteen positions on the second helix were designed (pink). The figure is reproduced from Yin et al.[51]

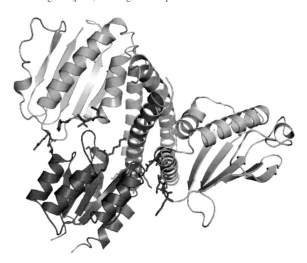

FIGURE 8.5 ZDOCK prediction for test case 1F51 using rebuilt residues prior to docking. 1SRR is colored magenta, 1IXM chain A is colored green, and 1IXM chain B is colored cyan. The rebuilt interface residues with missing atoms are displayed as red sticks.

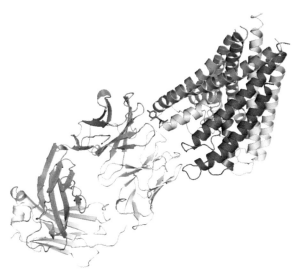

FIGURE 8.6 Structure prediction for test case 1K4C. 1JVM is colored in blue, cyan, magenta, and green and 1K4C is colored in salmon and gray. The rebuilt interface residue with missing atoms is displayed as sticks in red.

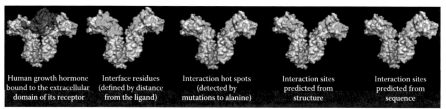

| Human growth hormone bound to the extracellular domain of its receptor | Interface residues (defined by distance from the ligand) | Interaction hot spots (detected by mutations to alanine) | Interaction sites predicted from structure | Interaction sites predicted from sequence |

FIGURE 9.1 Interaction sites and their prediction: The complex of human growth hormone and the extracellular domain of its receptor (left). When the hormone is removed, the interface residues on the receptors are revealed. Only a few of them are critical for stabilizing the complex. Two prediction methods, one that is based on structure and one that is based on sequence, were used to predict the interaction sites.

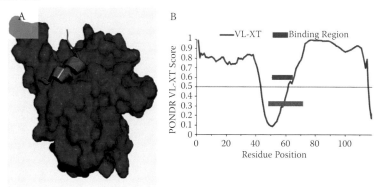

FIGURE 12.5 Example of binding regions and their positions relative to the regions of predicted order (PONDR® VL-XT score) and α-MoRF. (A) Eukaryotic initiation factor (blue) and the binding region of 4EBP1 (red). (B) The PONDR® VL-XT prediction for 4EBP1 with the binding region (blue bar) and the predicted α-MoRF region (pink bar) shown.

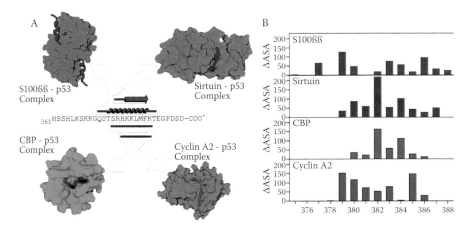

FIGURE 12.8 Sequence and structure comparison for the four overlapping complexes in the C-terminus of p53. (A) Primary, secondary, and quaternary structure of p53 complexes. (B) The ΔASA for rigid association between the components of complexes for each residue in the relevant sequence region of p53.

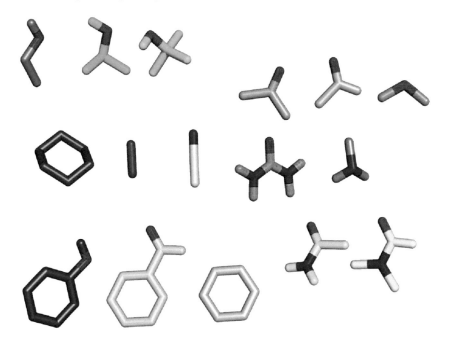

FIGURE 13.1 Small organic molecules used as probes in protein mapping. The probes are colored as follows: ethanol, sky blue; isopropanol, green; isobutanol, cyan; acetone, violet; acetaldehyde, olive; dimethyl ether, sand; cyclohexane, chocolate; ethane, purple; acetonitrile, pale green; urea, orange; methylamine, zinc; phenol, deep blue; benzaldehyde, gold; benzene, silver; acetamide, lemon; N,N-dimethylformamide, yellow. Oxygen atoms are colored red; nitrogen atoms are colored blue. (From Brenke, R. et al., 2009. Reprinted with permission from *Bioinformatics*.)

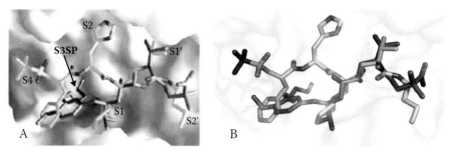

FIGURE 13.4 (A) Published figure from Novartis depicting the conformation of aliskiren with respect to a peptidomimetic. (From J. M. Wood et al., Structure-based design of aliskiren, a novel orally effective renin inhibitor, *Biochem Biophys Res Commun*, 308, 698–705, 2003. Reprinted with permission from *Biochemical and Biophysical Research Communications*.) (B) Resulting docked conformation of aliskiren, shown in purple, in the peptide-binding pocket of renin. The relative conformation of the same peptidomimetic from panel A is shown in green. The docked conformation of aliskiren is qualitatively consistent with the published figure. Key structural features, such as the occupation of the S_3^{SP} pocket, conjectured to be responsible for the affinity of aliskiren, are preserved in the docked conformation. (From M. R. Landon et al., Identification of hot spots within druggable binding sites of proteins by computational solvent mapping, *J Med Chem*, 50, 1231–1240, 2007. Reprinted with permission from the *Journal of Medicinal Chemistry*.)

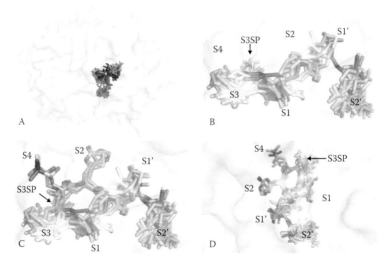

FIGURE 13.5 Mapping results for five structures of renin. (A) The first and second ranked consensus sites resulting from the mapping of the different structures are superimposed in the peptide binding pocket of renin, demonstrating the reproducibility of the results. Each color represents the results of a distinctive protein. (B) Closer examination of the consensus sites depicted in panel A, now all colored light blue and shown in relation to the docked conformation of aliskiren, supports the importance of the S_1, S_2, S_3, and S_3^{SP} subsites for ligand affinity. (C and D) The preferred binding mode of aliskiren as compared to the peptidomimetics, shown in green, is confirmed by the mapping results. The S_2 and S_4 subsites are bound preferentially by the peptidomimetics, but not by aliskiren or the mapping probes. (From M. R. Landon et al., Identification of hot spots within druggable binding sites of proteins by computational solvent mapping, *J Med Chem*, 50, 1231–1240, 2007. Reprinted with permission from the *Journal of Medicinal Chemistry*.)

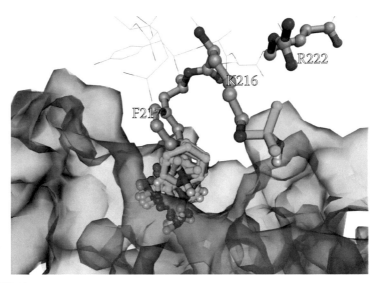

FIGURE 13.9 Detail of the *E. coli* RecA CS1 pocket with the consensus cluster shown in green and neighboring subunit shown in orange. The probes are shown in atomic colors with carbons colored green, oxygens colored red, and nitrogen colored blue.

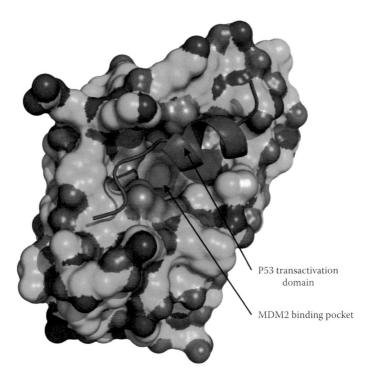

P53 transactivation domain

MDM2 binding pocket

FIGURE 14.3 X-ray structure of the MDM2–p53 fragment complex. MDM2 is displayed as a solid surface, whereas the p53 fragment is displayed as a cartoon colored in magenta. The MDM2 binding pocket is colored in orange. Figure generated with PyMol.

Wang, Chen, et al. 2006; Wang, Wong, et al. 2006; Dong, Wang, et al. 2007; Kufareva, Budagyan, et al. 2007; Negi, Schein, et al. 2007; Neuvirth, Heinemann, et al. 2007; Darnell, LeGault, et al. 2008; Murga, Ondrechen, et al. 2008).

Another category of methods relies on external sources of information, such as protein–protein interactions, in an attempt to identify sequence motifs that may overlap with the interaction sites (Sprinzak and Margalit 2001; Sprinzak, Altuvia, et al. 2006; Guo, Wu, et al. 2008). Similarly, it has been suggested to use protein–protein interaction data to search for positions that coevolve in interacting proteins and identify them as putative interaction sites (Pazos and Valencia 2002).

WHAT THE USER NEEDS TO KNOW

For the user, the critical differences between the methods are the required input data and the performance. These two questions are usually interdependent: A method that requires only sequence would be widely applicable but usually performs poorly. The performance is dramatically improved when also using evolutionary conservation, but then the method is not applicable to proteins with no or just a few known homologues (Koike and Takagi 2004; Res, Mihalek, et al. 2005; Ofran and Rost 2007a). Methods that rely also on protein–protein interaction data restrict applicability further but may improve performance. The best performing methods are those that rely also on 3-D structure, but they are also the most limited in their applicability.

ASSESSING PERFORMANCE

The assessment of the performance of any prediction method is not a trivial task. In the case of interaction site prediction, there are multiple difficulties. Since developers use different definitions for interaction sites, they essentially attempt to predict slightly different things. Hence, their results are not fully comparable. Developers report their own assessment of the method when they first introduce it, but different developers use different data sets to assess their performance and different statistical measures to report it (for further discussion of assessment in bioinformatics, see Baldi, Brunak, et al. 2000). Furthermore, they use different criteria for deeming a prediction successful. What fraction of an observed patch should be covered by the prediction to be considered a hit? What fraction of the predicted patch should be covered by the observed one? There is no standard answer to these questions.

Some attempts are made to establish a standard benchmark that will allow for objective, independent assessment of different methods (Zhou and Qin 2007). So far these attempts have only limited success (de Vries and Bonvin 2008). As per the statistical methods for assessing performance, the most commonly used measures are:

- $accuracy = \dfrac{TP+TN}{TP+FP+TN+FN}$

where TP, TN, FP, and FN are the number of predictions that are true positive, true negative, false positive, and false negative, respectively. This measure gives the same weight for successful prediction of interaction sites and of residues that are not interaction sites. This usually results in over optimistic assessment of the performance. The developers' self assessed values for accuracy are currently between 0.5 and 0.9.

- $precision = \dfrac{TP}{TP+FP}$

which measures what fraction of the positive predictions is correct (i.e., how often a predicted interaction site is also an experimentally observed one). Self-assessed values are between 0.4 and 0.9.

- $coverage = recall = sensitivity = \dfrac{TP}{TP+FN}$

which measures what fraction of the observed interaction sites were correctly identified by the method. Self-assessed values are between 0.05 and 0.7.

- $specificity = \dfrac{TN}{TN+FP}$

which measures the fraction of correct negative predictions.

- Matthews Correlation coefficient (MCC) =

$$\dfrac{TP \times TN - FP \times FN}{\sqrt{(TP+FP)(TP+FN)(TN+FP)(TN+FN)}}$$

which attempts to account for data sets that are not balanced but scores positive and negative predictions equally.

- Area under receiver operating curve (AUC), which measures the area under the curve of the graph one gets from plotting sensitivity versus (1 – specificity). However, this figure is known to be problematic for unbalanced data sets (when there are substantial differences between the size of the negative and the size of the positive data sets, like in the case of interaction sites versus other residues).

The relevant measure that should be considered when choosing a method depends on, to a large extent, the user's needs: Most users want to know how reliable the positive predictions are (namely, how reliable is the identification of interaction sites). Hence, for them the most relevant measures are precision and recall. In most cases there is a trade-off between these two measures: The user can choose parameters that increase precision on the account of recall and vice versa. Therefore, users should ask themselves what is more important for them: not to miss any putative interaction site (higher recall) or not to receive false positives (higher precision). For a review of the self-reported performance of current methods, see de Vries and Bonvin (2008).

HOT SPOTS

Several methods have reported high levels of precision, but most of them had fairly low levels of recall. That is, when they identify a residue as part of the interaction site, they are usually correct; however, they fail to identify many of the residues in the interface. This fact coincides with a fundamental observation about protein–protein interfaces in general: Only very few of the residues in protein–protein interfaces are absolutely essential for the interaction. In a typical 1200–2000 Å^2 interface, less than 5% of interface residues contribute more than 2 kcal/mol to binding. In small interfaces this can mean as few as one amino acid on each protein (Bogan and Thorn 1998). A common way to explore the importance of a residue for interaction is by mutating it, typically to alanine, and measuring the effect of this substitution on the interaction (Wells 1991; Morrison and Weiss 2001). Often this is done sequentially on a large scale in a procedure known as alanine scanning. Many experiments have demonstrated that most interface residues could be mutated without affecting the affinity of the protein to its partners (Clackson and Wells 1995; Thorn and Bogan 2001). Those few residues that, upon mutation, change the affinity are often assumed to be the most essential for the interaction and are deemed hot spots (Bogan and Thorn 1998). Identification of hot spots was also shown to be useful in docking (Halperin, Wolfson, et al. 2004). It has been suggested that the poor recall should be attributed to the fact that some methods actually predict hot spots rather than all interface residues (Ofran and Rost 2007b). Several new methods, databases, and analyses, therefore, attempt explicitly to identify hot spots rather than all interface residues (Ma, Wolfson, et al. 2001; Kortemme and Baker 2002; Ofran and Rost 2007b; Darnell, LeGault, et al. 2008; Guney, Tuncbag, et al. 2008). Figure 9.1 shows the complex of human growth hormone bound to the extracellular part of its domain. When removing the hormone, the interface is revealed: it covers 70 residues of the receptor, 35 on each chain. However, in an alanine scan only 10 of them—5 on each chain—were found to be critical to the stability of the complex. The structure of one chain of the homodimeric receptor was used to predict interaction sites, using a structure-based method called ProMate (Neuvirth, Raz, et al. 2004)). The sequence of the same

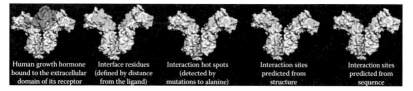

FIGURE 9.1   (SEE COLOR INSERT FOLLOWING PAGE 174.) Interaction sites and their prediction: The complex of human growth hormone and the extracellular domain of its receptor (left). When the hormone is removed, the interface residues on the receptors are revealed. Only a few of them are critical for stabilizing the complex. Two prediction methods, one that is based on structure and one that is based on sequence, were used to predict the interaction sites.

chain was fed to ISIS (Ofran and Rost 2007), a sequence-based method, to predict its interaction sites from sequence. It is interesting to note that the structure-based method identified not only all of the hot spots and some of the other interface residues, but also the site of the homodimeric interaction at the bottom of the receptor. The sequence-based method captured four of the five hot spots and one other interface residue.

DESIGN OF TOOLS

Virtually all recent methods are based on supervised machine learning. That is, they are based on a large training set classified into verified interaction sites and verified noninteraction sites. The data are fed into an algorithm that learns subtle—and possibly nonlinear—dependencies between various features, or descriptors, of a residue and its classification as an interaction site or a noninteraction site. The statistical model of dependencies can then be used to classify a new residue, based on its descriptors. Prediction methods differ in their choice of training sets, their choice of descriptors, and their choice of algorithms. Other than the distinction between patches and individual residues mentioned earlier, fundamental distinctions between prediction methods are:

- The type of interaction (permanent or transient, homodimeric or heterodimeric, specific to a family or a function, etc.). Table 9.1 lists types of protein–protein interactions predicted by different methods.
- The type of algorithm used for training (most common are parametric approaches, artificial neural networks [ANNs], support vector machines [SVMs], and Bayesian networks).
- The way the interface is defined (patch, contacting residues, hot spots, etc.).
- The data used for training (large or small data set, hand selected, or automatically generated).
- The descriptors that are used for classification (physicochemical characteristics, structural ones, evolutionary conservation, hydrophobicity). Table 9.2 lists various descriptors.

TABLE 9.2
Common Descriptors Used for Prediction of Interaction Sites

Descriptor	Comments
Features Derived from 3-D Structure	
Neighbor list: Residues in spatial vicinity to the residue in question	9–20 residues
B-factor	A crystallographic measure that approximates the flexibility of a residue.
Solvent accessibility (ASA)	Measured in Å^2.
Relative solvent accessibility	Measured as a fraction of the overall surface of the residue that is exposed to solvent.

TABLE 9.2 (CONTINUED)
Common Descriptors Used for Prediction of Interaction Sites (Continued)

Descriptor	Comments
Shape index/curvedness	
Secondary structure	Three state (helix, strand, loop) or more.
Sequence distance	The separation in sequence between residues within the same patch. Some results indicate that structurally contiguous residues that are not adjacent in sequence are more likely to form interaction sites.
Planarity	

Predicted/Approximate Structural Features

Predicted secondary structure	Methods relying only on sequence can use computational tools to
Predicted solvent accessibility	generate predicted solvent accessibility/secondary structure. This improves performance without limiting applicability to proteins with known 3-D structure.
Sequence neighbor list	Can be used instead of neighbor list to approximate the environment of the analyzed residue. Nine to fifteen residues around the residue in question. Four to seven on each side of the residue. Some structure-based methods use this in addition to neighbor list.

Evolutionary Features

Sequence profile	Extracted from a multiple sequence alignment, a profile reveals patterns of evolutionary conservation.
Conservation score	A quantification of the level of conservation of an individual position.
Conservation of physicochemical traits	If the position is not conserved, scoring conservation of traits such as charge, hydrophobicity, or size may improve prediction.

Physicochemical Features

Hydrophobicity	Several different scales are available.
Electrostatic potential	Measured for individual residue or for a patch. Requires 3-D structure.
Atom propensities	Serves as a way to sum physicochemical properties across residues in the patch.
Desolvation energy	Used mostly in predictions for rigid-body docking.

External Knowledge

Protein–protein interaction	Can be used to: (1) identify sequence or structural elements that are significantly overrepresented in interacting pairs, and (2) to assess coevolution of positions in interacting pairs.
Functional annotation of the protein	Enzyme–inhibitor and antigen–antibody have different types of interfaces than other complexes. Adding this information may improve prediction.

Different methods have different strengths and weaknesses. Therefore, a user that wisely integrates the output of several different prediction methods would most likely get predictions that are better than those of any single method. This could also be done automatically with a metaserver that automatically submits a query to different methods and weighs the results of each of them to produce a consensus prediction (Qin and Zhou 2007).

CONCLUSION

Between 2003 and 2009, dozens of new methods for the prediction of interaction sites were introduced. Their performance steadily improves and they are reliable enough to inform experiments. They are often used to choose targets for mutagenesis and for analyzing protein function and interaction. Three developments allow for the growth of these methods. First, the deluge of protein–protein interaction data, particularly structural data, provides sufficient data sets for the training of machine-learning algorithms. Second, elaborate study, both experimental and computational, revealed some of the principles of protein interaction and enabled a careful choice of descriptors for training, and finally, communication with computer scientists allows for the choice of state-of-the-art algorithms that improve performance further. These three factors are likely to be the keys for further improvements of the methods. The new data that are required will enable the training of more specialized methods that focus on specific types of interactions, such as interaction of membrane proteins, antibody–antigen interactions, interactions of enzymes and their targets, interactions that are mediated by water, and so forth. More data on each type of interaction will also allow for more detailed analysis of its traits and, therefore, for better choice of descriptors. Finally, based on these descriptors, computer scientists will be able to devise specific methods for better predictions.

REFERENCES

Archakov, A. I., V. M. Govorun, et al. (2003). "Protein–protein interactions as a target for drugs in proteomics." *Proteomics* **3**(4): 380–91.

Arkin, M. R. and J. A. Wells (2004). "Small-molecule inhibitors of protein–protein interactions: Progressing towards the dream." *Nat Rev Drug Discov* **3**(4): 301–17.

Armon, A., D. Graur, et al. (2001). "ConSurf: An algorithmic tool for the identification of functional regions in proteins by surface mapping of phylogenetic information." *J Mol Biol* **307**(1): 447–63.

Bahadur, R. P., P. Chakrabarti, et al. (2004). "A dissection of specific and non-specific protein–protein interfaces." *J Mol Biol* **336**(4): 943–55.

Baldi, P., S. Brunak, et al. (2000). "Assessing the accuracy of prediction algorithms for classification: An overview." *Bioinformatics* **16**(5): 412–24.

Balter, M. (1998). "Revealing HIV's T cell passkey." *Science* **280**(5371): 1833–4.

Blythe, M. J. and D. R. Flower (2005). "Benchmarking B cell epitope prediction: Underperformance of existing methods." *Protein Sci* **14**(1): 246–8.

Bogan, A. A. and K. S. Thorn (1998). "Anatomy of hot spots in protein interfaces." *J Mol Biol* **280**(1): 1–9.

Bordner, A. J. and R. Abagyan (2005). "Statistical analysis and prediction of protein–protein interfaces." *Proteins* **60**(3): 353–66.

Bradford, J. R., C. J. Needham, et al. (2006). "Insights into protein–protein interfaces using a Bayesian network prediction method." *J Mol Biol* **362**(2): 365–86.

Bradford, J. R. and D. R. Westhead (2005). "Improved prediction of protein–protein binding sites using a support vector machines approach." *Bioinformatics* **21**(8): 1487–94.

Burgoyne, N. J. and R. M. Jackson (2006). "Predicting protein interaction sites: Binding hot-spots in protein–protein and protein-ligand interfaces." *Bioinformatics* **22**(11): 1335–42.

Caffrey, D. R., S. Somaroo, et al. (2004). "Are protein–protein interfaces more conserved in sequence than the rest of the protein surface?" *Protein Sci* **13**(1): 190–202.

Chen, H. and H. X. Zhou (2005). "Prediction of interface residues in protein–protein complexes by a consensus neural network method: Test against NMR data." *Proteins* **61**(1): 21–35.

Chung, J. L., W. Wang, et al. (2006). "Exploiting sequence and structure homologs to identify protein–protein binding sites." *Proteins* **62**(3): 630–40.

Clackson, T. and J. A. Wells (1995). "A hot spot of binding energy in a hormone-receptor interface." *Science* **267**(5196): 383–6.

Darnell, S. J., L. LeGault, et al. (2008). "KFC Server: Interactive forecasting of protein interaction hot spots." *Nucleic Acids Res* **36**(Web Server issue): W265–9.

de Vries, S. J. and A. M. Bonvin (2008). "How proteins get in touch: Interface prediction in the study of biomolecular complexes." *Curr Protein Pept Sci* **9**(4): 394–406.

de Vries, S. J., A. D. van Dijk, et al. (2006). "WHISCY: What information does surface conservation yield? Application to data-driven docking." *Proteins* **63**(3): 479–89.

Dong, Q., X. Wang, et al. (2007). "Exploiting residue-level and profile-level interface propensities for usage in binding sites prediction of proteins." *BMC Bioinformatics* **8**: 147.

Fariselli, P., F. Pazos, et al. (2002). "Prediction of protein–protein interaction sites in heterocomplexes with neural networks." *Eur J Biochem* **269**(5): 1356–61.

Fernandez-Recio, J., M. Totrov, et al. (2004). "Identification of protein–protein interaction sites from docking energy landscapes." *J Mol Biol* **335**(3): 843–65.

Gallet, X., B. Charloteaux, et al. (2000). "A fast method to predict protein interaction sites from sequences." *J Mol Biol* **302**(4): 917–26.

Glaser, F., D. M. Steinberg, et al. (2001). "Residue frequencies and pairing preferences at protein–protein interfaces." *Proteins* **43**(2): 89–102.

Greenbaum, J. A., P. H. Andersen, et al. (2007). "Towards a consensus on datasets and evaluation metrics for developing B-cell epitope prediction tools." *J Mol Recognit* **20**(2): 75–82.

Guney, E., N. Tuncbag, et al. (2008). "HotSprint: Database of computational hot spots in protein interfaces." *Nucleic Acids Res* **36**(Database issue): D662–6.

Guo, J., X. Wu, et al. (2008). "Genome-wide inference of protein interaction sites: Lessons from the yeast high-quality negative protein–protein interaction dataset." *Nucleic Acids Res* **36**(6): 2002–11.

Halperin, I., H. Wolfson, et al. (2004). "Protein–protein interactions; coupling of structurally conserved residues and of hot spots across interfaces. Implications for docking." *Structure* **12**(6): 1027–38.

Haste Andersen, P., M. Nielsen, et al. (2006). "Prediction of residues in discontinuous B-cell epitopes using protein 3D structures." *Protein Sci* **15**(11): 2558–67.

Jackson, R. M. (1999). "Comparison of protein–protein interactions in serine protease-inhibitor and antibody-antigen complexes: Implications for the protein docking problem." *Protein Sci* **8**(3): 603–13.

Janin, J. (1995). "Protein–protein recognition." *Progress in Biophysics and Molecular Biology* **21**: 145–66.

Jones, S. and J. M. Thornton (1996). "Principles of protein–protein interactions." *Proc Natl Acad Sci USA* **93**(1): 13–20.

Jones, S. and J. M. Thornton (1997). "Prediction of protein–protein interaction sites using patch analysis." *Journal of Molecular Biology* **272**(1): 133–43.

Kabat, E. A. (1985). Immunoglobulin sequence data. The generation of antibody diversity and its genetic control. In *The Role of Data in Scientific Progress*, edited by P. S. Glaeser, pp. 97–102. North Holland: Elsevier.

Kerrien, S., Y. Alam-Faruque, et al. (2007). "IntAct: Open source resource for molecular interaction data." *Nucleic Acids Res* **35**(Database issue): D561–5.

Keskin, O., I. Bahar, et al. (1998). "Empirical solvent-mediated potentials hold for both intramolecular and inter-molecular inter-residue interactions." *Protein Sci* **7**(12): 2578–86.

Keskin, O., C. J. Tsai, et al. (2004). "A new, structurally nonredundant, diverse data set of protein–protein interfaces and its implications." *Protein Sci* **13**(4): 1043–55.

Kini, R. M. and H. J. Evans (1996). "Prediction of potential protein–protein interaction sites from amino acid sequence. Identification of a fibrin polymerization site." *FEBS Lett* **385**(1-2): 81–86.

Koblish, H. K., S. Zhao, et al. (2006). "Benzodiazepinedione inhibitors of the Hdm2:p53 complex suppress human tumor cell proliferation in vitro and sensitize tumors to doxorubicin in vivo." *Mol Cancer Ther* **5**(1): 160–9.

Koike, A. and T. Takagi (2004). "Prediction of protein–protein interaction sites using support vector machines." *Protein Eng Des Sel* **17**(2): 165–73.

Kortemme, T. and D. Baker (2002). "A simple physical model for binding energy hot spots in protein–protein complexes." *Proc Natl Acad Sci USA* **99**(22): 14116–21.

Kufareva, I., L. Budagyan, et al. (2007). "PIER: Protein interface recognition for structural proteomics." *Proteins* **67**(2): 400–17.

Kwong, P. D., R. Wyatt, et al. (1998). "Structure of an HIV gp120 envelope glycoprotein in complex with the CD4 receptor and a neutralizing human antibody." *Nature* **393**(6686): 648–59.

Laskowski, R. A., N. M. Luscombe, et al. (1996). "Protein clefts in molecular recognition and function." *Protein Sci* **5**(12): 2438–52.

Li, J. J., D. S. Huang, et al. (2006). "Identifying protein–protein interfacial residues in heterocomplexes using residue conservation scores." *Int J Biol Macromol* **38**(3-5): 241–7.

Liang, S., C. Zhang, et al. (2006). "Protein binding site prediction using an empirical scoring function." *Nucleic Acids Res* **34**(13): 3698–707.

Lo Conte, L., C. Chothia, et al. (1999). "The atomic structure of protein–protein recognition sites." *J Mol Biol* **285**(5): 2177–98.

Ma, B., T. Elkayam, et al. (2003). "Protein–protein interactions: Structurally conserved residues distinguish between binding sites and exposed protein surfaces." *Proc Natl Acad Sci USA* **100**(10): 5772–7.

Ma, B., H. J. Wolfson, et al. (2001). "Protein functional epitopes: Hot spots, dynamics and combinatorial libraries." *Curr Opin Struct Biol* **11**(3): 364–9.

McCoy, A. J., V. Chandana Epa, et al. (1997). "Electrostatic complementarity at protein/protein interfaces." *J Mol Biol* **268**(2): 570–84.

Mika, S. and B. Rost (2006). "Protein–protein interactions more conserved within species than across species." *PLoS Comput Biol* **2**(7): e79.

Morrison, K. L. and G. A. Weiss (2001). "Combinatorial alanine-scanning." *Curr Opin Chem Biol* **5**(3): 302–7.

Murga, L. F., M. J. Ondrechen, et al. (2008). "Prediction of interaction sites from apo 3D structures when the holo conformation is different." *Proteins* **72**(3): 980–92.

Negi, S. S., C. H. Schein, et al. (2007). "InterProSurf: A web server for predicting interacting sites on protein surfaces." *Bioinformatics* **23**(24): 3397–9.

Neuvirth, H., U. Heinemann, et al. (2007). "ProMateus—an open research approach to protein-binding sites analysis." *Nucleic Acids Res* **35**(Web Server issue): W543–8.

Neuvirth, H., R. Raz, et al. (2004). "ProMate: A structure based prediction program to identify the location of protein–protein binding sites." *J Mol Biol* **338**(1): 181–99.

Ofran, Y. and B. Rost (2003a). "Analysing six types of protein–protein interfaces." *J Mol Biol* **325**(2): 377–87.

Ofran, Y. and B. Rost (2003b). "Predicted protein–protein interaction sites from local sequence information." *FEBS Lett* **544**(1-3): 236–9.

Ofran, Y. and B. Rost (2007a). "ISIS: Interaction sites identified from sequence." *Bioinformatics* **23**(2): e13–6.

Ofran, Y. and B. Rost (2007b). "Protein–protein interaction hotspots carved into sequences." *PLoS Comput Biol* **3**(7): e119.

Ofran, Y., A. Schlessinger, et al. (2008). "Automated identification of complementarity determining regions reveals peculiar characteristics of CDRs and B-Cell epitopes." *J Immunol* **181**: 6230–35.

Pazos, F. and A. Valencia (2002). "In silico two-hybrid system for the selection of physically interacting protein pairs." *Proteins* **47**(2): 219–27.

Peters, K. P., J. Fauck, et al. (1996). "The automatic search for ligand binding sites in proteins of known three-dimensional structure using only geometric criteria." *J Mol Biol* **256**(1): 201–13.

Pettit, F. K., E. Bare, et al. (2007). "HotPatch: A statistical approach to finding biologically relevant features on protein surfaces." *J Mol Biol* **369**(3): 863–79.

Pettit, F. K. and J. U. Bowie (1999). "Protein surface roughness and small molecular binding sites." *J Mol Biol* **285**(4): 1377–82.

Qin, S. and H. X. Zhou (2007). "meta-PPISP: A meta web server for protein–protein interaction site prediction." *Bioinformatics* **23**(24): 3386–7.

Res, I., I. Mihalek, et al. (2005). "An evolution based classifier for prediction of protein interfaces without using protein structures." *Bioinformatics* **21**(10): 2496–501.

Rost, B. (1999). "Twilight zone of protein sequence alignments." *Protein Eng* **12**(2): 85–94.

Rost, B., J. Liu, et al. (2003). "Automatic prediction of protein function." *Cell Mol Life Sci* **60**(12): 2637–50.

Rudolph, J. (2007). "Inhibiting transient protein–protein interactions: Lessons from the Cdc25 protein tyrosine phosphatases." *Nat Rev Cancer* **7**(3): 202–11.

Sheinerman, F. B., R. Norel, et al. (2000). "Electrostatic aspects of protein–protein interactions." *Curr Opin Struct Biol* **10**(2): 153–9.

Smith, G. R. and M. J. Sternberg (2002). "Prediction of protein–protein interactions by docking methods." *Curr Opin Struct Biol* **12**(1): 28–35.

Sprinzak, E., Y. Altuvia, et al. (2006). "Characterization and prediction of protein–protein interactions within and between complexes." *Proc Natl Acad Sci USA* **103**(40): 14718–23.

Sprinzak, E. and H. Margalit (2001). "Correlated sequence-signatures as markers of protein–protein interaction." *J Mol Biol* **311**(4): 681–92.

Thorn, K. S. and A. A. Bogan (2001). "ASEdb: A database of alanine mutations and their effects on the free energy of binding in protein interactions." *Bioinformatics* **17**(3): 284–5.

Vassilev, L. T. (2004). "Small-molecule antagonists of p53-MDM2 binding: Research tools and potential therapeutics." *Cell Cycle* **3**(4): 419–21.

Vassilev, L. T., B. T. Vu, et al. (2004). "In vivo activation of the p53 pathway by small-molecule antagonists of MDM2." *Science* **303**(5659): 844–8.

Wang, B., P. Chen, et al. (2006). "Predicting protein interaction sites from residue spatial sequence profile and evolution rate." *FEBS Lett* **580**(2): 380–4.

Wang, B., H. S. Wong, et al. (2006). "Inferring protein–protein interacting sites using residue conservation and evolutionary information." *Protein Pept Lett* **13**(10): 999–1005.

Wells, J. A. (1991). "Systematic mutational analyses of protein–protein interfaces." *Methods Enzymol* **202**: 390–411.

Wells, J. A. and C. L. McClendon (2007). "Reaching for high-hanging fruit in drug discovery at protein–protein interfaces." *Nature* **450**(7172): 1001–9.

Yan, C., V. Honavar, et al. (2004). "Identification of interface residues in protease-inhibitor and antigen-antibody complexes: A support vector machine approach." *Neural Computing & Applications* **13**(3): 123–129.

Zhou, H. X. and S. Qin (2007). "Interaction-site prediction for protein complexes: A critical assessment." *Bioinformatics* **23**(17): 2203–9.

10 Predicting Molecular Interactions in Structural Proteomics

Irina Kufareva and Ruben Abagyan

CONTENTS

INTRODUCTION

As the number of files in the Protein Data Bank (PDB) exceeded 50,000 (representing around 10,000 protein domains at 95% level of sequence identity), it is becoming increasingly important to develop the understanding of the protein function and the next level of subcellular structural organization.[1,2] This, among other aspects, requires understanding of what other biological molecules or cellular structures interact with each domain, which residues are involved in this interaction (e.g., References 3 and 4), and what conformational changes accompany the binging. Structure-based computational approaches to these questions invariably face the issue of protein flexibility, which is further complicated by the existence of unstructured, partially structured, or conditionally structured interfaces.[5] While the dream of predictive millisecond-scale molecular dynamics serving as a "computational microscope" persists (K. Schulten, award lecture at the ISQBP meeting in Ascona, 2008; also Reference 6) and may even be getting more tangible as computers become faster, the ability to make reliable predictions on the basis of such trajectory is still lacking.

The task of predicting molecular interactions has three principal aspects:

A. *Predicting the interfaces* on a given molecule that are involved in intermolecular interactions. As a subtask one may include predicting a *class* (but hardly the identity) of the interaction partner (say, protein, peptide, membrane, a small substrate).[7] During the last years, computational methods making these kinds of predictions have improved dramatically and may be quite useful.

B. *Predicting the spatial arrangement* of two interacting molecules given the apostructures of both, aka docking. Existence of homologous interacting pairs with already solved three-dimensional (3D) structures greatly facilitates solving this problem. However, when such template complex structure is not available, obtaining a crystallographic quality model may be exceedingly difficult due to the *induced fit*.

C. *Predicting the identity* of molecules (including proteins) involved in direct transient specific interactions with each other. In the most general form, solving this problem requires precise, large-scale prediction of conformational ensembles and Gibbs free binding energies between all possible pairs of biological molecules, which is unrealistic even with the use of the best state-of-the-art computing resources.

From the biological standpoint, the three aspects should be considered in a different order, by increasing attention to details: C (what) to A (where) to B (how). We, however, order them by their computational complexity. For example, in context of protein–protein interactions, task A is tangible and applicable to thousands of proteins constituting entire structural genomes. Task B, in spite of the achieved limited success in protein docking,[8] largely remains an academic exercise. Task C appears to be impossible to solve due to the enormous complexity of biological systems and the imperfections in existing methods of free energy calculations.

While for protein–protein interactions only task A can be solved with reasonable effort and outcome, all three kinds of predictions are approaching widespread practical use in cases when the interacting partner is a small chemical. Recent advances in small molecule docking and related applications led to a number of successful solutions of tasks A and B in this context. Though more difficult than others, task C becomes quite tangible for druglike compounds and is represented by two kinds of screening:

C1. *Ligand screening*, that is, searching for a natural substrate or a new compound to specifically bind to the source protein.

C2. *Ligand specificity profiling,* that is, searching for the proteins in a subclass or even in the entire structural proteome that bind specifically to a given small molecule.

In this chapter, we present an overview of some methods for predicting the three aspects of molecular interactions. We will focus on targets where a good quality atomic resolution 3D model either has been determined experimentally or can be reliably built by homology (unfortunately, de novo predictors of the 3D structure from the amino acid sequence are still unreliable).[9,10] We will also focus on the *transient,* not permanent, interactions. In most cases, permanent binding partners are known in advance, and when this is not the case, they are more easily predictable.[11] We will present an analysis of the induced conformational changes upon binding that create the single biggest challenge for modelers of protein interactions, and describe several methods to overcome this difficulty. Our analysis and the optimization of the prediction methods relied on an ever-growing body of structural data and the improved methods of molecular mechanics with related energy functions.

CHARACTERIZING MOLECULAR INTERFACES

COMPREHENSIVE SETS OF TRANSIENT MOLECULAR INTERACTIONS IN 3D

Of more than 10,000 unique protein domains found in the 2008 release of the PDB,[12] only about 10% are represented in transient complexes with their biological protein partners. Selection and preparation of a sufficiently large collection of these complexes to be used as a training and validation set is a prerequisite for any study addressing the problem of protein interface prediction.[13–15] Unfortunately, artificial constructs, crystal packing, and other artifacts present a substantial challenge for both manual and automatic identification of true biological interactions. Although manual intervention during the set collection helps reduce the number of errors, it limits the size of the set and possibilities of timely updates. On the other hand, only a truly large-scale effort can lead to a statistically significant and diverse set without overrepresentation of large families of homologues.

We collected a set of as many as 858 protein domains participating in crystallized transient protein–protein complexes. The entire PDB was organized into families, one family per domain, with each family containing all publicly available good quality structures of the domain with its possible binding partners. To reduce the noise while preserving the automation, we only collected the domains represented by *multiple*, yet "partner-diverse" structures, and used consistency criteria to achieve the following goals:

- Transient complexes were distinguished from permanent ones based on comparison of PDB complex compositions across the family.
- Each domain was treated in context of its permanent biological unit. In ~20% of the set, the biological unit was found to be different from the

monomer. The permanent biological multimers were treated as a whole to avoid potential contamination of the data set with intersubunit (obligate, permanent) interfaces that never get exposed in biological environments.

- Each transient complex was guaranteed to have at least one *unbound* structure of its receptor domain.
- Multiple protein partners binding to the same or different sites on the protein surface were taken into account.
- Superimposition and structural comparison of the multiple structures provided means for characterization of the induced conformational changes.

For simplicity, we did not include in the set any protein domain that formed permanent heterotrimers or higher multimers, or any domain that was simultaneously bound to more than four distinct protein partners.

The collected set provides a fairly comprehensive representation of transient protein–protein interactions in the PDB. It covered all major classes of biological interactions such as enzyme–inhibitor, hormone–receptor, structural protein, and many types of regulatory interactions. However, antibody–antigen interactions were purposely excluded from the set, as well as all families featuring antibodies as the only type of interacting partner. Epitope prediction must be considered as a standalone task in computational biology. Being different from biological interfaces by both physicochemical properties and (typically) location, epitopes are only recognized by antibodies, naturally selected to target even most noninterface-like patches.

The family size ranged from 2 to 30 (median 6, mean 8.61) structures (Figure 10.1) and was limited by the requirement of using no more than 15 PDB entries and no more than two chains from each entry per protein domain. In a large fraction of cases (361 of 858, 42%), protein domains were found to interact with a variety of protein partners. Such interactions often involved nonoverlapping patches on the protein surface.

Using a similar approach, we also collected a set of ~800 protein domains that have been crystallized apo or in complexes with *small molecule ligands*. In the following, we present a comparative analysis of the two sets and a comprehensive description of induced fit changes.

PROPERTIES AND FLEXIBILITY OF TRANSIENT MOLECULAR INTERFACES

Protein surface patches involved in transient interaction with other proteins or small molecule ligands differ

- from the rest of the surface,
- from permanent multimer interfaces (e.g., Figure 10.2), and
- from each other (small molecule interface vs. protein interface)

by a number of properties. Properties such as relative residue frequencies, physical fields, hydrophobicity, size, charge, evolutionary rates, and so forth have statistically significant differences when compared between classes of protein surface patches (e.g., Reference 16). These properties can be used to predict molecular interfaces (task A).

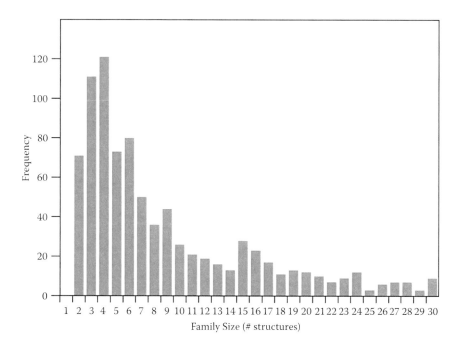

FIGURE 10.1 Eight hundred fifty-eight transient protein interfaces represented by two or more PDB entries. *Family size* refers to the number of PDB structures representing the same protein domain in apo form or in complex with transient protein partners. These families were used to evaluate the induced conformational changes at protein interfaces.

To predict complex geometries (task B) a different question gains primary importance. Since induced fit presents the major challenge for all docking algorithms, one needs a clear understanding of the nature and the degree of changes that can happen upon binding of a protein to a protein or a small molecule partner. Such studies were previously performed only for small sets of proteins.[17,18]

To collect the induced fit data, we used the sets of transient protein interactions in 3D described earlier. Given a family of complexes formed by a particular protein domain, we compared each complex with all other complexes of the same composition (same protein partner in case of protein interactions, same small molecule for protein–ligand interactions), complexes of other compositions, and unbound structures. The unbound structures were also compared to one another to assess the degree of changes stemming from natural protein flexibility rather than induced by binding partners.

For protein–protein interactions, the obtained data for 858 protein ensembles are presented in Figure 10.3. In the majority of the cases (77%), comparison of a bound form of a protein to its unbound form or a complex of different compositions shows a strong deviation (>1.5 Å) of at least one interface residue. On average, about one-fourth of interface residue backbones deviate above that threshold. Moreover, at least one interface side chain is displaced by more than 1.5 Å almost always (99%), and more than one-half of side chains strongly deviate on average. The corresponding

(a)

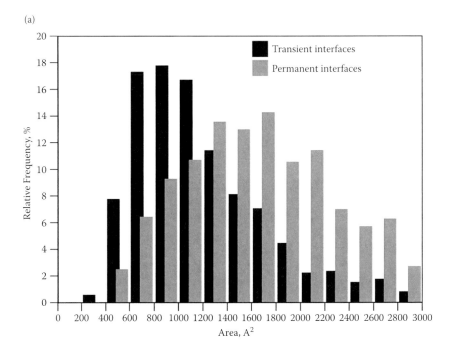

(b)

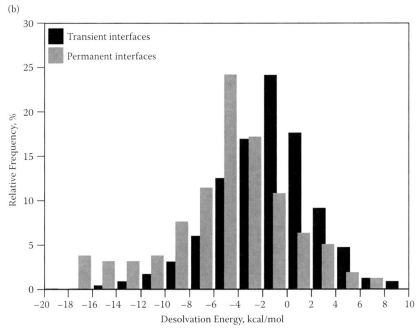

FIGURE 10.2 Desolvation properties of obligate and transient interfaces in the collected set of 858 protein domains involved in crystallized transient protein complexes in PDB. (a) Buried solvent accessible surface area, (b) desolvation energy. As shown, the transient interfaces are smaller in size, are associated with smaller desolvation penalty, and, therefore, are more difficult to predict.

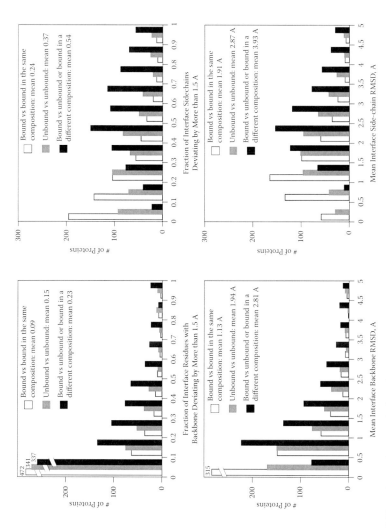

FIGURE 10.3 Flexibility of protein–protein interfaces and induced fit. On average, more than one-half of interface side chains are displaced by more than 1.5 Å when compared between different complex compositions (bound vs. unbound/bound to a different protein partner), giving an average interface side chain RMSD of ~4 Å. In contrast, when compared between complexes of the same composition (bound vs. bound to the same protein partner), the expected fraction of strongly deviating side chains is less than one-fourth, and the average interface side chain RMSD is below 2 Å. At least one interface residue backbone deviates by more than 1.5 Å in 78% of the cases, and by at least one side chain in 99% of the cases.

values observed between complexes of the same composition due to natural protein flexibility (white bars), or even between unbound structures (gray bars), are significantly lower. Some induced changes involved large-scale domain, termini, or loop movements and extended as far as 25–50 A (Figure 10.4).

In contrast, ligand-binding interfaces appear more stable. Being a little smaller in size (the number of residues involved in small-molecule binding is about two-thirds of an average protein interface size; Figure 10.5) they are usually more buried, which restricts potential movements of the interface side chains. Only about 4% of the interface residue backbones deviate above the threshold of 1.5 A, and about 18% of the side chains (1 to 2 side chains per interface; Figure 10.6).

In short, our analysis proved that in spite of comparable sizes on the interface, small molecules induce less conformational changes upon binding to their receptors than possible protein partners. Even though some exceptions to this rule exist (e.g., activation loop transitions in protein kinases),[19] the task of predicting interfaces, binding geometry, and even identifying the small molecule ligands appears more tangible compared with protein–protein interaction predictions.

PREDICTING PROTEIN–PROTEIN INTERACTIONS

PHYSICOCHEMICAL PROPERTIES OR EVOLUTIONARY PATTERNS?

Computational methods for protein interface prediction can be divided into two major classes: (1) methods incorporating evolutionary conservation information derived from multiple sequence alignments (MSA) and projected on a protein surface, and (2) those based solely on geometrical and physicochemical properties of the surface.

Methods on the first class rely on the broad evidence of interface residues mutating at slower rates than the rest of the protein surface.[20–23] In general, functionally important surface residues are expected to be conserved. Since interior residues responsible for efficient folding and stability also fall under this category, a strong conservation signal from protein interfaces is only observed when the residue conservation of the interface is compared with that of the surface. Modern multiple sequence alignment methods incorporating residue substitution matrices and phylogenetic trees[20,24–27] allow detection of even weak conservation signal. It was argued, however, that conservation score alone is not sufficient for accurate discrimination and can be misleading in several ways.[28–31] The high variability of alignment composition and extent, unbalanced subfamily representations, and local alignment errors need to be taken into account. The prediction greatly depends on the algorithm of deriving scores from the alignment. Even the most sophisticated algorithms break down on the proteins with no or few orthologs. Most important, many protein interfaces are not expected to be better conserved at all, either because of their function (e.g., the adaptable binding surfaces of the immune system proteins) or because they were formed late in evolution.[29]

Alignment-independent prediction methods rely on an assumption that protein interfaces are different from the rest of the surface by their physicochemical and geometrical properties. Although it was demonstrated that the composition of

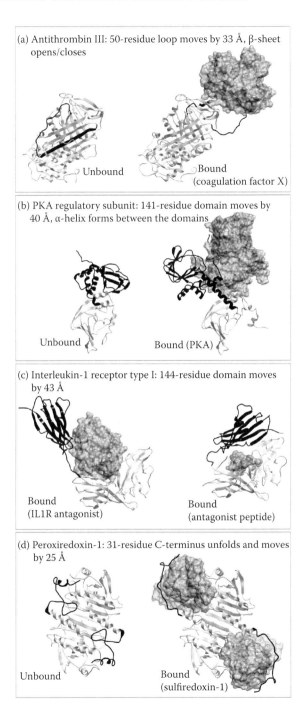

(a) Antithrombin III: 50-residue loop moves by 33 Å, β-sheet opens/closes

Unbound Bound (coagulation factor X)

(b) PKA regulatory subunit: 141-residue domain moves by 40 Å, α-helix forms between the domains

Unbound Bound (PKA)

(c) Interleukin-1 receptor type I: 144-residue domain moves by 43 Å

Bound (IL1R antagonist) Bound (antagonist peptide)

(d) Peroxiredoxin-1: 31-residue C-terminus unfolds and moves by 25 Å

Unbound Bound (sulfiredoxin-1)

FIGURE 10.4 Some examples of large conformational changes at protein–protein interfaces. Whereas the interaction surfaces could be predicted with the methods described in this chapter, the induced rearrangements and the docking pose could not.

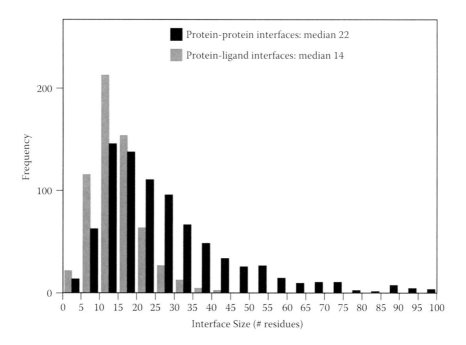

FIGURE 10.5 Protein–protein and protein–ligand interface sizes in residues. The average size of a ligand-binding interface constitutes about two-thirds of a protein-binding interface.

protein interface patches had statistically significant biases,[32–41] the attempts of using the differences for patch discrimination have encountered several difficulties. The physical properties of the interfaces are highly diverse[38,42,43] and vary between protein families and complex types. Even within a single interface the binding energy is not distributed evenly among residues; instead, there are so-called hot spots, which contribute most of the interaction energy, while the other interface residues are of relatively minor importance.[17,44,45] Finally, the extent and shape of a protein patch in which the small local biases accumulate into a statistically significant signal is not known in advance.

Despite the described difficulties, both approaches have been successfully applied to prediction of protein interfaces on isolated protein structures. The decision about choosing one of the two approaches in each particular case depends on the nature of the protein of interest and available resources. It should be taken into account that in realistic situations, the absence of knowledge about the interaction patch shape (which depends strongly on the partner), and the ambiguity of interface definition make 100% success rates unachievable. On the other hand, all methods provide a statistically significant prediction with high likelihood for the predicted interface residues to be really involved in protein interactions.

In the following we present three methods for prediction of protein interfaces on the surface of isolated proteins with available 3D structures. The first method, REVCOM, belongs to the class of alignment-dependent methods, and the other two, ODA and PIER, to that of the alignment-independent methods.

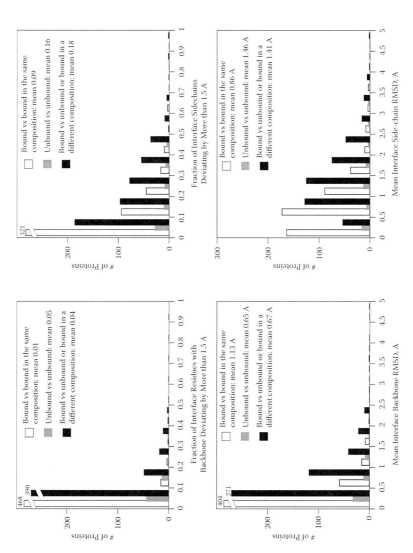

FIGURE 10.6 Flexibility of small molecule binding interfaces and induced fit. Less than one-fifth of interface side chains are displaced by more than 1.5 Å when compared between different complex compositions. At least one interface residue backbone deviates by more than 1.5 Å in only 33% of the cases, and by at least one side chain in 77% of the cases. The median number of strongly deviating side chains is one.

Protein Interface Prediction with REVCOM

Many residue conservation calculation algorithms only work well if a positionally accurate and compositionally balanced sequence alignment is used, and provide highly variable results otherwise. This variability, along with limited performance of simple conservation measures on weakly conserved interfaces, urged the development of a Bayesian method, robust evolutionary conservation measure (REVCOM),[16] that employs phylogenetic trees to calculate evolutionary rates. REVCOM improves the conservation prediction algorithm by making it more robust and less sensitive to (1) local alignment errors, (2) overrepresentation of sequences in some branches, and (3) occasional presence of unrelated sequences. The method was evaluated and compared with an entropy-based conservation measure on a set of 1494 protein interfaces. By REVCOM conservation measures, 62% of the analyzed protein interfaces were found to be more conserved than the remaining surface at the 5% significance level. A consistent method to incorporate alignment reliability was proposed and demonstrated to reduce arbitrary variation of calculated rates upon inclusion of distantly related or unrelated sequences into the alignment.

REVCOM measures were combined with residue-type distributions in a support vector machine (SVM)-based method for predicting protein interfaces on the structure of an isolated protein. The models were trained and cross-validated on a carefully selected set of biologically relevant protein–protein interfaces. Data for noninterface residues was not removed from the data set. Removing this data reduces the number of false positives in the cross-validation, which provides a biased measure of accuracy since the identity of noninterface residues is not known beforehand for an actual prediction.

The recall and precision achieved by the model on cross-validation were, respectively, 35% higher and 24% higher than expected from a random assignment. Ninety-seven percent of the predicted interface patches overlapped with the actual interface, even though on average only 22% of the surface residues were included in the predicted patch. The receiver operating characteristic (ROC) curve for the fivefold cross-validation on the complete dimer set is shown in Figure 10.7. This curve shows the tradeoff between sensitivity and specificity for the prediction.

Optimal Docking Area (ODA)

Extensive experimental and theoretical kinetic studies (reviewed in Reference 46) indicate that specific protein association is often preceded by formation of the encounter complex, which is primarily driven by electrostatics and desolvation. More specific interactions, such as hydrogen bonding and salt bridges, form later and account for the specificity of the final orientation. The important role of desolvation was discussed and used in several applications, for example, to discriminate between docking solutions.[47]

The relative contributions of electrostatic and hydrophobic forces to complex formation vary widely among different complexes.[48] In particular, statistical analyses of known protein–protein complex structures have clearly shown the hydrophobic character of protein interfaces in obligate complexes.[33,40,49] Moreover, it has been

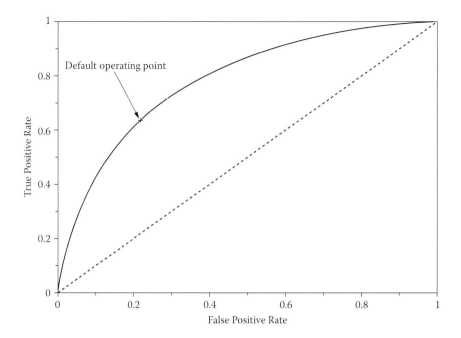

FIGURE 10.7   Receiver operator characteristic (ROC) curve for fivefold cross-validation of the REVCOM-based protein interface prediction method on the set of 632 dimer interfaces. The area under the curve is 0.79. False positive rate = (false positives)/(false positives + true negatives) and true positive rate = (true positives)/(true positives + false negatives). The default SVM operating point with a decision value cutoff is marked.

demonstrated that large hydrophobic patches correlate with obligate protein interfaces.[50] However, in transient complexes, the residue preferences at the interface are less pronounced and are not strong enough to unambiguously predict the interface location.

The optimal docking area (ODA)[11] method represented an attempt to characterize the desolvation properties of protein surfaces and further analyze their role in transient complex formation.

Different subsets of adjacent surface residues were evaluated in order to map low desolvation areas on protein surfaces. In contrast to the common approach that involves dividing the protein surface into equal-area patches, the ODA method generated a series of patches of increasing size with a common center and searched for the ODA, that is, the patch with the lowest desolvation energy. The surface points that generated the ODAs with significant low-energy values were used to define a region over the protein surface most likely to be involved in interaction with other proteins. Instead of accounting for trivial hydrophobicity, the patch surface energy was evaluated based on atomic solvation parameters previously derived from octanol/water transfer experiments and adjusted for protein–protein binding.[51] The method was applied and shown to successfully identify *nonobligate* protein interaction sites.

Protein Interface Recognition (PIER)

Inspired by the success of ODA, the protein interface recognition (PIER)[52] method applied machine learning techniques for further optimization of atom desolvation parameters in the context of protein–protein associations. This led to an alignment-independent method for protein interface identification with improved reliability, accuracy, and speed.

Interface prediction by PIER starts with generation of surface patches in the spirit of ODA; however, for each protein the patch generation radius was fixed and given by the formula

$$d = \sqrt{27.235 + 0.018 \times ASA_{400}^{1000}} \, ,$$

where ASA_{400}^{1000} is the accessible surface area (ASA) of the isolated molecule, trimmed to fit in the range of [400,1000] Å^2. For proteins whose total ASA exceeded 10,000 Å^2, using this equation produced the distance of 14 Å, and resulted in surface patches with average ASA between 900 and 1000 Å^2. The deviations in the values of ASA between different patches reflected the curvature and packing of the surface atoms within the patch. For each patch (P), 12 patch descriptors were calculated. These descriptors simply represented a total ASA of 12 subresidue atomic groups, whose representation was previously found to be significantly different between interface and noninterface patches. The PIER value for the patch was calculated as a linear combination of the obtained descriptors and further transferred to individual residues within the patch. Based on the per-residue PIER decision value, the residues are predicted to be either interface or noninterface.

The proposed alignment-independent method demonstrated improved performance over the previously published methods. On a diverse benchmark of 748 proteins known to be involved in homo- and heterodimeric interactions, permanent as well as transient, the overall precision at the residue level was 60% at the recall threshold of 50%. The method was also tested on other benchmarks. Using the method, we identified potential new interfaces and corrected mislabeled oligomeric states. Several predictions with PIER are presented in Figure 10.8.

A cross-validated partial least squares (PLS) regression algorithm[53] provides a natural environment to incorporate and evaluate the relative contribution of new surface descriptors including those derived from sequence alignments. In particular, we found that when added to the set of PIER descriptors, the evolutionary signal contributed as little as 7%–10%, with the rest (90%–93%) being provided by atomic group composition descriptors. Adding evolutionary signals only marginally influenced the prediction performance; moreover, for certain classes of proteins, using conservation scores actually resulted in deteriorated prediction. This exercise demonstrated that while both alignment-dependent and alignment-independent approaches maybe successful, combining them does not improve the success rate significantly.

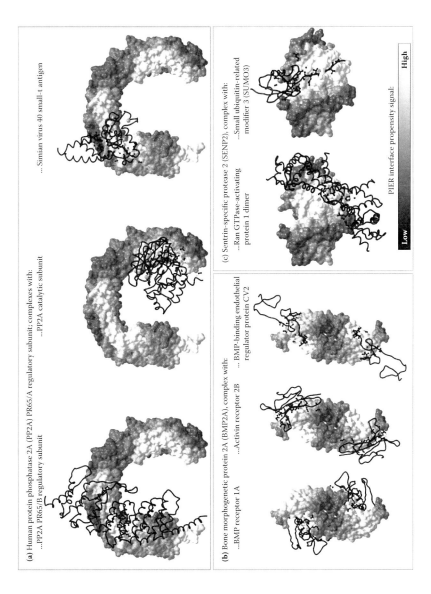

FIGURE 10.8   Precise identification of multiple interfaces (white patches) on the surface of an isolated protein with PIER.

Predicting Membrane Interfaces: MODA

Reversible recruitment of soluble proteins to cellular membranes represents another class of important biological intermolecular interactions and underlines many cellular processes and events. In many cases, this recruitment is not mediated by explicit covalently attached membrane anchors and is based solely on the surface properties. A number of computational methods predicting peripheral membrane interactions are based on homology with known membrane-targeting modules. They, however, fail to encompass the entire biological space and yield many false positives, since conservation of the structural motif does not always mean conservation of the function (i.e., PH domains and C2 domains).[54,55] By combining the PIER approach with the analysis of protein surface curvature and local electrostatic potential, we developed a fast computational method, which, given a 3D structure of a protein, predicts its membrane propensity and identifies the membrane contact elements on its surface. The method was named MODA,[56] or membrane ODA. MODA prediction of the protein–membrane interfaces on known peripheral proteins correlated well with the experimental data; it achieved the precision of 64% at 50% recall at the residue level (7% precision expected from random prediction). Moreover, using the method, we were able to identify several novel potential peripheral proteins, which were consequently validated using NMR spectroscopy and micelle titration (Figure 10.9).

Protein Docking

Predicting the geometry of association of two proteins *known to bind each other* is also a legitimate, albeit more rare and much more difficult task than predicting protein interaction patches.[57] Its applicability is limited to the cases where both the identity and three-dimensional structure of the second partner are known. Since the beginning of the Critical Assessment of PRediction of Interactions (CAPRI) competitions,[58] two main problems make the protein docking task unreliable: (1) an unpredictable *degree* of induced rearrangements or restructuring upon complex formation, and (2) poorly predictable *specifics* of conformational changes even for small degrees of induced fit (e.g., Reference 59).

The large degree of induced rearrangements may include (1) conditional structuring of the previously unstructured loop or region (e.g., the activation loop of protein kinases,[19,60,61] or loop 6 of TIM α/β barrels); (2) domain swapping (e.g., References 62 and 63); (3) large displacements of secondary structure elements, especially N- or C-terminal (e.g., helix 12 in nuclear receptors, opening of two parallel β-strands in serpins); and (4) large relative movements of protein domains, for example, the "swivelling" in pyruvate phosphate dikinase, Ca-dependent domain movements in calmodulin, and integrin domains. The changes may have some functional context, for example, being associated with protein activation like in serpins or kinases. Sadly, no computational method so far has claimed any success in predicting those changes without prior knowledge of the answer.

For cases where the restructuring of domains, parts, and the interface atoms is minor, the protein docking has been a half-solved problem (e.g., References 8 and 64). If the changes are limited to minor backbone shifts and rearrangements of a

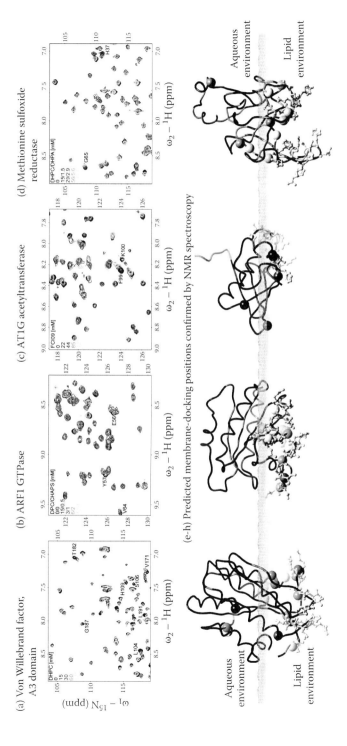

FIGURE 10.9 Discovery of novel peripheral membrane proteins and identification of membrane interfaces with MODA. MODA-positive residues are shown in sticks; backbone amides with chemical shifts are shown as Spheres.

fraction of the interface side chains, in 40% to 80% of the cases (depending on a problem set) the top scoring pose will have most of the residue–residue contacts correctly predicted and the interface patch can be around 2 A RMSD from the crystallographic solution. Of course the caveats include the fact that the correct solution can be either completely missed because of a large rearrangement, or not ranked as a top solution due to inability of the refinement to find all smaller changes necessary for the native interface and scoring function to recognize the near-native solution.[65]

The first successful combination ab initio of all-atom protein–protein docking and interface refinement that led to a single solution was published in *Nature Structural Biology* in 1994.[66] The uncomplexed lysozyme was docked to the HyHel5 antibody without any prior knowledge about the epitope location. The procedure used all-atom representation and worked in two natural phases. The initial positional sampling was performed by searching from multiple starting positions with the pseudo-Brownian jumps followed by minor relaxations of torsional angles. At the refinement phase, the 30 lowest-energy conformations from all the searches were then refined by a *global optimization* of the interface side chains and positional variables[67] with relaxed backbone using an extended set of energy terms with desolvation and side-chain entropy. The near-native solution found was surprisingly close to the crystallographic structure (root mean square deviation [RMSD] of 1.57 A for all backbone atoms of lysozyme) and had a considerably lower energy (by 20 kcal/mol) than any other solution after refinement.

Although the first phase of the docking procedure can be performed on various resolution models using a variety of search methods (most notably, multidimensional fast Fourier transform, or FFT) and scoring functions[68] (also reviewed in References 47, 64, 69, and 70), it became clear after 1994 that the refinement with a detailed atomic model and detailed energy function was necessary to both improve the ability to recognize the native solution and improve the quality of the pose. The 1994 refinement protocol using internal coordinate sampling was essentially preserved in the later modifications of the Internal Coordinate Mechanics–based protocols leading to the high accuracy predictions in the first two CAPRI meetings[71,72] and was later "rediscovered."[73] Refinement procedures using molecular dynamics were also proven to be successful.[74–76]

Overall, the unpredictable and unexpected rearrangements of protein segments and domains, their partially unstructured character,[5] as well as a relative paucity of the available high quality and complete (i.e., "dockable") three-dimensional models of the interacting components prevent protein docking from becoming a routine practical tool of molecular biology comparable to x-ray crystallography.

PROTEIN INTERACTIONS WITH SMALL MOLECULES

Small molecules are much easier to deal with. First, their interfaces with proteins are guaranteed to be small in contrast to proteins where a large variation of interface area is observed. Consequently, specific ligands cannot afford to avoid a tight fit to one of the ligand's conformers, which, in turn, makes the prediction of ligand binding sites for nanomolar ligands a feasible task (see References 77 and 78 for a list).

Furthermore, for any new ligand it is relatively easy to predict how it docks into a "preformed" pocket because the number of degrees of freedom including the positional ones is small enough for a reliable sampling and structure prediction.

POCKETOME BASED ON GAUSSIAN CONVOLUTION OF THE LENNARD-JONES POTENTIAL

Since strong, small ligands require a tight fit whether the binding has polar, charge, hydrophobic, or mixed nature, the most general form of interatomic interaction potential, van der Waals potential, may be predictive of a small-molecule interface on a protein. Straightforwardly computed with the Lennard-Jones equation, it is not predictive; however, its aggregate value allows clear distinction between an area occupied by a typical small molecule and the rest of the protein surface.[79] The cumulative value can be mathematically defined as a convolution of the potential with a Gaussian kernel. We tried different radii and found that the radius of 2.6A results in the best performance of the method.

The generated potential can be contoured at a certain level to give an envelope that resembles the envelope of strong binders in most cases. The predicted envelope is particularly valuable for open and extended pockets with many binding surfaces and does not require anything but a set of coordinates. We used this method to generate the bounding box for unbiased ligand–receptor cross-docking.[80]

The method is fast and can be applied to a currently known structural proteome of an organism. For example, we applied the Gaussian convolution pocket prediction method to a small set of proteins from the malarial parasite, *Plasmodium falciparum*, but it can be scaled up to much larger sets of proteins.[81] The site prediction methods can also be adjusted to a particular task and involve other principles, from geometrical to evolutionary (see Reference 82 for a review).

CROSS-DOCKING OF A LIGAND TO A SINGLE RECEPTOR CONFORMATION WITH THE ICM ALGORITHM OVERCOMES LIMITED INDUCED FIT

As described earlier, in most cases, ligand binding induces only small conformational changes. This opens possibilities for successful prediction of binding poses of ligands in the pockets that have not been co-crystallized with them (the so-called *cross-docking* problem). To quantify the extent that rigid-body receptor docking can overcome the extent on the induced fit, we performed cross-docking simulations for 1000 ligands and 300 proteins from different complex structures.[88] It was shown that for about one-half of the cases (46%) such cross-docking with the Internal Coordinate Mechanics (ICM) docking algorithm[83] predicts a correct near-native geometry as the top-scoring solution. To overcome the induced fit obstacle for the remaining cases, one of the induced fit protocols needs to be applied.

If multiple conformations of a pocket are known in advance, a simple protocol, called MRC (multiple receptor conformation) docking can be used.[84] This procedure can also be accelerated using a so-called 4D approach in which the receptor conformer becomes a variable in the docking procedure.[88]

ADVANCED APPROACHES TO INDUCED FIT IN LIGAND DOCKING

Often only one conformer of the receptor is known. It was shown that to dock a ligand correctly to an incorrect pocket, one can simply delete the uncertain parts in a certain way and rely on the rest of the pocket to position the ligand correctly. The idea came from the "omission" modeling proven to be successful for side chain prediction.[85] The uncertain parts can later be refined *around* the identified ligand position. The scan alanines and refine (SCARE) algorithm[80] replaces *pairs* of neighboring side chains by alanines and docks the ligand to each gapped version of the pocket. Unlike in Reference 86, the selection of the residue pairs by SCARE is performed in an unbiased systematic fashion, which results in a general algorithm applicable on a scale of complete proteomes. All docked positions are scored, refined with original side chains and flexible backbone, and rescored. In the optimal version of the protocol, pairs of residues were replaced by alanines and only one best scoring conformation was selected from each "gapped" pocket for refinement. The optimal SCARE protocol identifies a near-native conformation (under 2 A RMSD) as the lowest rank for 80% of pairs if the docking bounding box is defined by the predicted pocket envelope, and for as many as 90% of the pairs if the bounding box is derived from the known answer with a 5 A margin.

LIGAND SCREENING AND PROFILING

As shown earlier, predicting the correct binding pose of a ligand required some form of treatment of the induced fit. Those relevant receptor models provide necessary starting points for (1) identification of a small-molecule binder to a protein of interest in a large database of chemicals (ligand screening), or (2) identification of possible protein targets for a single, given small molecule (ligand profiling).

The DOLPHIN protocol[89] gives a specific recipe for predicting the changes associated with binding of so-called type II kinase inhibitors to their target kinases. Similar to the SCARE protocol, the type II compatible model of a target kinase is built by omission of the part of the structure with consequent ligand docking and full-atom complex refinement. Using the ICM binding score[87] that has been previously derived from a multireceptor screening benchmark as a compromise between approximated Gibbs free energy of binding and numerical errors, we screened a large database of kinase ligands and showed that the correct type II ligands for the modified kinase are selected in the top 1.5%–3.5% of the database. Further on, based on experimental data, we derived the kinase-specific systematic free energy contributions originating from the different abundance of the relevant conformer and other protein features. By combining these energy contributions with the calculated binding energies, we could identify the kinase specificity profile of individual type II ligands. The ligand specificity profiling approach can be extended to the whole structural interactome as the relevant models and protein offsets become available.

CONCLUSIONS

The structural interactome is the next great challenge for structural proteomics that clearly cannot be solved by crystallography alone and requires computational structural methods because of enormous combinatorics of possible interactions. In this chapter we reviewed (1) the interaction interface prediction methods; (2) the induced conformational changes for different types of interacting partners; and (3) the ability of the available methods to predict the binding pose, its relative score or binding energy, and binding specificity. The recent methods for small-molecule ligands, including pocket prediction, receptor flexible docking (SCARE, 4D, and MRC protocols), and improved scoring, are paving the way to predictive structural chemogenomics.

For protein–protein (or membrane) interactions, the most practical kinds of structure-based predictions include predicting interfaces with other proteins (without the prediction of their identity) and with a membrane. Predicting the geometry, identity, or strength of protein–protein interactions de novo from unbound structures, without evolutionary heuristics, seems to be much more problematic because of a much larger role and scale of unpredictable, induced conformational changes. A comprehensive benchmark described in this review may help to develop and test better methods for predicting large-scale conformational changes upon interaction.

ACKNOWLEDGMENTS

We thank Maxim Totrov and Eugene Raush for valuable discussions and technical help. The project was partially funded by National Institutes of Health/National Institute of General Medical Sciences grants 5-R01-GM071872-02 and 1-R01-GM074832-01A1.

REFERENCES

1. Janin J. 2007. Structural genomics: Winning the second half of the game. *Structure* *15*(11):1347–1349.
2. Vakser IA. 2008. PSI has to live and become PCI: Protein Complex Initiative. *Structure* *16*(1):1–3.
3. Ma B, Nussinov R. 2007. Trp/Met/Phe hot spots in protein–protein interactions: Potential targets in drug design. *Curr Top Med Chem* *7*(10):999–1005.
4. Cohen M, Reichmann D, Neuvirth H, Schreiber G. 2008. Similar chemistry, but different bond preferences in inter- versus intra-protein interactions. *Proteins* *72*(2):741–753.
5. Radivojac P, Iakoucheva LM, Oldfield CJ, Obradovic Z, Uversky VN, Dunker AK. 2007. Intrinsic disorder and functional proteomics. *Biophys J* *92*(5):1439–1456.
6. Sotomayor M, Schulten K. 2007. Single-molecule experiments in vitro and in silico. *Science* *316*(5828):1144–1148.
7. Vajda S, Guarnieri F. 2006. Characterization of protein–ligand interaction sites using experimental and computational methods. *Curr Opin Drug Discov Devel* *9*(3):354–362.
8. Vakser IA, Kundrotas P. 2008. Predicting 3D structures of protein–protein complexes. *Curr Pharm Biotechnol* *9*(2):57–66.
9. Kryshtafovych A, Fidelis K, Moult J. 2007. Progress from CASP6 to CASP7. *Proteins* *69*(Suppl 8):194–207.

10. Moult J. 2008. Comparative modeling in structural genomics. *Structure 16*(1):14–16.

11. Fernandez-Recio J, Totrov M, Skorodumov C, Abagyan R. 2005. Optimal docking area: A new method for predicting protein–protein interaction sites. *Proteins 58*(1):134–143.

12. Berman HM, Westbrook J, Feng Z, Gilliland G, Bhat TN, Weissig H, Shindyalov IN, Bourne PE. 2000. The Protein Data Bank. *Nucleic Acids Res 28*(1):235–242.

13. Fernández-Recio J, Totrov M, Abagyan R. 2002. Soft protein–protein docking in internal coordinates. *Protein Sci 11*(2):280–291.

14. Hwang H, Pierce B, Mintseris J, Janin J, Weng Z. 2008. Protein–protein docking benchmark version 3.0. *Proteins 73*(3):705–709.

15. Gao Y, Douguet D, Tovchigrechko A, Vakser IA. 2007. DOCKGROUND system of databases for protein recognition studies: Unbound structures for docking. *Proteins 69*(4):845–851.

16. Bordner AJ, Abagyan R. 2005. REVCOM: A robust Bayesian method for evolutionary rate estimation. *Bioinformatics 21*(10):2315–2321.

17. Rajamani D, Thiel S, Vajda S, Camacho CJ. 2004. Anchor residues in protein–protein interactions. *Proc Natl Acad Sci USA 101*(31):11287–11292.

18. Smith GR, Sternberg MJ, Bates PA. 2005. The relationship between the flexibility of proteins and their conformational states on forming protein–protein complexes with an application to protein–protein docking. *J Mol Biol 347*(5):1077–1101.

19. Schindler T, Bornmann W, Pellicena P, Miller WT, Clarkson B, Kuriyan J. 2000. Structural mechanism for STI-571 inhibition of abelson tyrosine kinase. *Science 289*(5486):1938–1942.

20. Lichtarge O, Bourne HR, Cohen FE. 1996. An evolutionary trace method defines binding surfaces common to protein families. *J Mol Biol 257*(2):342–358.

21. Halperin I, Wolfson H, Nussinov R. 2004. Protein–protein interactions: coupling of structurally conserved residues and of hot spots across interfaces. Implications for docking. *Structure (Camb) 12*(6):1027–1038.

22. Ma B, Elkayam T, Wolfson H, Nussinov R. 2003. Protein–protein interactions: Structurally conserved residues distinguish between binding sites and exposed protein surfaces. *Proc Natl Acad Sci USA 100*(10):5772–5777.

23. Valencia A, Pazos F. 2003. Prediction of protein–protein interactions from evolutionary information. *Methods Biochem Anal* 44:411–426.

24. Lichtarge O, Sowa ME. 2002. Evolutionary predictions of binding surfaces and interactions. *Curr Opin Struct Biol 12*(1):21–27.

25. Armon A, Graur D, Ben-Tal N. 2001. ConSurf: An algorithmic tool for the identification of functional regions in proteins by surface mapping of phylogenetic information. *J Mol Biol 307*(1):447–463.

26. Glaser F, Pupko T, Paz I, Bell RE, Bechor-Shental D, Martz E, Ben-Tal N. 2003. ConSurf: Identification of functional regions in proteins by surface-mapping of phylogenetic information. *Bioinformatics 19*(1):163–164.

27. Pupko T, Bell RE, Mayrose I, Glaser F, Ben-Tal N. 2002. Rate4Site: An algorithmic tool for the identification of functional regions in proteins by surface mapping of evolutionary determinants within their homologues *Bioinformatics 18*(Suppl 1):71–77.

28. Aloy P, Russell RB. 2002. Interrogating protein interaction networks through structural biology. *Proc Natl Acad Sci USA 99*(9):5896–5901.

29. Bradford JR, Westhead DR. 2003. Asymmetric mutation rates at enzyme-inhibitor interfaces: Implications for the protein–protein docking problem. *Protein Sci 12*(9):2099–2103.

30. Caffrey DR, Somaroo S, Hughes JD, Mintseris J, Huang ES. 2004. Are protein–protein interfaces more conserved in sequence than the rest of the protein surface? *Protein Sci 13*(1):190–202.

31. Mintseris J, Weng Z. 2005. Structure, function, and evolution of transient and obligate protein–protein interactions. *Proc Natl Acad Sci USA 102*(31):10930–10935.
32. Bahadur RP, Chakrabarti P, Rodier F, Janin J. 2004. A dissection of specific and non-specific protein–protein interfaces. *J Mol Biol 336*(4):943–955.
33. Jones S, Thornton JM. 1997. Analysis of protein-protein interaction sites using surface patches. *J Mol Biol 272*(1):121–132.
34. Larsen TA, Olson AJ, Goodsell DS. 1998. Morphology of protein–protein interfaces. *Structure 6*(4):421–427.
35. Conte LL, Chothia C, Janin J. 1999. The atomic structure of protein–protein recognition sites. *J Mol Biol 285*(5):2177–2198.
36. MacCallum RM, Martin AC, Thornton JM. 1996. Antibody–antigen interactions: Contact analysis and binding site topography. *J Mol Biol 262*(5):732–745.
37. Nooren IM, Thornton JM. 2003. Structural characterisation and functional significance of transient protein-protein interactions. *J Mol Biol 325*(5):991–1018.
38. Ofran Y, Rost B. Analysing six types of protein–protein interfaces. 2003. *J Mol Biol 325*(2):377–387.
39. Rodier F, Bahadur RP, Chakrabarti P, Janin J. 2005. Hydration of protein–protein interfaces. *Proteins 60*(1):36–45.
40. Tsai CJ, Lin SL, Wolfson HJ, Nussinov R. 1997. Studies of protein–protein interfaces: A statistical analysis of the hydrophobic effect. *Protein Sci 16*(1):53–64.
41. Young L, Jernigan RL, Covell DG. 1994. A role for surface hydrophobicity in protein–protein recognition. *Protein Sci 3*(5):717–729.
42. Jones S, Thornton JM. 1996. Principles of protein–protein interactions. *Proc Natl Acad Sci USA 93*(1):13–20.
43. Nooren IM, Thornton JM. 2003. Diversity of protein–protein interactions. *EMBO J 22*(14):3486–3492.
44. Bogan AA, Thorn KS. 1998. Anatomy of hot spots in protein interfaces. *J Mol Biol 280*(1):1–9.
45. Clackson T, Wells JA. 1995. A hot spot of binding energy in a hormone–receptor interface. *Science 267*(5196):383–386.
46. Gabdoulline RR, Wade RC. 2002. Biomolecular diffusional association. *Curr Opin Struct Biol 12*(2):204–213.
47. Camacho CJ, Vajda S. 2002. Protein–protein association kinetics and protein docking. *Curr Opin Struct Biol 12*(1):36–40.
48. Sheinerman FB, Honig B. 2002. On the role of electrostatic interactions in the design of protein–protein interfaces. *J Mol Biol 318*(1):161–177.
49. Jones S, Thornton JM. 1997. Prediction of protein–protein interaction sites using patch analysis. *J Mol Biol 272*(1):133–143.
50. Lijnzaad P, Argos P. 1997. Hydrophobic patches on protein subunit interfaces: Characteristics and prediction. *Proteins 28*(3):333–343.
51. Fernández-Recio J, Totrov M, Abagyan R. 2004. Identification of protein–protein interaction sites from docking energy landscapes. *J Mol Biol 335*(3):843–865.
52. Kufareva I, Budagyan L, Raush E, Totrov M, Abagyan R. 2007. PIER: Protein interface recognition for structural proteomics. *Proteins 67*(2):400–417.
53. Geladi P, Kowalski B. 1986. Partial least squares regression: A tutorial. *Analytica Chimica Acta 185*:1–17.
54. Lemmon MA. 2007. Pleckstrin homology (PH) domains and phosphoinositides. *Biochem Soc Symp 74*:81–93.
55. Lemmon MA. 2008. Membrane recognition by phospholipid-binding domains. *Nat Rev Mol Cell Biol 9*(2):99–111.

56. Kufareva I, Dancea F, Kiran MR, Polonskaya Z, Overduin M, Abagyan R. 2008. Computational identification of novel peripheral protein-membrane interactions. *FASEB J* 22:811–815.

57. Villoutreix BO, Bastard K, Sperandio O, Fahraeus R, Poyet J-L, Calvo F, Deprez B, Miteva MA. 2008. In silico–in vitro screening of protein–protein interactions: Towards the next generation of therapeutics. *Curr Pharm Biotechnol* 9(2):103–122.

58. Wodak SJ, Méndez R. 2004. Prediction of protein–protein interactions: The CAPRI experiment, its evaluation and implications. *Curr Opin Struct Biol* 14(2):242–249.

59. Lensink MF, Mendez R. 2008. Recognition-induced conformational changes in protein–protein docking. *Curr Pharm Biotechnol* 9(2):77–86.

60. Liu Y, Gray NS. 2006. Rational design of inhibitors that bind to inactive kinase conformations. *Nat Chem Biol* 2(7):358–364.

61. Mol CD, Lim KB, Sridhar V, Zou H, Chien EYT, Sang B-C, Nowakowski J, Kassel DB, Cronin CN, McRee DE. 2003. Structure of a c-kit product complex reveals the basis for kinase transactivation. *J Biol Chem* 278(34):31461–31464.

62. Sinha N, Tsai CJ, Nussinov R. 2001. A proposed structural model for amyloid fibril elongation: Domain swapping forms an interdigitating beta-structure polymer. *Protein Eng* 14(2):93–103.

63. Posy S, Shapiro L, Honig B. 2008. Sequence and structural determinants of strand swapping in cadherin domains: Do all cadherins bind through the same adhesive interface? *J Mol Biol* 378(4):952–966.

64. Ritchie DW. 2008. Recent progress and future directions in protein–protein docking. *Curr Protein Pept Sci* 9(1):1–15.

65. Fernandez-Recio J, Abagyan R, Totrov M. 2005. Improving CAPRI predictions: Optimized desolvation for rigid-body docking. *Proteins* 60(2):308–313.

66. Totrov M, Abagyan R. 1994. Detailed ab initio prediction of lysozyme-antibody complex with 1.6 A accuracy. *Nat Struct Biol* 1(4):259–263.

67. Abagyan R, Totrov M. 1994. Biased probability Monte Carlo conformational searches and electrostatic calculations for peptides and proteins. *J Mol Biol* 235(3):983–1002.

68. Ritchie DW, Kozakov D, Vajda S. 2008. Accelerating and focusing protein–protein docking correlations using multi-dimensional rotational FFT generating functions. *Bioinformatics* 24(17):1865–1873.

69. Halperin I, Ma B, Wolfson H, Nussinov R. 2002. Principles of docking: An overview of search algorithms and a guide to scoring functions. *Proteins* 47(4):409–443.

70. Smith GR, Sternberg MJ. 2002. Prediction of protein–protein interactions by docking methods. *Curr Opin Struct Biol* 12(1):28–35.

71. Méndez R, Leplae R, Maria LD, Wodak SJ. 2003. Assessment of blind predictions of protein–protein interactions: Current status of docking methods. *Proteins* 52(1):51–67.

72. Méndez R, Leplae R, Lensink MF, Wodak SJ. 2005. Assessment of CAPRI predictions in rounds 3-5 shows progress in docking procedures. *Proteins* 60(2):150–169.

73. Gray JJ, Moughon S, Wang C, Schueler-Furman O, Kuhlman B, Rohl CA, Baker D. 2003. Protein–protein docking with simultaneous optimization of rigid-body displacement and side-chain conformations. *J Mol Biol* 331(1):281–299.

74. Heifetz A, Pal S, Smith GR. 2007. Protein–protein docking: Progress in CAPRI rounds 6-12 using a combination of methods: The introduction of steered solvated molecular dynamics. *Proteins* 69(4):816–822.

75. Qin S, Zhou H-X. 2007. A holistic approach to protein docking. *Proteins* 69(4):743–749.

76. Motiejunas D, Gabdoulline R, Wang T, Feldman-Salit A, Johann T, Winn PJ, Wade RC. 2008. Protein-protein docking by simulating the process of association subject to biochemical constraints. *Proteins* 71(4):1955–1969.

77. Sotriffer C, Klebe G. 2002. Identification and mapping of small-molecule binding sites in proteins: Computational tools for structure-based drug design. *Farmaco* 57(3):243–251.
78. Villoutreix BO, Renault N, Lagorce D, Sperandio O, Montes M, Miteva MA. 2007. Free resources to assist structure-based virtual ligand screening experiments. *Curr Protein Pept Sci* 8(4):381–411.
79. An J, Totrov M, Abagyan R. 2005. Pocketome via comprehensive identification and classification of ligand binding envelopes. *Mol Cell Proteomics* 4(6):752–761.
80. Bottegoni G, Kufareva I, Totrov M, Abagyan R. 2008. A new method for ligand docking to flexible receptors by dual alanine scanning and refinement (SCARE). *J Comput Aided Mol Des* 22(5):311–325.
81. Nicola G, Smith CA, Abagyan R. 2008. New method for the assessment of all drug-like pockets across a structural genome. *J Comput Biol* 15(3):231–240.
82. Laurie ATR, Jackson RM. 2006. Methods for the prediction of protein-ligand binding sites for structure-based drug design and virtual ligand screening. *Curr Protein Pept Sci* 7(5):395–406.
83. Totrov M, Abagyan R. 1997. Flexible protein-ligand docking by global energy optimization in internal coordinates. *Proteins* 29(Suppl 1):215–220.
84. Totrov M, Abagyan R. 2008. Flexible ligand docking to multiple receptor conformations: A practical alternative. *Curr Opin Struct Biol* 18(2):178–184.
85. Eisenmenger F, Argos P, Abagyan R. 1993. A method to configure protein side-chains from the main-chain trace in homology modelling. *J Mol Biol* 231(3):849–860.
86. Sherman W, Day T, Jacobson MP, Friesner RA, Farid R. 2006. Novel procedure for modeling ligand/receptor induced fit effects. *J Med Chem* 49(2):534–553.
87. Schapira M, Totrov M, Abagyan R. 1999. Prediction of the binding energy for small molecules, peptides and proteins. *J Mol Recog* 12(3):177–190.
88. Bottegoni G, Kufareva I, Totrov M, Abagyan R. 2009. Four-dimensional docking: a fast and accurate account of discrete receptor flexibility in ligand docking. *Journal of Medicinal Chemistry* 52(2):397–406.
89. Kufareva I, Abagyan R. 2008. Type-II kinase inhibitor docking, screening, and profiling using modified structures of active kinase states. *Journal of Medicinal Chemistry* 51(24):7921–7932.

11 Rearrangements and Expansion of the Domain Content in Proteins Frequently Increase the Protein Connectivity in the Protein–Protein Interaction Network

Inbar Cohen-Gihon, Roded Sharan, and Ruth Nussinov

CONTENTS

OVERVIEW

Protein–protein recognition occurs via interacting domains. Protein domains are conserved components within proteins, usually folding independently of other parts of the protein. This property allows modularity in the protein domain architecture. In most cases domains have a distinct function and many domains mediate protein–protein interactions. Combining the modular nature of domain rearrangement and the relative independence of their functionality allows proteins to acquire additional roles by the acquisition of new domains. Indeed, organisms tend to manipulate existing domain architectures to create new proteins, rather than creating new architectures ab initio. The acquired domain can mediate a new protein–protein interaction and therefore increase the protein connectivity in the protein–protein interaction network. This increased connectivity within and between the biological systems of an organism frequently enhances its complexity. Thus, the rate of domain rearrangements rises with the increase in organisms' complexity during evolution. Here we survey various ways to create new protein domain architectures and the impact of creating new proteins by domain rearrangements on protein connectivity and organisms' complexity. Within this framework, we highlight the role of domains as mediating protein–protein interactions.

INTRODUCTION

Protein domains are highly conserved sequences with a distinct structure, function, and/or common ancestry. A definition of domain may be either sequence or structure based; from a structural point of view, a domain is defined as an independently folding unit, which is both compact and stable. On the other hand, domains are commonly defined as protein regions where their sequence is highly conserved during evolution [1]. The average length of domains ranges from 100 to 250 amino acids [2]. Families of domains contain domains that probably share a common ancestor. Although the structure and sequence definitions of domains are different in their essence, their domain assignment to families is often similar [3]. Small proteins are composed of a single domain, whereas larger, and most of the eukaryotic proteins, are multidomains [4]. Domains, sometimes termed modules, compose proteins in a modular manner. Although the potential number of domain combinations is huge, there is a limited collection of domain types that are duplicated, diverged, and integrated in other proteins in various ways. This usually involves the following scenario: (1) a genomic sequence that codes for one or more domains is duplicated; (2) the duplicated area selectively diverges by mutations, deletions, or insertions; and, sometimes, (3) a recombination or fusion with other genes occurs [5]. Most of the new proteins were created by expansion of existing architectures rather than the creation of new ones ab initio. Domain shuffling creates new functions, enables proteins to form more interactions with other proteins or DNA, and generally increases the organisms' functionality. In this chapter, we describe the mechanisms by which domain rearrangements occur in the genome; review the main findings in the field of complexity by protein domain rearrangements during evolution; and highlight the role of co-occurring domains in protein–protein interactions.

DOMAINS HAVE A KEY ROLE IN PROTEIN–PROTEIN INTERACTIONS

The modular nature of the formation of multidomain proteins allows proteins to acquire new properties without disrupting the ones they already have. One of the most important properties a protein can acquire is the ability to interact with other proteins. Many domains are known to mediate particular protein–protein interactions and a relatively large fraction of these are modular interaction domains. These domains usually fold such that their N and C termini are adjacent in space [6]. In most cases, a protein interaction domain recognizes a particular sequence of residues. For example, the well-known SH3 domain, which mediates a broad range of protein–protein interaction in various organisms, recognizes the motif PXXP, where a proline is followed by any two residues followed by an additional proline. The SH2 domain, which has a role in many tyrosine kinase signaling proteins, requires a phosphotyrosine site in its ligands [7]. These domains are examples of well-known mediators of protein–protein interactions and are found in more than one hundred copies in the human genome [6]. Itzhaki et al. [8] investigated cellular protein–protein interaction networks from their domain–domain interactions. They found that there is a repertoire of conserved domain pairs that is responsible for a variety of interactions in the cell and conclude that different organisms use the same interaction building blocks to perform protein–protein interactions. Park et al. [9] investigated domain–domain interactions as related to their structural families. They found that interactions can be classified based on their domain families and that there are different interaction types for functional, structural, and regulatory purposes. In addition, they showed that domains from the same family tend to interact, both within and between polypeptide chains. This type of functional coupling of domains can be identified by their co-occurrence within the same protein complex as well [10].

CREATING NEW DOMAIN ARCHITECTURES

As noted earlier, the repertoire of protein domains is limited, and novel proteins are created by combining existing proteins with additional domains. The creation of new gene structures during evolution involves one or more molecular mechanisms, in which duplication of a whole gene or a part of it plays a crucial role. Several genetic mechanisms may lead to the formation of a new domain architecture and new functional genes. Possible domain rearrangements are exemplified in Figure 11.1.

GENE AND DOMAIN DUPLICATION

An important mechanism to create functional innovation during evolution is the duplication of either an entire gene or particular gene segments. Duplication of an entire gene may result in one of the following possibilities: (1) the duplicate acquires mutations and becomes nonfunctional; and (2) one of the duplicates accumulates mutations without disrupting the activity in the cell. When a contributing mutation occurs, new functions are added to the organism (neofunctionalization), and (3) both duplicates lose some of their functions and together maintain the original

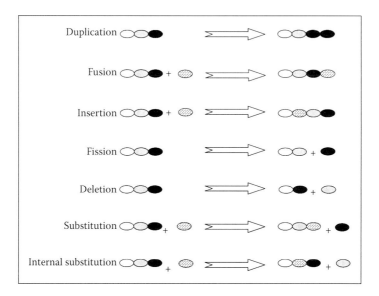

FIGURE 11.1   Possible scenario for evolution of protein architecture.

function (subfunctionalization) [11]. The latter becomes possible mainly because of the modular nature of proteins, that is, their domain compositions, because the particular function of a mutated domain in one of the duplicates is backed up by its corresponding domain in the other duplicate. However, sometimes only parts of the gene are duplicated. This type of duplication is initiated by duplication of the area in the gene that encodes one or more particular domain. The duplicated domain may retain its original function, resulting in a duplicated specific function of the protein, or mutate, diverge, and acquire a new or modified function, resulting in a specialized protein composed of an old combination of domains and a new specialized, diverged domain. In some cases, either the unmodified or the modified domain may undergo recombination with other genes. Occurrence of gene and domain duplicates in the genome may take place via domain shuffling through paralogous recombinations and/or the formation of fused proteins by the union of different domains [12]. In some cases, domain pairs or triplets tend to be duplicated together and therefore appear jointly more than expected by chance [13]. It is believed that there is a single origin of domain combinations, that is, similar combinations that are found in different species have a common ancestor [14].

EXON AND DOMAIN SHUFFLING

It has been shown that there is a correlation between the borders of exons and protein domains in many eukaryotic genomes [15]. In many globular proteins, a more or less exact correlation exists between the exon of the gene and the domain architecture of its protein product. In most of the cases, domain duplication at the protein level implies that exon duplication occurred at the DNA level [16]. Exon shuffling, known also as domain shuffling, is a well-known mechanism in creating new genes. In this process, two or more exons, either from different genes or a duplicated exon, are brought together to create a new exon–intron arrangement. This may occur either during recombination or by reverse transcription exon insertions [17]. For example, the construction of many trypsin-like proteases is believed to be related to exon shuffling [18]. These proteins have a common ancestor for a particular part, and between the signal peptide and the zymogen activation domain of the ancestral protease, some other regulatory modules were inserted. It was assumed that exon shuffling is the mechanism that enables a mobility of domains. Exon shuffling plays a central role in the evolution of protein modularity. Most of these proteins are extracellular, for example, extracellular parts of receptors or components of the extracellular matrix. Modular proteins that are believed to be formed by exon shuffling are unique to metazoans and equivalents were not found in prokaryotes or in plants [17]. Furthermore, it has been shown that about 19% of the exons in eukaryotic genes are assumed to be created by exon shuffling [19].

Fusions, Fissions, and Domain Accretion

Particular types of recombinations may result in either a fusion of two proteins or in the split of a single gene into two or more smaller proteins. Fusions of either domains and proteins or two entire proteins are extremely important for the creation of new domain architectures. The union of two components participating in the same biological process has two main advantages. First, there is an increased efficiency of the united gene on top of the activation of two independent genes. Furthermore, as opposed to the original genes that had separate regulatory regions, the fused gene is found under a single regulatory system, which promises coexpression of the gene products [20,21]. Figure 11.2 illustrates the two scenarios. Marcotte et al. [20] showed that it is possible to infer protein–protein interactions from genome sequences if the two apparently interacting proteins have homologs in other genomes, where they are fused into a single protein chain. This technique has been termed the Rosetta stone method [20]. Rosetta stone proteins have been detected in a variety of organisms. It has been shown that in many cases, microbial genes that are found fused in other genomes are of the same functional category [22]. About half of the fused genes are

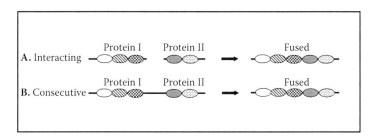

FIGURE 11.2 Detecting protein–protein interaction from genome sequences (the Rosetta stone model). (A) Fusion of two proteins that interact with each other. (B) Fusion of two consecutive genes. Both fusions increase the efficiency of the united protein and promise coexpression of the gene products.

found in their standalone form in close proximity on the chromosome, usually as a part of a microbial operon. This may be an intermediate stage of fusion, where the genes are separated from each other, yet are coregulated [23]. During the course of evolution, fusion of small protein components into larger protein units have contributed much more functional novelties in organisms than the split of proteins into smaller components, and it has been found that fusions are about four times more common than fission events [24].

Domain accretion is the event characterized by the addition of a new DNA fragment encoding a distinct domain to an existing combination of domains. The contribution of such an addition of domain is in the supply of a new functional unit to the protein, which results in an increase in the protein and proteome complexity. This phenomenon is much more common in the evolution of eukaryotes than in that of prokaryotes. Moreover, there is an increasing tendency of domain accretion in metazoans in general and in primates in particular. In fact, there is a twofold increase in domain accretion events in the human genome as compared to invertebrate genomes. In the human genome, domain accretion is most prominent in genes involved in transcription regulation and chromatin remodeling [25].

RETROPOSITION, TRANSPOSABLE ELEMENTS, AND AB INITIO CREATION OF DOMAINS

Transposable elements can be divided into two types: those that require an RNA intermediate, which is reverse transcribed back to the DNA, and those that move by a cut and paste mechanism. The former, an important mechanism for creating gene duplicates, is the reverse transcription of mRNA back into the genome. This process is termed retroposition and is mediated by reverse transcriptase [26] and the resulting cDNA is integrated back to the DNA usually at a random target [27]. Since these fragments are retroposed without their regulatory regions, the functional novelty usually arises from the addition of a retroposed coding region into an existing protein. In mammals and especially in the human genome, long interspersed nuclear elements (LINEs) serve as a source for reverse transcriptase activity. They have a major role in the retrotransposition of promoter and exons, especially of nuclear genes [27–29]. The second type of transposable elements is those that are removed from one locus on the genome and immediately relocated in other place. The most famous family of transposable elements is the Alu sequences [30]. Transposable elements are abundant in mammals and especially in primates and are believed to have a large contribution to protein diversity [31].

An ab initio creation of an entire protein almost does not exist. There are two main reasons why evolution tends to create new genes by the expansion of existing ones. First, duplication, divergence, and recombination of existing genes are much faster than de novo initiation of domains. Second, it is very difficult to overcome the error-correction machineries in the transcription and translation processes to obtain a new domain. However, in a few cases there is de novo formation of the functional domain, which usually occurs when a previously intronic sequence is converted into a coding exon [32].

NETWORK REPRESENTATION OF PROTEIN DOMAINS

Due to the modularity of the protein domain world, it is straightforward to use graph theoretical tools to explore domain composition of proteins. The domain content of proteins either can be related by retaining the exact sequential order of the domains in the protein or by consisting of a similar "bag of domains," that is, without considering the order. In most graphical representations of domains, the order of domains within the protein chain is not considered, and the result is the so-called co-occurring domain graph or referred to as a domain overlap graph. In such a network, nodes correspond to domains, and edges represent two domains that are found within the same protein. Wuchty and Almaas [33] found that in many co-occurrence domain networks there is a giant component that contains the vast majority of the nodes. With regard to the evolutionary aspect, they found that although the numbers of domains across several genomes are similar, the sizes of highly connected domain subgraphs grow with evolution. Przytycka et al. [34] used this graph representation of domains to study the dynamics of the acquisition of new domains and persistence of domain combinations during evolution. They found that in most cases similar domain architectures have a common ancestor and that the topology of most domain co-occurrence networks cannot be explained by a possible mechanism of preferential attachment. The co-occurrence domain graph can also be used to investigate functionality. Ye and Godzik [35] divided this network into clusters and found that clusters tend to contain domains with similar functions. Domain composition of proteins can also be represented on a bipartite graph, where nodes correspond to proteins and domains, and edges connect between proteins and their constituent domains. Cohen-Gihon et al. [36] applied graph theoretical tools to this type of network and found co-occurring domain sets (CDSs) in the *Saccharomyces cerevisiae* proteins more than expected by chance. They found that these sets tend to contain ancient domains that are conserved from bacteria or archaea and that the proteins containing them are highly functionally coherent and enriched in protein–protein interactions. As opposed to graph representation that does not consider the order of domains, Bornberg-Bauer et al. [37] have used directed graphs to represent the order of domains within proteins. On such a graph, nodes correspond to protein domains and a directed edge connects between two consecutive domains in their N-to-C terminal order. They found that such graphs are sparsely connected, except for some regions that are highly clustered. These clusters tend to include functionally related domains, most of them related to signal transduction and cell adhesion.

INCREASING DOMAIN COMPLEXITY IN EUKARYOTES AND METAZOA

The formation of novel proteins by various domain rearrangements is in most cases the result of either one of the mechanisms described earlier or combinations of several of them, and is likely to contribute considerably to the increasing complexity of organisms. Complexity of organisms is mainly reflected by the ability to create interactions among an organism's components. Such components may be protein–protein and protein–DNA interactions at the molecular level, or interactions between cells, tissues, and organs at higher levels. Thus, the contribution of a novel protein to its organism complexity can be estimated by its new connectivity, where highly connected proteins are more likely to increase complexity [38]. The acquisition of a new domain into an existing protein may immediately enable the protein to mediate new interactions and therefore increase its connectivity. As a result, it is not surprising that the evolution of eukaryotes in general and multicellular organisms in particular is correlated with an increased rate of protein domain manipulations. In the following paragraphs, we survey recent findings regarding the correlation between various protein domain manipulations and the increasing complexity during evolution.

The origin of eukaryotes was an extremely important milestone in evolution. As opposed to prokaryotes, eukaryotic genomes encode completely novel systems such as the organization of chromosomes, distinct growth phases in cell cycle, RNA processing, ubiquitin-based posttranslational modifications, endoplasmic reticulum, and obviously the emergence of the nucleus. The entire set of eukaryotic innovations may be classified into three main evolutionary scenarios [39]:

1. Prokaryotic domains that have recruited modified functions in eukaryotes but with no significant biochemical changes. An example is the actin- and tubulin-based eukaryotic cytoskeleton that has emerged from the prokaryotic cytoskeletal proteins MreB and FtsZ [40].
2. Prokaryotic domains that were expanded into new superfamilies in eukaryotes, for example, the ancient helix-turn-helix fold produced several eukaryotic superfamilies of peptide interacting adaptors [41].
3. Completely new domains that originated after the emergence of the eukaryotes, for example, the POZ domain that mediates interactions in transcription and Ub signaling [52].

Apart from this classification, it was found that eukaryotic proteins significantly tend to acquire additional domains, increasing with these accretions their connectivity in protein networks [25]. In addition, eukaryotic genomes have more multidomain architectures than prokaryotic ones, ranging from 65% to 80% multidomain architectures in eukaryotes and 40% to 65% in prokaryotes, depending on the detecting methods [4,42].

The number of different domains encoded by a genome increases with the organism's complexity and the number of unique domain architectures is correlated with the organism's complexity [1,39,43]. Whereas in single-celled organisms only a small fraction of proteins contain tandem domains, the corresponding fraction in multicellular organisms is doubled and rises to 11% on average. Fong et al. [44] studied the evolution of domain architectures in eukaryotes by reconstructing ancestral protein architectures. They found that organisms have gained domains mainly by adding new domains to the protein termini, rather than substituting internal domains within the protein chain. Bjorklund et al. [45] defined a domain distance measure to investigate the evolution of protein architectures in eukaryotes. They also found that domains tend to be added at the protein termini and that insertions and deletions are more common than internal repetitions of domains.

It has been shown that the expansion in protein families resulting in a greater number of genes is not simply correlated with organism complexity, measured by its number of cell types. The expansion of only a small fraction of protein families is correlated with the organism's complexity [46]. Furthermore, they found that some expansions in protein families are progressive, that is, related to the increasing biological complexity of the organism, while some are conservative and related to a successful adaptation of the organism to its environment. Ciccarelli et al. [47] have shown that in some cases, a selection against duplication may result in creative changes of the protein architecture. They found a family of genes that were under a strong selection against duplication across metazoans apart from primates. The latter family had undergone a series of segmental duplications, inversions, translocations, exon loss, and domain accretion that lead to a novel function. Ekman et al. [48] studied the evolution of multidomain architectures and the emergence of new domains in eukaryotic in general and in metazoan species in particular. They found that the rate of creation of new domain architectures was accelerated in the metazoan clade, compared to other species. However, there was no significant increase in the formation of new domains in metazoan, emphasizing the earlier description of creation of new domain architectures by shuffling of existing ones. Tordai et al. [49] showed that intronic recombinations facilitated the shuffling of some domains in metazoa, contributing to the increased functionality of many metazoan systems. Recently, it has been found that the origin of metazoans was accompanied by an abundant domain shuffling, mainly in cell adhesion proteins and in signaling systems [50]. In higher eukaryotes, such as mouse and human, it has been shown that tissue-specific proteins are composed significantly of more domains than housekeeping ones, and that they particularly contain more metazoan-specific domains [51].

CONCLUSIONS

In the postgenomic era, the ever-growing genome and protein sequence data and their domain assignments facilitate the exploration of domain architectures of proteins and their implication for organisms' evolution; yet, despite the large amount of data, deciphering the actual and precise processes in which biological systems acquire new functions and improve existing ones by adding, duplicating, or removing domains is still an extremely difficult task. Our survey provides a summary of what is currently known on domain rearrangement during evolution and its contribution to creating new protein–protein interactions. However, this broad field is far from being fully covered and there is still much we can learn on the evolution of complex biological systems by their domain content.

ACKNOWLEDGMENTS

Inbar Cohen-Gihon is a fellow of the Edmond J. Safra Bioinformatics Program and of the Ela Kodesz Research and Scholarship Fund at Tel Aviv University. This project has been funded in whole or in part with federal funds from the National Cancer Institute, National Institutes of Health (NIH), under contract number N01-CO-12400. The content of this publication does not necessarily reflect the views or policies of the Department of Health and Human Services, nor does mention of trade names, commercial products, or organizations imply endorsement by the U.S. government. This research was supported (in part) by the Intramural Research Program of the NIH, National Cancer Institute, Center for Cancer Research.

REFERENCES

1. Koonin EV, Wolf YI, Karev GP. The structure of the protein universe and genome evolution. *Nature* 2002, **420**(6912):218–223.
2. Wheelan SJ, Marchler-Bauer A, Bryant SH. Domain size distributions can predict domain boundaries. *Bioinformatics* 2000, **16**(7):613–618.
3. Elofsson A, Sonnhammer EL. A comparison of sequence and structure protein domain families as a basis for structural genomics. *Bioinformatics* 1999, **15**(6):480–500.
4. Apic G, Gough J, Teichmann SA. Domain combinations in archaeal, eubacterial and eukaryotic proteomes. *J Mol Biol* 2001, **310**(2):311–325.
5. Chothia C, Gough J, Vogel C, Teichmann SA. Evolution of the protein repertoire. *Science* 2003, **300**(5626):1701–1703.
6. Pawson T, Nash P. Assembly of cell regulatory systems through protein interaction domains. *Science* 2003, **300**(5618):445–452.
7. Pawson T, Scott JD. Signaling through scaffold, anchoring, and adaptor proteins. *Science* 1997, **278**(5346):2075–2080.
8. Itzhaki Z, Akiva E, Altuvia Y, Margalit H. Evolutionary conservation of domain–domain interactions. *Genome Biol* 2006, **7**(12):R125.
9. Park J, Lappe M, Teichmann SA. Mapping protein family interactions: Intramolecular and intermolecular protein family interaction repertoires in the PDB and yeast. *J Mol Biol* 2001, **307**(3):929–938.
10. Betel D, Isserlin R, Hogue CW. Analysis of domain correlations in yeast protein complexes. *Bioinformatics* 2004, **20**(Suppl. 1):i55–i62.
11. He X, Zhang J. Rapid subfunctionalization accompanied by prolonged and substantial neofunctionalization in duplicate gene evolution. *Genetics* 2005, **169**(2):1157–1164.
12. Eichler EE. Recent duplication, domain accretion and the dynamic mutation of the human genome. *Trends Genet* 2001, **17**(11):661–669.
13. Vogel C, Berzuini C, Bashton M, Gough J, Teichmann SA. Supradomains: Evolutionary units larger than single protein domains. *J Mol Biol* 2004, **336**(3):809–823.
14. Vogel C, Teichmann SA, Pereira-Leal J. The relationship between domain duplication and recombination. *J Mol Biol* 2005, **346**(1):355–365.
15. Liu M, Grigoriev A. Protein domains correlate strongly with exons in multiple eukaryotic genomes: Evidence of exon shuffling? *Trends Genet* 2004, **20**(9):399–403.
16. Graur DL, Li WH. *Fundamentals of Molecular Evolution*. Sunderland, MA: Sinauer Associates, 1999.
17. Patthy L. Exon shuffling and other ways of module exchange. *Matrix Biol* 1996, **15**(5):301–310; discussion 311–302.
18. Patthy L. Modular design of proteases of coagulation, fibrinolysis, and complement activation: Implications for protein engineering and structure-function studies. *Methods Enzymol* 1993, **222**:10–21.
19. Long M, Betran E, Thornton K, Wang W. The origin of new genes: Glimpses from the young and old. *Nat Rev Genet* 2003, **4**(11):865–875.
20. Marcotte EM, Pellegrini M, Ng HL, Rice DW, Yeates TO, Eisenberg D. Detecting protein function and protein–protein interactions from genome sequences. *Science* 1999, **285**(5428):751–753.

21. Enright AJ, Iliopoulos I, Kyrpides NC, Ouzounis CA. Protein interaction maps for complete genomes based on gene fusion events. *Nature* 1999, **402**(6757):86–90.

22. Yanai I, Derti A, DeLisi C. Genes linked by fusion events are generally of the same functional category: A systematic analysis of 30 microbial genomes. *Proc Natl Acad Sci USA* 2001, **98**(14):7940–7945.

23. Yanai I, Wolf YI, Koonin EV. Evolution of gene fusions: Horizontal transfer versus independent events. *Genome Biol* 2002, **3**(5):research0024.

24. Kummerfeld SK, Teichmann SA. Relative rates of gene fusion and fission in multidomain proteins. *Trends Genet* 2005, **21**(1):25–30.

25. Koonin EV, Aravind L, Kondrashov AS. The impact of comparative genomics on our understanding of evolution. *Cell* 2000, **101**(6):573–576.

26. Weiner AM, Deininger PL, Efstratiadis A. Nonviral retroposons: Genes, pseudogenes, and transposable elements generated by the reverse flow of genetic information. *Annu Rev Biochem* 1986, **55**:631–661.

27. Brosius J. The contribution of RNAs and retroposition to evolutionary novelties. *Genetica* 2003, **118**(2-3):99–116.

28. Esnault C, Maestre J, Heidmann T. Human LINE retrotransposons generate processed pseudogenes. *Nat Genet* 2000, **24**(4):363–367.

29. Moran JV, DeBerardinis RJ, Kazazian HH, Jr. Exon shuffling by L1 retrotransposition. *Science* 1999, **283**(5407):1530–1534.

30. Makalowski W, Mitchell GA, Labuda D. Alu sequences in the coding regions of mRNA: A source of protein variability. *Trends Genet* 1994, **10**(6):188–193.

31. Lorenc A, Makalowski W. Transposable elements and vertebrate protein diversity. *Genetica* 2003, **118**(2-3):183–191.

32. Nurminsky DI, Nurminskaya MV, De Aguiar D, Hartl DL. Selective sweep of a newly evolved sperm-specific gene in Drosophila. *Nature* 1998, **396**(6711):572–575.

33. Wuchty S, Almaas E. Evolutionary cores of domain co-occurrence networks. *BMC Evol Biol* 2005, **5**(1):24.

34. Przytycka T, Davis G, Song N, Durand D. Graph theoretical insights into evolution of multidomain proteins. *J Comput Biol* 2006, **13**(2):351–363.

35. Ye Y, Godzik A. Comparative analysis of protein domain organization. *Genome Res* 2004, **14**(3):343–353.

36. Cohen-Gihon I, Nussinov R, Sharan R. Comprehensive analysis of co-occurring domain sets in yeast proteins. *BMC Genomics* 2007, **8**:161.

37. Bornberg-Bauer E, Beaussart F, Kummerfeld SK, Teichmann SA, Weiner J, III. The evolution of domain arrangements in proteins and interaction networks. *Cell Mol Life Sci* 2005, **62**(4):435–445.

38. Patthy L. Modular assembly of genes and the evolution of new functions. *Genetica* 2003, **118**(2-3):217–231.

39. Aravind L, Iyer LM, Koonin EV. Comparative genomics and structural biology of the molecular innovations of eukaryotes. *Curr Opin Struct Biol* 2006, **16**(3):409–419.

40. Lowe J, van den Ent F, Amos LA. Molecules of the bacterial cytoskeleton. *Annu Rev Biophys Biomol Struct* 2004, **33**:177–198.

41. Aravind L, Anantharaman V, Balaji S, Babu MM, Iyer LM. The many faces of the helix-turn-helix domain: Transcription regulation and beyond. *FEMS Microbiol Rev* 2005, **29**(2):231–262.

42. Ekman D, Bjorklund AK, Frey-Skott J, Elofsson A. Multi-domain proteins in the three kingdoms of life: Orphan domains and other unassigned regions. *J Mol Biol* 2005, **348**(1):231–243.

43. Lander ES, Linton LM, Birren B, Nusbaum C, Zody MC, Baldwin J, Devon K, Dewar K, Doyle M, FitzHugh W, et al. Initial sequencing and analysis of the human genome. *Nature* 2001, **409**(6822):860–921.

44. Fong JH, Geer LY, Panchenko AR, Bryant SH. Modeling the evolution of protein domain architectures using maximum parsimony. *J Mol Biol* 2007, **366**(1):307–315.

45. Bjorklund AK, Ekman D, Light S, Frey-Skott J, Elofsson A. Domain rearrangements in protein evolution. *J Mol Biol* 2005, **353**(4):911–923.

46. Vogel C, Chothia C. Protein family expansions and biological complexity. *PLoS Comput Biol* 2006, **2**(5):e48.

47. Ciccarelli FD, von Mering C, Suyama M, Harrington ED, Izaurralde E, Bork P. Complex genomic rearrangements lead to novel primate gene function. *Genome Res* 2005, **15**(3):343–351.

48. Ekman D, Bjorklund AK, Elofsson A. Quantification of the elevated rate of domain rearrangements in metazoa. *J Mol Biol* 2007, **372**(5):1337–1348.

49. Tordai H, Nagy A, Farkas K, Banyai L, Patthy L. Modules, multidomain proteins and organismic complexity. *FEBS J* 2005, **272**(19):5064–5078.

50. King N, Westbrook MJ, Young SL, Kuo A, Abedin M, Chapman J, Fairclough S, Hellsten U, Isogai Y, Letunic I, et al. The genome of the choanoflagellate Monosiga brevicollis and the origin of metazoans. *Nature* 2008, **451**(7180):783–788.

51. Cohen-Gihon I, Lancet D, Yanai I. Modular genes with metazoan-specific domains have increased tissue specificity. *Trends Genet* 2005, **21**(4):210–213.

52. Stogios PJ, GS Downs et al. Sequence and structural analysis of BTB domain proteins *Genome Biol* 2005, **6**(10):R82.

12 Intrinsically Disordered Proteins and Their Recognition Functions

Vladimir N. Uversky, Monika Fuxreiter, Christopher J. Oldfield, A. Keith Dunker, and Peter Tompa

CONTENTS

OVERVIEW

The existence of intrinsically disordered proteins (IDPs) that lack well-folded states yet fulfill key biological roles has challenged the classical structure–function

223

paradigm. Their recognition led to the onset of an entire field that aims to elaborate the structural characteristics and mechanism of action of IDPs. Based on bioinformatics predictions, IDPs are abundant in different genomes and carry out mostly regulatory functions that are often related to molecular recognition. Biophysical data provide evidence that most IDPs do not behave like fully random coils, but rather exhibit a limited structural organization either at a secondary or tertiary structure level. Residual structures of IDPs, which are often distinguished in interactions with their partners, are described by related concepts of preformed elements, molecular recognition features, primary contact sites, or linear motifs. All these structural elements act as recognition motifs that facilitate formation of productive contacts with the target and result in specific binding modes. IDPs often adopt a partly or fully folded state in their bound form, yet their interactions and the nature of the interfaces are distinct from that of globular proteins. Although many details are yet to be revealed, the basic concepts of IDP action transform our view on the structural basis of protein functionality.

INTRODUCTION: THE CONCEPT OF PROTEIN INTRINSIC DISORDER

Proteins are major components of living cells and play crucial roles in the maintenance of life. Aberrations in protein function may result in pathological conditions and lead to disease. A unique three-dimensional (3-D) structure was long believed to be an obligatory prerequisite to protein function. This belief became a cornerstone in molecular biology and is supported by several thousand crystal structures. According to the classical structure–function paradigm, the primary amino acid sequence determines the protein's unique 3-D structure, which, in turn, determines its specific function. Although counterexamples were scattered in the literature for more than 70 years, evidence accumulated only recently that many protein regions and even entire proteins might lack stable tertiary and/or secondary structure in solution yet possess crucial biological functions [1–20]. This transition occurred mostly due to the efforts of four research groups, which almost simultaneously and completely independently came to the important conclusion that naturally flexible proteins, instead of being just rare exceptions, represent a new and very broad class of proteins [1,2,4,10]. This important conclusion was reached commencing from absolutely different starting points and was based on rather different areas of expertise: bioinformatics (Dr. A. K. Dunker's group), nuclear magnetic resonance (NMR) spectroscopy (Dr. P. E. Wright's group), multiparametric protein folding/misfolding studies (Dr. V. N. Uversky's group), and protein structural characterization (Dr. P. Tompa's group). Since the publication of key studies and reviews describing this new concept, the literature on this class of proteins is virtually exploding. The recognition that protein function may not require a well-folded state has led to the development of the protein trinity model [21], according to which function may originate from three distinct states: ordered, molten globule, and random coil, and transitions between them. This model was subsequently expanded to include a fourth state (pre-molten globule) and transitions between all four states [8].

These naturally flexible proteins/regions are known by different names, including intrinsically disordered [5], natively denatured [22], natively unfolded [23], intrinsically unstructured [2], and natively disordered proteins [18]. This clearly indicates that there is not a consensus in this field regarding nomenclature, which suggests the need for an ontology of incompletely folded proteins and regions. In this chapter, all types of such proteins are called *intrinsically disordered proteins* (IDP), which means that they exist as a structural ensemble, either at the secondary or at the tertiary level. The terms *natively unfolded* or *intrinsically unstructured* will be used to indicate subclasses of IDPs either random-coil-like and premolten globular forms. Intrinsically disordered proteins or regions (IDRs) exist as a dynamic ensemble of conformations with significantly varying backbone torsion angles, without a single equilibrium structure. In contrast, ordered proteins can be characterized by a distinguished conformation, where backbone torsion angles are confined to small regions of the Ramachandran map. Globular proteins may undergo large-scale conformational changes, which occur in a cooperative manner. Such cooperativity cannot operate in IDPs, due to the lack of well-defined tertiary structure interactions.

IDPs and IDRs differ from structured globular proteins and domains with regard to many attributes, including amino acid composition, sequence complexity, hydrophobicity, charge, flexibility, and type and rate of amino acid substitutions over evolutionary time. For example, IDPs are significantly depleted in a number of so-called order-promoting residues, including bulky hydrophobic (Ile, Leu, and Val) and aromatic amino acid residues (Trp, Tyr, and Phe), which would normally form the hydrophobic core of a folded globular protein, and IDPs/IDRs also possess low content of Cys and Asn residues. On the other hand, IDPs were shown to be substantially enriched in so-called disorder-promoting amino acids: Ala, Arg, Gly, Gln, Ser, Pro, Glu, and Lys [5,24–26]. This is illustrated in Figure 12.1, which represents the relative content of each amino acid in two IDP data sets in reference to ordered proteins [18,27]. Data are extracted from the DisProt database [28] that currently assembles 520 experimentally verified disordered proteins.

Based on the above-mentioned differences between IDPs and globular proteins, numerous disorder predictors have been developed, including the family of PONDR® (Predictor of Naturally Disordered Regions) algorithms [24,29], charge-hydropathy plot (CH-plot) [4], cumulative distribution function analysis (CDF-analysis) [16], NORSp [30], GlobPlot [31,32], FoldIndex© [33], IUPred [34], and DisoPred [35–37]. Several IDP predictors have been compared in recent publications [16,26,38–43]. In fact, comparing several predictors on an individual protein of interest or on a protein data set can provide additional insight regarding the predicted disorder [44,45].

Application of various disorder predictors to different proteomes revealed that IDPs and IDRs are highly abundant in nature, and the overall amount of disorder in proteins increases from bacteria to archaea to eukaryota, with over half of the eukaryotic proteins containing long-predicted IDRs [3,16,37]. The increasing abundance of IDRs in higher organisms is likely due to a change in the cellular requirements for certain protein functions, particularly regulatory functions/cellular signaling. In support of this hypothesis, the majority of known signaling proteins and transcription factors were predicted to contain long ID regions [6,46].

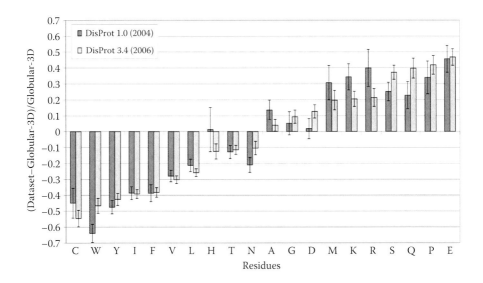

FIGURE 12.1 Amino acid composition of intrinsically disordered regions of 10 residues or longer from the DisProt database, relative to the set of globular proteins computed as (Disordered–Ordered)/(Ordered). Dark gray indicates DisProt 1.0 (152 proteins), while light gray indicates DisProt 3.4 (460 proteins). Amino acid compositions were calculated per disordered regions and then averaged. The arrangement of the amino acids is by peak height for the DisProt 3.4 release. Confidence intervals were estimated using per-protein bootstrapping with 10,000 iterations. Negative peaks correspond to the amino acids in which the disordered segments are depleted, and positive peaks indicate the amino acids in which IDRs/IDPs are enriched in comparison with the set of ordered proteins.

In this review we describe functions and molecular mechanisms of IDPs/IDRs, with specific focus on recognition, and attempt to create links with the structural properties of these proteins. These concepts serve as a basis for a novel structure/disorder–function paradigm.

FUNCTIONS OF DISORDERED PROTEINS

FUNCTIONAL CHARACTERIZATION OF IDPs

Early bioinformatics analysis revealed that IDPs/IDRs carry out pivotal biological functions [2,4,5,8–10,21,47]. They are commonly involved in signaling and regulatory pathways, via specific protein–protein, protein–nucleic acid, and protein–ligand interactions [2,4,5,7–11,13,14,17–19,21,48–52]. Sites of posttranslational modifications, such as acetylation, hydroxylation, ubiquitination, methylation, phosphorylation, and those of proteolytic attack are frequently associated with regions of intrinsic disorder [50]. The functional diversity provided by IDPs and IDRs was suggested to complement functions of ordered proteins and ordered protein regions [50–52]. Specific functions of IDPs can be grouped into four broad classes: (1) molecular recognition; (2) molecular assembly; (3) protein modification; and (4) entropic chain activities with 28 separate functions initially identified [7].

In fact, signaling and regulation are among the most important functions of IDPs [6,15,17]. Qualitatively, it seems reasonable that highly mobile proteins would provide a better basis for signaling and recognition. For example, disordered regions can bind partners with both high specificity and low affinity [53]. This means that the regulatory interactions can be specific and also can be easily dispersed. Obviously this represents a keystone of signaling—turning a signal off is as important as turning it on [5]. Another crucial property of ID proteins for their function in signaling networks is binding diversity, that is, their ability to interact with many partners [54,55]. This opens a unique opportunity for one regulatory region or one regulatory protein to bind to various targets [47,55] or the ability of several regulatory regions/proteins to bind to one protein [47]. An interesting consequence of this feature is the polymorphism of IDPs/IDRs in the bound state; they might have completely different geometries in the complex form, depending on the nature of the bound partner [55–57].

Promiscuity of IDPs, that is, their capability to interact with various partners, can also be utilized in organizing complex protein–protein interaction networks. The architecture of these networks can be described by a highly heterogeneous scale-free topology, in which proteins have widely different connectivities, with some of them having only one connection (ends), whereas others possess tens, hundreds, and possibly thousands of links (hubs) [58]. It has been shown that there are two general ways of how a hub can have multiple interactions: it either can be intrinsically disordered and serve as an anchor or it can act as a stable globular scaffold interacting with intrinsically disordered regions of its targets [17]. The general importance of intrinsic disorder in hubs has been corroborated by several studies [59–62]. The importance of disorder in the partners of structured hub proteins was also shown for the partners of 14-3-3 [55,63] and calmodulin [62].

Recently, the functional annotations in the SwissProt database were analyzed from a structured-versus-disordered point of view [50–52] using function-related keywords that are associated with at least 20 proteins. By performing predictions for the function-associated and properly chosen random sets, the disorder- and order-associated functions as well as structurally ambiguous functions were identified. Out of 710 functions (keywords), 310 were found to be order-associated, 238 disorder-associated, and 170 structurally ambiguous [50–52]. If functional keywords are grouped into 11 Gene Ontology (GO) categories, order-associated functions fall into only 7 categories, while disorder-associated functions cover essentially all functional categories [50]. This observation might imply that the functional repertoire is larger for disordered proteins compared to that for structured proteins [18].

FUNCTIONAL CLASSIFICATION OF IDPS

Functions of IDPs can also be classified by their actual molecular mechanism of action. In this respect, their functions can either stem directly from their disordered state or be related to molecular recognition, when they either transiently or permanently bind other macromolecule(s) or small ligand(s). By this criterion, IDPs were classified into five [10] and later into six [19] categories. Recently, the recognition of disorder in prion proteins suggested their addition to this scheme [64]. These classes (Table 12.1) cover all distinct modes of IDP/IDR actions described thus far [28].

Entropic Functions

Entropic chains are not involved in partner recognition. Their function directly results from disorder. By their actual mode of action, entropic chains can function as entropic springs, bristles/spacers, linkers, and clocks, and in all these subcategories they either influence the localization of attached domains or generate force against movements/structural changes [7]. For the best characterized examples, one should refer to entropic gating in the nuclear pore complex by disordered regions of nucleoporins (NUPs) [65], the entropic spacer/bristle domains of microtubule-associated

---

TABLE 12.1

Classification Scheme of IDPs Based on Their Molecular Modes of Action

Protein	Partner	Function
	Entropic Chains	
Nup2p FG repeat region	na	Gating in NPC
K channel N-terminal region	na	Timing of gate inactivation
	Display Sites	
CREB KID	PKA	Phosphorylation site
Cyclin B N-terminal domain	E3 ubiquitin ligase	Ubiquitination site
	Chaperones	
ERD 10/14	(e.g.) luciferase	Prevention of aggregation
hnRNP A1	(e.g.) DNA	Strand reannealing
	Effectors	
p27Kip1	CycA-Cdk2	Inhibition of cell cycle
Securin	Separase	Inhibition of anaphase
	Assemblers	
RNAP II CTD	mRNA maturation factors	Regulation of mRNA maturation
CREB	p300/CBP	Initiation of transcription
	Scavengers	
Casein	Calcium phosphate	Stabilization of calcium phosphate in milk
Salivary PRPs	Tannin	Neutralization of plant tannins
	Prions	
Ure2p		Utilization of urea under nitrogen
Sup35p	NusA, mRNA	Suppression of stop codon, translation read through

Note: Two examples within each category are given, specifying the binding partner (if applicable) and the actual cellular function of the protein.

---

proteins [66], and the entropic spring PEVK region of titin, which is responsible for passive tension in resting muscle [67].

Functions by Transient Binding

Display sites serve as the sites of posttranslational modifications because enzymatic modifications are facilitated in flexible and structurally adaptable regions in proteins. This feature has been demonstrated by limited proteolysis, which tends to occur in loop and linker regions in globular proteins [68]. Phosphorylation [69], ubiquitination [70], and deacetylation [71] also correlate with local disorder. Disorder prediction in proteins with short recognition elements (linear motifs [72]) corroborates this idea, and in a systematic study, linear motifs were found to preferentially reside in a locally disordered sequential environment [73].

Chaperones can also require disorder for transient binding, and IDPs were frequently identified as protein and RNA chaperones [13]. RNA chaperones have a very high proportion of disorder (40% of their residues fall into long-disordered regions), but protein chaperones are also among the most disordered proteins (15% of their residues are located within long disordered regions). To elucidate the involvement of disordered regions in the mechanism of chaperone action, an "entropy transfer" model of disordered chaperones was suggested [13].

Functions by Permanent Binding

Effectors usually modify the activity of their partner enzyme [10]. Some of the best characterized IDPs, such as p27^{Kip1} (the inhibitor of Cdks) [54,74], securin (the inhibitor of separase) [75], and calpastatin (the inhibitor of calpain) [76,77], can be found in this category. Effectors may sometimes have the potential to both inhibit and activate their partners, as suggested in the case of p27^{Kip1} [78], or the C fragment of DHPR II-III loop [79] that was captured by the concept of moonlighting [49].

Assemblers either target the activity of attached domains or assemble multiprotein complexes. Due to their open and extended structural state, IDPs may also function by permanent partner binding such as accessory proteins [10]. The evidence for this function is the high level of disorder in a significant number of scaffold proteins [80], such as BRCA1 and Ste5 [81,82], the correlation of disorder with hub function in the interactome [60–62], and the increase of average disorder with the number of partners in multiprotein complexes [83].

Scavengers store and/or neutralize small ligand molecules by permanent binding. Casein(s) in milk, for example, store calcium phosphate seeds and enable a high total calcium-phosphate concentration [84].

Prions present a novel functional category of IDPs that was not included in previous classification schemes [9,10]. Prions are thought of traditionally as pathogens, mostly because of their involvement in mad cow diseases [85]. The autocatalytic conformational conversion characteristic of prions, however, also occurs in the normal physiological functions of proteins of yeast [86] and even higher organisms [87,88]. These prion proteins have disordered Q/N-rich prion domains [64] that are capable of autocatalytic transition to an altered conformation, with potentially advantageous phenotypic consequences on the organism that harbors them.

TRANSIENT STRUCTURAL ELEMENTS IN IDPS

STRUCTURAL CHARACTERIZATION OF IDPS

A wide range of physicochemical methods suitable to characterize partly folded protein conformations was applied to elaborate the structural organization of IDPs/IDRs. The list of these techniques includes x-ray crystallography [89], NMR spectroscopy [18,90–93], near-ultraviolet (UV) circular dichroism (CD) [94], far-UV CD [4,95–97], optical rotatory dispersion (ORD) [4,95], Fourier transform infrared spectroscopy (FTIR) [4], Raman spectroscopy and Raman optical activity [98], different fluorescence techniques [99,100], numerous hydrodynamic techniques (including gel filtration, viscometry, small angle x-ray scattering [SAXS], small angle neutron scattering [SANS], sedimentation, and dynamic and static light scattering) [99,100], rate of proteolytic degradation [68,101–105], mobility in sodium dodecyl sulfate (SDS) gel electrophoresis [10,106], conformational stability [99,107–110], hydrogen deuterium (H/D) exchange [100], immunochemical methods [111,112], interaction with molecular chaperones [99], electron microscopy or atomic force microscopy [99,100,113], and the charge state analysis of electrospray ionization mass spectrometry [114]. For more detailed reviews on methods used to detect IDPs and IDRs, see References 8, 18, 91, and 100.

Based on the results obtained by a series of biophysical methods, conformational behavior and structural features of IDPs/IDRs resemble those of the nonnative states of "normal" globular proteins, which may exist in at least four different conformations: ordered, molten globule, premolten globule, and coil-like [11,20,110,115]. The main structural features of extended IDPs are [5,8,9,11,18,116]: (1) a specific amino acid sequence with low overall hydrophobicity and high net charge; (2) hydrodynamic properties typical of a random coil in poor solvent, or premolten globule-like conformation; (3) low content of ordered secondary structure; (4) the absence of a tightly packed core; (5) high conformational flexibility; (6) a "turn out" response to the environmental changes, with the structural complexity increase at extreme pH or high temperature; and (7) an ability to partially fold in the presence of specific binding partners. Such behavior is detected by numerous hydrodynamic techniques (gel filtration, viscometry, SAXS, SANS, sedimentation, dynamic and static light scattering), far-UV CD, ORD, FTIR, and NMR spectroscopy.

IDPs/IDRs can be classified into two groups: (1) collapsed disorder, where intrinsic disorder is present in a molten globule form; and (2) extended disorder, where intrinsic disorder is present in a form of random coil or premolten globule under physiological conditions *in vitro* [5,11,18]. Native molten globules, being highly compact, possess a well-developed secondary structure, whereas native coils and native premolten globules are extremely flexible, essentially noncompact (extended), and have little or no ordered secondary structure under physiological conditions. Because of the lack of the hydrophobic core and the presence of only marginal levels of residual secondary structure, native coils and native premolten globules are grouped together in a class of natively unfolded or intrinsically unstructured proteins. It has been proposed that premolten globules exhibit the behavior of squeezed macromolecular coils, as water is a poor solvent for a polypeptide chain [8,9,11,117].

Recently, it has been shown that water at ambient temperatures is a poor solvent for generic polypeptide backbones [118–120]. As a result, archetypal IDP sequences such as polyglutamine and glycine-serine block copolypeptides, despite the lack of hydrophobic residues, prefer ensembles of collapsed structures in aqueous environment [118–120]. These collapsed random coils are similar to, if not identical with, the premolten globule forms described earlier.

The idea of a systematic structural analysis of natively unfolded proteins seems to be redundant—one expects to be consistently observing the properties of an unfolded polypeptide chain. However, the situation is not so simple even for a "normal" globular protein in the presence of large concentrations of strong denaturants, such as 8 M urea or 6 M GdmCl; that is, under conditions where these proteins lose the majority of their specific structure, become essentially unfolded. In fact, it has been repeatedly shown that even under these conditions the polypeptide chains might contain some residual structure, suggesting that a polypeptide chain does not reach a random coil conformation [121–127]. Natively unfolded proteins were observed to possess a noticeable residual structure. In fact, a systematic analysis of several dozens of natively unfolded proteins revealed that, according to their structural properties and conformational behavior, they can be grouped into two structurally different categories: native coils and native premolten globules. Figure 12.2 illustrates a striking difference between these two protein classes, representing a "double wavelength" plot, $[\theta]_{222}$ versus $[\theta]_{200}$ dependence. In this plot, one group of natively unfolded proteins was characterized by far-UV CD spectra characteristic of almost completely unfolded polypeptide chains (with $[\theta]_{200} = -(18,900 \pm 2800)$ deg·cm^2·dmol^{-1} and $[\theta]_{222} = -(1700 \pm 700)$ deg·cm^2·dmol^{-1}), whereas another group possessed spectra typical of

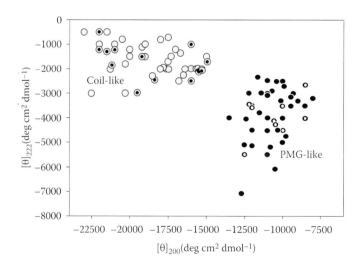

FIGURE 12.2 Analysis of far-UV CD spectra in terms of double wavelength plot, $[\theta]_{222}$ versus $[\theta]_{200}$, allows discrimination of the natively unfolded, coil-like proteins (gray circles) and premolten globule-like proteins (black circles). Premolten globules and coils with experimentally measured hydrodynamic parameters available are marked by white-dotted and black-dotted symbols, respectively.

the premolten globule state of globular proteins, being consistent with the existence of some residual secondary structure (with $[\theta]_{200} = -(10,700 \pm 1300)$ deg·cm^2·dmol^{-1} and $[\theta]_{222} = -(3900 \pm 1100)$ deg·cm^2·dmol^{-1}) [8,9,11,116].

Detailed NMR analysis of several IDPs has provided deeper understanding of their fine structural details. In general, because the amino acid residues within coil-like and premolten globule-like IDPs share a similar chemical environment, being all prevalently exposed to the solvent, they share similar NMR frequencies, thus resulting in a strong overlap of the ^{1}H resonances. Therefore, while bidimensional NMR spectra of folded proteins display a broad distribution of NMR frequencies, NMR spectra of IDPs typically possess a very low spread of the resonance frequencies of amide protons [90,92,93]. However, the recent advances in the multidimensional NMR based on the uniformly ^{15}N- and ^{13}C-labeled proteins have enabled studies of highly disordered proteins. This is because the chemical shifts of the ^{15}N and ^{13}C resonances are well dispersed even for completely unfolded polypeptide chains, allowing unambiguous resonance assignments [90,92,93]. Based on the assigned backbone resonances, various NMR parameters can be used to characterize residual structure in unfolded and partly folded states. For example, deviations of the chemical shifts from random coil values, especially for ^{13}C-, ^{15}N-, and ^{1}H-, can be used to calculate the relative population of dihedral angles in the α or β regions [90,92,93]. The chemical shifts of the C- nuclei from those expected for a fully random coil ensemble of conformations are known as secondary chemical shifts. Residues with positive C- secondary chemical shifts have preferences for helical regions of the ϕ–ψ space, whereas residues that populate more extended conformations show the opposite pattern. Additionally, abnormally low (relative to random coil values) amide proton temperature coefficients and low hydrogen exchange rates indicate the existence of residual structures within IDPs [90,92,93].

For example, a high-resolution NMR analysis of α-synuclein revealed that although it is largely unfolded in a solution in agreement with other techniques, such as far-UV CD, SAXS, FTIR, gel filtration, and dynamic light scattering, it exhibits a region between residues 6 and 37 with a preference for helical conformation [128]. Interestingly, the helical propensity found in this N-terminal region was strongly attenuated by an A30P mutation associated with the early onset of Parkinson's disease, whereas another familial mutation, A53T, was shown to leave this region unperturbed, exerting a more modest and local influence on structural propensity [129]. Furthermore, it has been shown that the residual structure in another synuclein family member, β-synuclein (which lacks residues 62–72), differed significantly from that of α-synuclein. In fact, β-synuclein possessed a lower predisposition toward helical structure in the second half of its N-terminal domain, and a higher preference for extended structures in its C-terminal tail [130]. Importantly, as it will be discussed in subsequent sections, some regions with residual structure might be directly related to IDP functionality.

TRANSIENT STRUCTURAL ELEMENTS AND RECOGNITION BY IDPS

Biophysical characterization of IDPs/IDRs mentioned in the previous section made it evident that although these proteins/segments do not possess a single, well-folded state, they often exhibit transient, residual organization either at the secondary or tertiary structure level. In the following sections, these transient elements will be classified and their connection to binding functions of IDPs will be described.

Preformed Structural Elements

IDPs often undergo disorder-to-order transition upon binding to their partners, which raises the question whether the structure adopted in the bound form is enforced by the partner molecule or reflects inherent conformational preferences of IDPs. In other words, the binding-coupled folding of IDPs may be induced by the template or, alternatively, a conformational selection scenario may be occurring. To answer this question, the structures of 26 IDPs complexed with their globular partners were analyzed for the predictability of their secondary structural elements [131]. Three algorithms, GOR, ALB, and PROF, were used, which all suggested that the accuracy of predicting secondary structural elements in IDPs is higher than that in either their partner proteins or randomized sequences (Figure 12.3). This observation suggests that IDPs may have rather strong conformational preferences for their bound conformations, that is, they probably use elements for recognition that are partially/transiently preformed in the solution state (Figure 12.4). This preference is strongest for helices and is weakest for extended structures. Although the insight from studying complexes under steady-state conditions is not fully conclusive on the role of these preformed

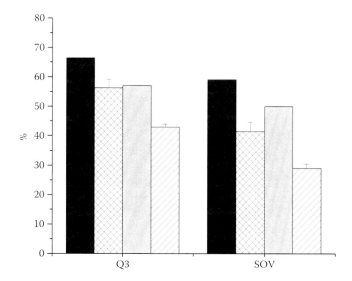

FIGURE 12.3 Predictability of the secondary structure of IDPs in the bound state. A selection of 26 IDP structures in complex with their partners was analyzed for the predictability of the secondary structure attained in the complex from sequence. Prediction accuracies calculated by the ALB method for full-length IDPs (black columns), randomized sequences of IDPs (cross-hatched columns), sequences of globular partners (gray columns), and randomized sequences of partners (hatched columns) are displayed by two measures, Q3 (position-based accuracy) and SOV (segmental overlap). These show that the intrinsic structural preferences of IDPs are strongly correlated with their conformation attained in the bound form.

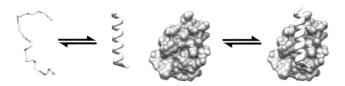

FIGURE 12.4 Illustration of disorder-to-order transition upon binding. This example shows the binding of a disordered region of Bad (ribbon) to Bcl-X$_L$ (globular structure) (PDB: 2bzw).

elements in binding, in certain cases a similar structure in the free and bound states was actually observed when IDPs have been characterized in the solution state by NMR. Such correlation is apparent in the case of the kinase inducible domain (KID) of cAMP response element binding (CREB) [132,133], p21Cip1/p27^{Kip2} [54,74], p53 [134], FlgM [135,136], PKI alpha [137], thymosin beta4 [138], Bad (PDB: 2bzw), and measles virus nucleoprotein [139]. Such preformed elements might also serve as initial contact points of interaction, but this issue requires further study. A very similar concept (intrinsically folded structural units, IFSUs) has also been suggested, based on studying the molecular function and binding of p27^{Kip1} [140].

Molecular Recognition Elements/Features

Intrinsic conformational preferences of IDPs/IDRs can be utilized to predict the sites for the disorder-to-order transitions [141]. In fact, it had been noticed long ago that PONDR VL-XT sometimes gives short regions of predicted order bounded by regions of predicted disorder [142]. This is illustrated in Figure 12.5, which represents a structure of the complex between the 4E binding protein 1 (4EBP1) and the eukaryotic translation initiation factor 4E [143] (Figure 12.5A) and the PONDR

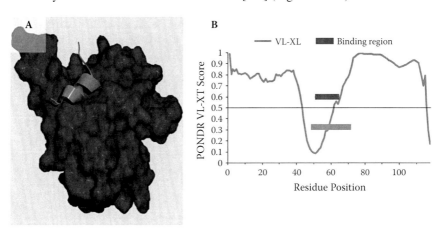

FIGURE 12.5   (SEE COLOR INSERT FOLLOWING PAGE 174.) Example of binding regions and their positions relative to the regions of predicted order (PONDR® VL-XT score) and α-MoRF. (A) Eukaryotic initiation factor (blue) and the binding region of 4EBP1 (red). (B) The PONDR® VL-XT prediction for 4EBP1 with the binding region (blue bar) and the predicted α-MoRF region (pink bar) shown.

VL-XT prediction for 4EBP1 (Figure 12.5B). In nonbound form, 4EBP1 was shown to be completely disordered by NMR [144], but its short stretch underwent a disorder-to-order transition upon interaction with the binding partner [143]. Figure 12.5B shows that there is a sharp dip in the PONDR VL-XT plot in the area of the binding region. This drop is flanked by long regions of predicted disorder.

Additional work has validated the use of these distinctive downward spikes in PONDR VL-XT curves to locate functional binding regions [141] within IDRs. These elements consist of a short region that undergoes coupled binding and folding within a longer region of disorder. We call these features molecular recognition elements (MoREs). Later, these regions were renamed as molecular recognition features (MoRFs) to emphasize their unique polymorphic nature. An algorithm has been elaborated [141] for identification of regions having a propensity for α-helix-forming molecular recognition features (α-MoRFs) based on a discriminant function that indicates such regions while giving a low false-positive error rate on a collection of structured proteins. Application of this predictor to databases of genomics and functionally annotated proteins indicates that α-MoRFs are likely to play important roles in protein–protein interactions involved in signaling events [141].

This first α-MoRF identifier was developed using a training data set of a limited size (a set of 13 proteins containing 15 potential α-MoRFs). Recently, the prediction algorithm was improved by (1) including additional α-MoRF examples and their cross species homologues in the positive training set; (2) carefully extracting monomer structure chains from the Protein Data Bank (PDB) as the negative training set; (3) including attributes from recently developed disorder predictors, secondary structure predictions, and amino acid indices; and (4) constructing neural network based predictors and performing validation [145]. The sensitivity, specificity, and accuracy of the resulting predictor, α-MoRF-PredII, were 0.87 ± 0.10, 0.87 ± 0.11, and 0.87 ± 0.08, respectively, over 10 cross-validations [145].

Systematic studies of PDB entries revealed that protein complexes deposited in this database often comprise a short protein segment bound to a larger globular protein [146]. Analysis of literature data showed that some of these short peptides, being specifically folded into α-helix, β-hairpin, β-strand, polyproline II helix, or irregular structure, and so forth, within their complexes with globular partners, are intrinsically disordered prior to the corresponding complex formation. Thus, all these regions can be considered as illustrative members of the subset of protein–protein interactions involving disorder-to-order transitions during the complex formation and, hence, can be considered as MoRFs. In a recent study, short polypeptide chains with lengths between 10 and 70 residues that are bound to a globular partner (with chains ≥100 residues) were extracted from PDB [146]. This process resulted in a data set comprising 372 nonredundant protein chains (9093 residues). The secondary structure assignment showed that 27% of this data set consisted of α-helical residues, 12% were β-sheet residues, and approximately 48% of the residues had an irregular conformation. The remaining 13% of the residues were found to be disordered as they were characterized by missing coordinate information in their respective PDB files [146]. This data set can be used for further studies to find sequence attributes for discriminating these different MoRFs from one another and from ordered proteins.

Primary Contact Sites

Primary contact sites (PCSs) are short recognition motifs that are kinetically distinguished in binding. PCSs were derived from the observation that the large-scale structural reorganization concomitant to binding of IDPs is usually realized very rapidly. In fact, structural disorder is thought to confer the advantage of rapid binding [147]. Hence, it was reasoned that certain regions within the disordered ensemble are more exposed than others, and thereby may serve as the first sites of contact with the partner. This idea was experimentally checked in the case of two IDPs: calpastatin and MAP2 [148].

It was suggested that limited proteolysis at extremely low concentrations of proteases may preferentially affect regions of an IDP that are exposed relative to others. At very low concentrations, proteases of narrow (trypsin, chymotrypsin, and plasmin) or broad (subtilisin and proteinase K) substrate specificity preferentially cleave both proteins in regions thought to make the first contact with the partner. In calpastatin, subdomains A, B, and C, and in MAP2c, the central Pro-rich region (PRR), were identified. This not fully random structural behavior was further probed by CD spectroscopy and NMR relaxation spectroscopy. In the case of calpastatin, the CD spectra and hydration of the two halves are not additive, which suggests long-range tertiary interactions within the protein. In MAP2c, no such tertiary interactions could be identified, but exposure of the PCSs could be accounted for by local structural constraints. Urea- and temperature-dependence of the CD spectrum of its central PRR pointed to the presence of PP-II helix conformation in this region, which is rather stretched out and keeps the interaction site exposed.

Some additional observations in the literature are also in line with the concept of PCSs. For example, rapid assembly of large membrane-bound complexes of highly repetitive and disordered membrane-associated proteins, such as AP180, epsin1, and auxilin in exocytosis [149,150], is essential for proper execution of the membrane fusion. It was suggested that a large capture radius of specific, exposed recognition elements enabled by the disordered nature of these proteins provides the key ingredient of this mechanism [151]. A structural study on p53, the tumor-suppressor transcription factor, also provides relevant information on the concept of PCSs. p53 has a long, disordered N-terminal transactivator domain (TAD), which has two binding sites, one for the E3 ubiquitin ligase MDM2, and the other for the 70 kDa subunit of replication protein A (RPA70). Paramagnetic relaxation enhancement (PRE) experiments identified distance constraints in TAD [152], and suggested that TAD is rather compact and dynamic, with the two binding motifs separated by an average distance of 10–15 Å. Prior to binding, a more extended conformation of the ensemble must be populated to expose binding for sites for either MDM2 or RPA70, in agreement with the PCS concept [152].

Short Linear Motifs

Analyses of sequences that were observed to mediate specific protein–protein interactions, whether they result in a stable complex or enzymatic modification, suggested that the element of recognition is often a short motif of discernible conservation, also denoted as a "consensus" sequence. The generality of this relation has led to the concept of linear motifs (LMs; also denoted as eukaryotic linear motifs [ELMs] and short linear motifs [SLiMs]). Such elements were first implicated in recognition by kinases or binding sites of SH3 domains [153]. Linear motifs are usually defined by a short consensus pattern, with conserved residues that are interspersed with rather freely exchangeable, variable positions. The first set of residues serves as specificity determinants, whereas the second likely acts as spacers. Due to their evolutionary variability, LMs may constitute dynamic switchlike elements, frequently generated and erased in evolution. The typical length of LMs is between 5 and 25 residues, and their specificity is determined by a few conserved residues, while the embedding sequence environment is hardly constrained. Due to the resulting limited information content, LMs are much more difficult to identify by sequence comparisons than domains. Traditional BLAST searches cannot positively identify LMs, but special algorithms that focus on nonglobular regions and that combine large-scale interaction data can tackle this problem (DILIMOT [154] and SLiMDisc [155]).

LMs have been collected in the ELM database available through the ELM server [72], which contains about 800 ELM examples that belong to more than 100 ELM classes. LMs were suggested to fall into locally disordered regions [32,72], which was corroborated in a systematic analysis by bioinformatics predictors [73]. It was found that disordered 20-residue-long flanking segments contribute to the plasticity of LMs (Figure 12.6A). LMs also have a peculiar amino acid composition, in that they resemble the characteristic composition of IDPs (Figure 12.6B), but they are enriched in both certain hydrophobic (Trp, Leu, Phe, and Tyr) and charged (Arg and Asp) residues. Furthermore, LMs are depleted in Gly and Ala, perhaps to limit the tendencies for both flexibility and secondary structure. In addition, Pro dominates in both LMs and their flanking segments, probably due to its direct involvement in

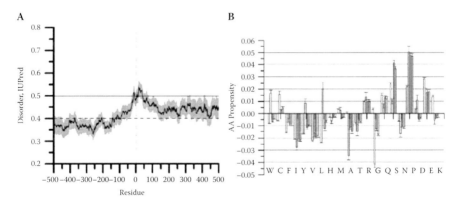

FIGURE 12.6 Disorder profiles and amino acid compositions of linear motifs (LMs, also denoted in the literature as ELMs and SLiMs). (A) Disorder profiles by the IUPred algorithm were computed and averaged. A thin horizontal line at 0.5 shows the threshold of disorder, whereas a dashed line at 0.4 shows the average score for experimentally verified disordered proteins in DisProt. Standard error of mean (SEM) values are displayed in light gray. (B) Amino acid propensities of IDPs of the DisProt database (white), LMs (light gray), 20-residue-long LM flanking segments (dark gray), and LMs plus 20-residue-long flanking segments (medium gray) are shown in reference to the composition of globular proteins.

interactions and also due to providing the extended secondary structural motif, PP-II helix.

We also noted that the amino acid composition of the specificity determinant (restricted) and variable (nonrestricted) sites markedly differ, which explains the above-noted peculiarity of amino acid propensities. Conserved positions are occupied by either hydrophobic and rigid, or charged and flexible residues, whereas nonrestricted positions abound in flexible residues, similarly to IDPs in general. The only exception is Pro, which prefers both restricted positions and flanking regions, indicating its dual role as a contact residue and promoter of an open structure. In all, the unique amino acid composition suggests a mixed nature of LMs, with a few specificity-determinant residues strongly favoring order, grafted on a completely disordered carrier sequence flanking and intervening the region critical for interaction.

INTERACTION OF IDPS WITH THEIR PARTNERS

INDUCED FOLDING ACCOMPANYING BINDING

Alterations in environmental or cellular conditions, or interactions with other proteins, nucleic acids, membranes, or small molecules induce function-associated conformational changes or disorder-to-order transitions in IDPs. One of the most important IDP features is their unique capability to fold under a variety of conditions [2,4–10,14,15,17,20,156], including functional folding triggered by the interaction with specific binding partners. This process is schematically illustrated by Figure 12.4 showing how the disordered region folds into an α-helical segment upon interaction with the binding partner. Such behavior was described for several individual IDPs involved in protein–protein interactions, including p27[Kip1] [74,157] and p53 [134], the conformation of which in the free state presage their structure in the folded, bound state. Some benefits of IDP binding are actually due to the binding-coupled folding process as the large decrease in conformational entropy accompanying the disorder-to-order transition decreases binding affinity and may uncouple it from specificity. Hence, highly specific interactions become easily reversible that can be exploited in signaling and regulation. We also have to note, however, that IDPs can remain substantially disordered [57] in the bound state.

A well-documented case of the astonishing conformational plasticity of IDPs was described for the highly abundant presynaptic brain protein, α-synuclein. In fact this protein is known to adopt a series of absolutely different conformations depending on its environment. For example, this protein may either stay substantially unfolded, or adopt a partially folded conformation, or fold into α-helix or β-structure species, both in monomeric and oligomeric forms. Furthermore, it might form several morphologically different types of aggregates, including oligomers (spheres or doughnuts), amorphous aggregates, and amyloid-like fibrils depending on the protein environment. In other words, α-synuclein has an exceptional ability to fold in a template-dependent manner [158–160].

When IDPs fold, their folding energy landscapes are markedly different from the funnel-like surface of globular proteins [161–163]. In the case of IDPs, out of a multiplicity of energy minima that coexist in the free state, different ones can be selected, depending on the partner. Furthermore, formation of a well-defined structure is initiated by tertiary structure interactions between an IDP and its partner; that is, the folding funnel is constructed by two terms that include the intermolecular interaction energies in addition to the intramolecular conformation energies.

Due to the noncooperative nature of IDP folding, their interaction interfaces are also different from those of complexes of globular proteins. An elegant approach to predict whether a given complex was formed from the association of ordered proteins or by the binding coupled folding of IDPs has been elaborated by Nussinov and collaborators [164]. According to this approach, a plot of the normalized monomer area (NMA) versus the normalized interface area (NIA) nicely separates complexes formed from structured proteins as compared to complexes formed from unfolded proteins by coupled binding and folding. That is, associations of structured proteins exhibit small NMAs and NIAs, and so lie near the origin of the NMA–NIA plot. On the other hand, complexes formed by coupled binding and folding have much larger NMAs and NIAs, and so are spread out and lie far from the origin of the NMA-versus-NIA plot. As a result, a linear boundary can be calculated to separate the two groups [164]. If a double NMA–NIA plot is constructed, one for each partner of a complex, the interacting pairs can be divided into three groups: (1) both partners are structured; (2) one partner is structured and the second partner is disordered; and (3) both partners are intrinsically disordered [55]. This approach was recently utilized to analyze structures of 13 complexes between various regions of p53 and its unique binding partners [55] (Figure 12.7). This analysis clearly showed the importance of disorder-to-order transitions for many of the structurally characterized interactions involving the p53 protein. For 10 of these partners, the interactions are mediated

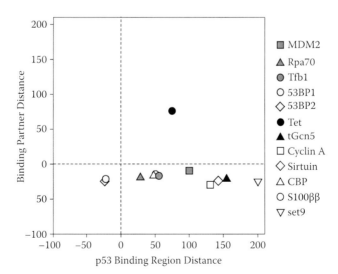

FIGURE 12.7 Double normalized monomer area (NMA) versus normalized interface area (NIA) plot for p53 complexes. The distance measured discriminates between complexes formed as a result of a disorder-to-order transition upon binding (positive values) and complexes formed by partners that were ordered prior to complex formation (negative values).

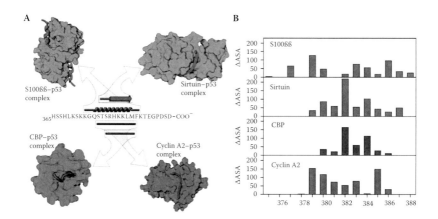

FIGURE 12.8 (SEE COLOR INSERT FOLLOWING PAGE 174.) Sequence and structure comparison for the four overlapping complexes in the C-terminus of p53. (A) Primary, secondary, and quaternary structure of p53 complexes. (B) The ΔASA for rigid association between the components of complexes for each residue in the relevant sequence region of p53.

by regions experimentally characterized as intrinsically disordered, where PONDR VL-XT detects the majority of these binding regions as short predictions of order within a longer prediction of disorder. These structures are complexes between p53 and cyclin A [165], sirtuin [166], CBP [167], S100ββ [168], set9 [169], tGcn5 [170], Rpa70 [171], Mdm2 [172], Tfb1 [173], and itself [174]. The remaining three interactions are mediated by the structured DBD, namely, between p53 and the following partners: DNA [175], 53BP1 [176], and 53BP2 [177].

Four complexes involve a single region of the p53 sequence bound to four different partners. This represents a unique feature of IDPs linked to multiple specificities. The mentioned p53 region is the 374–388 fragment, which was shown to interact with cyclin A [165], sirtuin [166], CBP [167], and S100ββ [168]. Interestingly, the four complexes displayed all three major secondary structure types: a region became a helix upon binding to S100ββ, a sheet upon binding to sirtuin, and a coil with two distinct backbone trajectories upon binding to CBP and cyclin A2 [55] (Figure 12.8). Next, the involvement of various p53 residues in the mentioned interactions were analyzed to test the hypothesis that p53 utilized different residues for the interactions with different partners [55]. To this end, the buried surface area for each residue in each interaction was quantified by calculating the ΔASA. Figure 12.8B shows that different amino acid interaction profiles are seen for each of the interactions, suggesting that the same residues are used to different extents in the four interfaces.

Molecular Architecture of the Interfaces of IDPs

The chemical nature of the interface of IDP complexes is unique in amino acid composition and geometry, as shown by a few analyses of IDP interaction sites. Gunasekaran and colleagues analyzed the interfaces of 10 two-state complexes, 44 ribosomal proteins, and 5 complexes of bona fide disordered proteins [164]. This study has been extended by Meszaros and colleagues, who have assembled a data set of 39 complexes of experimentally verified IDPs, and analyzed a variety of their features [178]. In a related study, Vacic and colleagues have presented a detailed analysis of 258 MoRFs, shown to correlate with disorder [27]. Although definitions of the subjects of the studies differed, the underlying features of interfaces were found to be very similar.

The comparison of the global geometrical features of the interfaces of disordered and globular proteins suggested that the distributions of size are not much different, with the area of IDP interfaces being slightly smaller but falling into about the same range as those of ordered complexes (1141 ± 110 Å²) [27], maybe with some bias against very large interfaces (>3000 Å²) [27,164,178]. Characteristic differences were seen when surfaces and interfaces were normalized to chain length; however, IDPs have a much larger per-residue surface area and a much larger per-residue interface area than globular proteins (Figure 12.9A). When their ratio is calculated, it is clear that IDPs not only have relatively larger surfaces, they also use its larger portion for interaction with their partner, sometimes 50% of the whole, as opposed to only 5%–15% for most ordered proteins [164,178].

The number of continuous segments the interface is made up of also differs significantly. In the case of globular proteins, distinct segments of the chain come into proximity to form a binding site, and thus their binding surfaces are more fragmented.

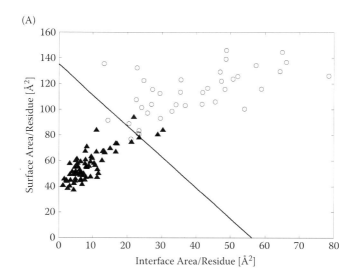

FIGURE 12.9 Surface and interface area, and segmentation of interfaces of IDPs. (A) IDPs use a large fraction of their surfaces for binding. The total surface area per residue is given as a function of the interface area per residue for the smaller chain of ordered complexes (triangles), and for disordered proteins in complex with an ordered protein (circles). (B) IDPs (dark gray) tend to use less segments to make up the binding site than globular proteins (light gray). The distribution of interfaces with the given number of noncontinuous sequence segments is shown. A database of 72 ordered complexes and 37 complexes of disordered proteins has been used in the study.

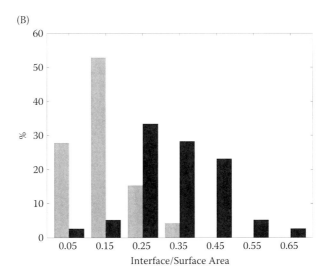

(B)

FIGURE 12.9   (Continued)

In fact, ordered proteins hardly ever use a single segment for binding, and their segmentation number occasionally even exceeds the value of 10 [178]. In contrast, in about 60% of cases, the IDP interface corresponds to a single continuous segment only (Figure 12.9B), and they contain more than three separate segments in a single case of a Sigma28-FlgM complex (PDB: 1sc5) only.

The ratio of buried/exposed area of IDPs is much smaller on average both for polar and hydrophobic residues, which suggests that they keep their hydrophobic residues exposed for contact with the partner, instead of generating a hydrophobic core [178]. Actually, IDPs have a larger fraction of their hydrophobic residues exposed than that of globular proteins (IDPs: 40%–90%, ordered proteins: 15%–50%) [178]. The interfaces of IDPs are not only more hydrophobic than their surfaces in general, but they are also more hydrophobic than the buried regions of the protein [27,164,178]. Exposure of hydrophobic amino acids and/or this compositional bias also follows from the analysis of ELMs [73], two-state complexes [164], and MoRFs [27].

IDP interfaces are also unique with respect to the types of contacts they usually make: They rely much more on hydrophobic–hydrophobic contacts (IDPs: 33%, ordered proteins: 22%), whereas ordered proteins utilize polar–polar contacts significantly more often (IDPS: 27%, ordered proteins: 33%) [164,178]. This difference might arise because IDPs require more enthalpic stabilization to counteract their decrease in configurational entropy. IDP interfaces are tighter, that is, structurally more complementary, which may be related from their binding by induced folding upon binding, which enables a better adaptation to the structure of the partner.

A critical point of their interactions is whether the foregoing differences manifest themselves in differences of the interaction energies of the two types of complexes. Whereas no comprehensive experimental analysis has been carried out to this end, we have shown that the pairwise interresidue interaction energy of IDPs can be estimated by low-resolution force fields deduced from globular structures. The ensuing algorithm, IUPred [34], can distinguish disordered from ordered proteins by showing characteristically less potential interresidue interaction energy for IDPs. By applying the same approach for analyzing the interfaces [178], it was found that ordered proteins tend to realize more stabilizing interactions within their polypeptide chains, whereas IDPs derive more stabilization from the interaction with the partner. This tilts their structural balance toward the folded state in the presence of the partner and explains why IDPs do not fold in isolation.

CONCLUDING REMARKS

A wealth of experimental data demonstrates that essential biological, mostly regulatory functions, such as signal transduction or transcription regulation are intertwined with the presence of IDPs/IDRs. Recognition by IDPs offers various advantages for biomolecular associations, such as fast kinetics and specificity without excessive binding strength. All these properties are encoded in the lack of a single, distinguished equilibrium structure as well as the noncooperative mode of conformational rearrangements of IDPs. Based on biophysical evidence, IDPs span a wide range of structural categories from random coils to molten globules. Various links between the residual structures of IDPs and their functions have been established. These recognition motifs can either provide thermodynamic benefits by reducing the entropic penalty of binding or can be kinetically advantageous by accelerating the formation of productive contacts via transiently exposed specific interaction sites. Current efforts aim to elaborate details of the binding mechanisms and also to predict putative binding sites from the primary sequence. Although much has to be learned on the nonconventional modes of action of IDPs, the basis of a new disorder–function paradigm has already been laid down.

REFERENCES

1. Romero, P., Z. Obradovic, C. R. Kissinger, J. E. Villafranca, E. Garner, S. Guilliot, and A. K. Dunker, Thousands of proteins likely to have long disordered regions. *Pac Symp Biocomput*, 1998. 437–48.
2. Wright, P. E. and H. J. Dyson, Intrinsically unstructured proteins: Re-assessing the protein structure-function paradigm. *J Mol Biol*, 1999. **293**(2): 321–31.
3. Dunker, A. K., Z. Obradovic, P. Romero, E. C. Garner, and C. J. Brown, Intrinsic protein disorder in complete genomes. *Genome Inform Ser Workshop Genome Inform*, 2000. **11**: 161–71.
4. Uversky, V. N., J. R. Gillespie, and A. L. Fink, Why are "natively unfolded" proteins unstructured under physiologic conditions? *Proteins*, 2000. **41**(3): 415–27.
5. Dunker, A. K., J. D. Lawson, C. J. Brown, R. M. Williams, P. Romero, J. S. Oh, C. J. Oldfield, A. M. Campen, C. M. Ratliff, K. W. Hipps, J. Ausio, M. S. Nissen, R. Reeves, C. Kang, C. R. Kissinger, R. W. Bailey, M. D. Griswold, W. Chiu, E. C. Garner, and Z. Obradovic, Intrinsically disordered protein. *J Mol Graph Model*, 2001. **19**(1): 26–59.
6. Iakoucheva, L. M., C. J. Brown, J. D. Lawson, Z. Obradovic, and A. K. Dunker, Intrinsic disorder in cell-signaling and cancer-associated proteins. *J Mol Biol*, 2002. **323**(3): 573–84.

7. Dunker, A .K., C. J. Brown, J. D. Lawson, L. M. Iakoucheva, and Z. Obradovic, Intrinsic disorder and protein function. *Biochemistry*, 2002. **41**(21): 6573–82.

8. Uversky, V. N., Natively unfolded proteins: A point where biology waits for physics. *Protein Sci*, 2002. **11**(4): 739–56.

9. Uversky, V. N., What does it mean to be natively unfolded? *Eur J Biochem*, 2002. **269**(1): 2–12.

10. Tompa, P., Intrinsically unstructured proteins. *Trends Biochem Sci*, 2002. **27**(10): 527–33.

11. Uversky, V. N., Protein folding revisited. A polypeptide chain at the folding-misfolding-nonfolding cross-roads: Which way to go? *Cell Mol Life Sci*, 2003. **60**(9): 1852–71.

12. Tompa, P., Intrinsically unstructured proteins evolve by repeat expansion. *Bioessays*, 2003. **25**(9): 847–55.

13. Tompa, P. and P. Csermely, The role of structural disorder in the function of RNA and protein chaperones. *FASEB J*, 2004. **18**(11): 1169–75.

14. Dyson, H. J. and P. E. Wright, Intrinsically unstructured proteins and their functions. *Nat Rev Mol Cell Biol*, 2005. **6**(3): 197–208.

15. Uversky, V. N., C. J. Oldfield, and A. K. Dunker, Showing your ID: Intrinsic disorder as an ID for recognition, regulation and cell signaling. *J Mol Recognit*, 2005. **18**(5): 343–84.

16. Oldfield, C. J., Y. Cheng, M. S. Cortese, C. J. Brown, V. N. Uversky, and A. K. Dunker, Comparing and combining predictors of mostly disordered proteins. *Biochemistry*, 2005. **44**(6): 1989–2000.

17. Dunker, A. K., M. S. Cortese, P. Romero, L. M. Iakoucheva, and V. N. Uversky, Flexible nets: The roles of intrinsic disorder in protein interaction networks. *FEBS J*, 2005. **272**(20): 5129–48.

18. Daughdrill, G. W., G. J. Pielak, V. N. Uversky, M.S. Cortese, and A.K. Dunker, Natively disordered proteins, in *Handbook of Protein Folding*, J. Buchner and T. Kiefhaber, Editors. 2005, Weinheim, Germany: Wiley-VCH, Verlag GmbH & Co., pp. 271–353.

19. Tompa, P., The interplay between structure and function in intrinsically unstructured proteins. *FEBS Lett*, 2005. **579**(15): 3346–54.

20. Fink, A. L., Natively unfolded proteins. *Curr Opin Struct Biol*, 2005. **15**(1): 35–41.

21. Dunker, A. K. and Z. Obradovic, The protein trinity—linking function and disorder. *Nat Biotechnol*, 2001. **19**(9): 805–6.

22. Schweers, O., E. Schonbrunn-Hanebeck, A. Marx, and E. Mandelkow, Structural studies of tau protein and Alzheimer paired helical filaments show no evidence for beta-structure. *J Biol Chem*, 1994. **269**(39): 24290–7.

23. Weinreb, P. H., W. Zhen, A. W. Poon, K. A. Conway, and P. T. Lansbury, Jr., NACP, a protein implicated in Alzheimer's disease and learning, is natively unfolded. *Biochemistry*, 1996. **35**(43): 13709–15.

24. Romero, P., Z. Obradovic, X. Li, E. C. Garner, C. J. Brown, and A. K. Dunker, Sequence complexity of disordered protein. *Proteins*, 2001. **42**(1): 38–48.

25. Williams, R. M., Z. Obradovi, V. Mathura, W. Braun, E. C. Garner, J. Young, S. Takayama, C. J. Brown, and A. K. Dunker, The protein non-folding problem: Amino acid determinants of intrinsic order and disorder. *Pac Symp Biocomput*, 2001: 89–100.

26. Radivojac, P., L. M. Iakoucheva, C. J. Oldfield, Z. Obradovic, V. N. Uversky, and A. K. Dunker, Intrinsic disorder and functional proteomics. *Biophys J*, 2007. **92**(5): 1439–56.

27. Vacic, V., V. N. Uversky, A. K. Dunker, and S. Lonardi, Composition Profiler: A tool for discovery and visualization of amino acid composition differences. *BMC Bioinformatics*, 2007. **8**: 211.

28. Sickmeier, M., J. A. Hamilton, T. LeGall, V. Vacic, M. S. Cortese, A. Tantos, B. Szabo, P. Tompa, J. Chen, V. N. Uversky, Z. Obradovic, and A. K. Dunker, DisProt: The Database of Disordered Proteins. *Nucleic Acids Res*, 2007. **35**(Database issue): D786–93.

29. Li, X., P. Romero, M. Rani, A. K. Dunker, and Z. Obradovic, Predicting protein disorder for N-, C-, and internal regions. *Genome Inform Ser Workshop Genome Inform*, 1999. **10**: 30–40.

30. Liu, J. and B. Rost, NORSp: Predictions of long regions without regular secondary structure. *Nucleic Acids Res*, 2003. **31**(13): 3833–5.

31. Linding, R., L. J. Jensen, F. Diella, P. Bork, T. J. Gibson, and R. B. Russell, Protein disorder prediction: Implications for structural proteomics. *Structure (Camb)*, 2003. **11**(11): 1453–9.

32. Linding, R., R. B. Russell, V. Neduva, and T. J. Gibson, GlobPlot: Exploring protein sequences for globularity and disorder. *Nucleic Acids Res*, 2003. **31**(13): 3701–8.

33. Prilusky, J., C. E. Felder, T. Zeev-Ben-Mordehai, E. H. Rydberg, O. Man, J. S. Beckmann, I. Silman, and J. L. Sussman, FoldIndex: A simple tool to predict whether a given protein sequence is intrinsically unfolded. *Bioinformatics*, 2005. **21**(16): 3435–8.

34. Dosztanyi, Z., V. Csizmok, P. Tompa, and I. Simon, IUPred: Web server for the prediction of intrinsically unstructured regions of proteins based on estimated energy content. *Bioinformatics*, 2005. **21**(16): 3433–4.

35. Jones, D. T. and J. J. Ward, Prediction of disordered regions in proteins from position specific score matrices. *Proteins*, 2003. **53**(Suppl 6): 573–8.

36. Ward, J. J., L. J. McGuffin, K. Bryson, B. F. Buxton, and D. T. Jones, The DISOPRED server for the prediction of protein disorder. *Bioinformatics*, 2004. **20**(13): 2138–9.

37. Ward, J. J., J. S. Sodhi, L. J. McGuffin, B. F. Buxton, and D. T. Jones, Prediction and functional analysis of native disorder in proteins from the three kingdoms of life. *J Mol Biol*, 2004. **337**(3): 635–45.

38. Melamud, E. and J. Moult, Evaluation of disorder predictions in CASP5. *Proteins*, 2003. **53**(Suppl 6): 561–5.

39. Jin, Y. and R. L. Dunbrack, Jr., Assessment of disorder predictions in CASP6. *Proteins*, 2005. **61**(Suppl 7): 167–75.

40. Peng, K., S. Vucetic, P. Radivojac, C. J. Brown, A. K. Dunker, and Z. Obradovic, Optimizing long intrinsic disorder predictors with protein evolutionary information. *J Bioinform Comput Biol*, 2005. **3**(1): 35–60.

41. Peng, K., P. Radivojac, S. Vucetic, A. K. Dunker, and Z. Obradovic, Length-dependent prediction of protein intrinsic disorder. *BMC Bioinformatics*, 2006. **7**: 208.

42. Ferron, F., S. Longhi, B. Canard, and D. Karlin, A practical overview of protein disorder prediction methods. *Proteins*, 2006. **65**(1): 1–14.

43. Bordoli, L., F. Kiefer, and T. Schwede, Assessment of disorder predictions in CASP7. *Proteins*, 2007. **69**(Suppl 8): 129–36.

44. Uversky, V. N., A. Roman, C. J. Oldfield, and A. K. Dunker, Protein intrinsic disorder and human papillomaviruses: Increased amount of disorder in E6 and E7 oncoproteins from high risk HPVs. *J Proteome Res*, 2006. **5**(8): 1829–42.

45. Mohan, A., W. J. Sullivan, Jr., P. Radivojac, A. K. Dunker, and V. N. Uversky, Intrinsic disorder in pathogenic and non-pathogenic microbes: Discovering and analyzing the unfoldomes of early-branching eukaryotes. *Mol Biosyst*, 2008. **4**(4): 328–40.

46. Liu, J., N. B. Perumal, C. J. Oldfield, E. W. Su, V. N. Uversky, and A. K. Dunker, Intrinsic disorder in transcription factors. *Biochemistry*, 2006. **45**(22): 6873–88.

47. Dunker, A. K., E. Garner, S. Guilliot, P. Romero, K. Albrecht, J. Hart, Z. Obradovic, C. Kissinger, and J. E. Villafranca, Protein disorder and the evolution of molecular recognition: Theory, predictions and observations. *Pac Symp Biocomput*, 1998: 473–84.

48. Dunker, A. K., C. J. Brown, and Z. Obradovic, Identification and functions of usefully disordered proteins. *Adv Protein Chem*, 2002. **62**: 25–49.

49. Tompa, P., C. Szasz, and L. Buday, Structural disorder throws new light on moonlighting. *Trends Biochem Sci*, 2005. **30**(9): 484–9.

50. Xie, H., S. Vucetic, L. M. Iakoucheva, C. J. Oldfield, A. K. Dunker, V. N. Uversky, and Z. Obradovic, Functional anthology of intrinsic disorder. 1. Biological processes and functions of proteins with long disordered regions. *J Proteome Res*, 2007. **6**(5): 1882–98.

51. Vucetic, S., H. Xie, L. M. Iakoucheva, C. J. Oldfield, A. K. Dunker, Z. Obradovic, and V. N. Uversky, Functional anthology of intrinsic disorder. 2. Cellular components, domains, technical terms, developmental processes, and coding sequence diversities correlated with long disordered regions. *J Proteome Res*, 2007. **6**(5): 1899–916.

52. Xie, H., S. Vucetic, L. M. Iakoucheva, C. J. Oldfield, A. K. Dunker, Z. Obradovic, and V. N. Uversky, Functional anthology of intrinsic disorder. 3. Ligands, post-translational modifications, and diseases associated with intrinsically disordered proteins. *J Proteome Res*, 2007. **6**(5): 1917–32.

53. Schulz, G.E., Nucleotide binding proteins, in *Molecular Mechanism of Biological Recognition*, M. Balaban, Editor. 1979, New York: Elsevier/North-Holland Biomedical Press, pp. 79–94.

54. Kriwacki, R. W., L. Hengst, L. Tennant, S. I. Reed, and P. E. Wright, Structural studies of p21Waf1/Cip1/Sdi1 in the free and Cdk2-bound state: Conformational disorder mediates binding diversity. *Proc Natl Acad Sci USA*, 1996. **93**(21): 11504–9.

55. Oldfield, C. J., J. Meng, J. Y. Yang, M. Q. Yang, V. N. Uversky, and A. K. Dunker, Flexible nets: Disorder and induced fit in the associations of p53 and 14-3-3 with their partners. *BMC Genomics*, 2008. **9**(Suppl 1): S1.

56. Dajani, R., E. Fraser, S. M. Roe, M. Yeo, V. M. Good, V. Thompson, T. C. Dale, and L. H. Pearl, Structural basis for recruitment of glycogen synthase kinase 3beta to the axin-APC scaffold complex. *Embo J*, 2003. **22**(3): 494–501.

57. Tompa, P. and M. Fuxreiter, Fuzzy complexes: Polymorphism and structural disorder in protein–protein interactions. *Trends Biochem Sci*, 2008. **33**(1): 2–8.

58. Jeong, H., S. P. Mason, A. L. Barabasi, and Z. N. Oltvai, Lethality and centrality in protein networks. *Nature*, 2001. **411**(6833): 41–2.

59. Ekman, D., S. Light, A. K. Bjorklund, and A. Elofsson, What properties characterize the hub proteins of the protein–protein interaction network of Saccharomyces cerevisiae? *Genome Biol*, 2006. **7**(6): R45.

60. Dosztanyi, Z., J. Chen, A. K. Dunker, I. Simon, and P. Tompa, Disorder and sequence repeats in hub proteins and their implications for network evolution. *J Proteome Res*, 2006. **5**(11): 2985–95.

61. Patil, A. and H. Nakamura, Disordered domains and high surface charge confer hubs with the ability to interact with multiple proteins in interaction networks. *FEBS Lett*, 2006. **580**(8): 2041–5.

62. Haynes, C., C. J. Oldfield, F. Ji, N. Klitgord, M. E. Cusick, P. Radivojac, V. N. Uversky, M. Vidal, and L. M. Iakoucheva, Intrinsic disorder is a common feature of hub proteins from four eukaryotic interactomes. *PLoS Comput Biol*, 2006. **2**(8): e100.

63. Bustos, D. M. and A. A. Iglesias, Intrinsic disorder is a key characteristic in partners that bind 14-3-3 proteins. *Proteins*, 2006. **63**(1): 35–42.

64. Pierce, M. M., U. Baxa, A. C. Steven, A. Bax, and R. B. Wickner, Is the prion domain of soluble Ure2p unstructured? *Biochemistry*, 2005. **44**(1): 321–8.

65. Elbaum, M., Materials science. Polymers in the pore. *Science*, 2006. **314**(5800): 766–7.

66. Mukhopadhyay, R. and J. H. Hoh, AFM force measurements on microtubule-associated proteins: The projection domain exerts a long-range repulsive force. *FEBS Lett*, 2001. **505**(3): 374–8.

67. Trombitas, K., M. Greaser, S. Labeit, J. P. Jin, M. Kellermayer, M. Helmes, and H. Granzier, Titin extensibility in situ: Entropic elasticity of permanently folded and permanently unfolded molecular segments. *J Cell Biol*, 1998. **140**(4): 853–9.

68. Fontana, A., P. P. de Laureto, B. Spolaore, E. Frare, P. Picotti, and M. Zambonin, Probing protein structure by limited proteolysis. *Acta Biochim Pol*, 2004. **51**(2): 299–321.

69. Iakoucheva, L. M., P. Radivojac, C. J. Brown, T. R. O'Connor, J. G. Sikes, Z. Obradovic, and A. K. Dunker, The importance of intrinsic disorder for protein phosphorylation. *Nucleic Acids Res*, 2004. **32**(3): 1037–49.

70. Cox, C. J., K. Dutta, E. T. Petri, W. C. Hwang, Y. Lin, S. M. Pascal, and R. Basavappa, The regions of securin and cyclin B proteins recognized by the ubiquitination machinery are natively unfolded. *FEBS Lett*, 2002. **527**(1-3): 303–8.

71. Khan, A. N. and P. N. Lewis, Unstructured conformations are a substrate requirement for the Sir2 family of NAD-dependent protein deacetylases. *J Biol Chem*, 2005. **280**(43): 36073–8.

72. Puntervoll, P., R. Linding, C. Gemund, S. Chabanis-Davidson, M. Mattingsdal, S. Cameron, D.M. Martin, G. Ausiello, B. Brannetti, A. Costantini, F. Ferre, V. Maselli, A. Via, G. Cesareni, F. Diella, G. Superti-Furga, L. Wyrwicz, C. Ramu, C. McGuigan, R. Gudavalli, I. Letunic, P. Bork, L. Rychlewski, B. Kuster, M. Helmer-Citterich, W.N. Hunter, R. Aasland, and T.J. Gibson, ELM server: A new resource for investigating short functional sites in modular eukaryotic proteins. *Nucleic Acids Res*, 2003. **31**(13): 3625–30.

73. Fuxreiter, M., P. Tompa, and I. Simon, Structural disorder imparts plasticity on linear motifs. *Bioinformatics*, 2007. **23**: 950–6.

74. Lacy, E.R., I. Filippov, W. S. Lewis, S. Otieno, L. Xiao, S. Weiss, L. Hengst, and R. W. Kriwacki, p27 binds cyclin-CDK complexes through a sequential mechanism involving binding-induced protein folding. *Nat Struct Mol Biol*, 2004. **11**(4): 358–64.

75. Waizenegger, I., J. F. Gimenez-Abian, D. Wernic, and J. M. Peters, Regulation of human separase by securin binding and autocleavage. *Curr Biol*, 2002. **12**(16): 1368–78.

76. Kiss, R., Z. Bozoky, D. Kovacs, G. Rona, P. Friedrich, P. Dvortsak, P. Tompa, and A. Perczel, Calcium-induced tripartite binding of intrinsically disordered calpastatin to its cognate enzyme, calpain. *FEBS Lett*, 2008. **582**(15):2149–54.

77. Kiss, R., D. Kovacs, P. Tompa, and A. Perczel, Local structural preferences of calpastatin, the intrinsically unstructured protein inhibitor of calpain. *Biochemistry*, 2008. **47**(26): 6936–45.

78. Olashaw, N., T. K. Bagui, and W. J. Pledger, Cell cycle control: A complex issue. *Cell Cycle*, 2004. **3**(3): 263–4.

79. Haarmann, C. S., D. Green, M. G. Casarotto, D. R. Laver, and A. F. Dulhunty, The random-coil 'C' fragment of the dihydropyridine receptor II-III loop can activate or inhibit native skeletal ryanodine receptors. *Biochem J*, 2003. **372**(Pt 2): 305–16.

80. Cortese, M. S., V. N. Uversky, and A. K. Dunker, Intrinsic disorder in scaffold proteins: Getting more from less. *Prog Biophys Mol Biol*, 2008. **98**(1):85–106.

81. Bhattacharyya, R. P., A. Remenyi, M. C. Good, C. J. Bashor, A. M. Falick, and W. A. Lim, The Ste5 scaffold allosterically modulates signaling output of the yeast mating pathway. *Science*, 2006. **311**(5762): 822–6.

82. Mark, W. Y., J. C. Liao, Y. Lu, A. Ayed, R. Laister, B. Szymczyna, A. Chakrabartty, and C. H. Arrowsmith, Characterization of segments from the central region of BRCA1: An intrinsically disordered scaffold for multiple protein–protein and protein–DNA interactions? *J Mol Biol*, 2005. **345**(2): 275–87.

83. Hegyi, H., E. Schad, and P. Tompa, Structural disorder promotes assembly of protein complexes. *BMC Struct Biol*, 2007. **7**: 65.

84. Holt, C., N. M. Wahlgren, and T. Drakenberg, Ability of a beta-casein phosphopeptide to modulate the precipitation of calcium phosphate by forming amorphous dicalcium phosphate nanoclusters. *Biochem J*, 1996. **314**(Pt 3): 1035–9.

85. Prusiner, S. B., Prions. *Proc Natl Acad Sci USA*, 1998. **95**(23): 13363–83.

86. Tuite, M. F. and N. Koloteva-Levin, Propagating prions in fungi and mammals. *Mol Cell*, 2004. **14**(5): 541–52.

87. Fowler, D. M., A. V. Koulov, W. E. Balch, and J. W. Kelly, Functional amyloid—from bacteria to humans. *Trends Biochem Sci*, 2007. **32**(5): 217–24.

88. Si, K., S. Lindquist, and E. R. Kandel, A neuronal isoform of the aplysia CPEB has prion-like properties. *Cell*, 2003. **115**(7): 879–91.

89. Ringe, D. and G. A. Petsko, Study of protein dynamics by X-ray diffraction. *Methods Enzymol.*, 1986. **131**: 389–433.

90. Dyson, H. J. and P. E. Wright, Insights into the structure and dynamics of unfolded proteins from nuclear magnetic resonance. *Adv Protein Chem*, 2002. **62**: 311–40.

91. Bracken, C., L. M. Iakoucheva, P. R. Romero, and A. K. Dunker, Combining prediction, computation and experiment for the characterization of protein disorder. *Curr Opin Struct Biol*, 2004. **14**(5): 570–6.

92. Dyson, H. J. and P. E. Wright, Unfolded proteins and protein folding studied by NMR. *Chem Rev*, 2004. **104**(8): 3607–22.

93. Dyson, H. J. and P. E. Wright, Elucidation of the protein folding landscape by NMR. *Methods Enzymol*, 2005. **394**: 299–321.

94. Fasman, G.D., *Circular Dichroism and the Conformational Analysis of Biomolecules.* 1996, New York: Plenum Press.

95. Adler, A. J., N. J. Greenfield, and G. D. Fasman, Circular dichroism and optical rotatory dispersion of proteins and polypeptides. *Methods Enzymol*, 1973. **27**: 675–735.

96. Provencher, S. W. and J. Glockner, Estimation of globular protein secondary structure from circular dichroism. *Biochemistry*, 1981. **20**(1): 33–7.

97. Woody, R. W., Circular dichroism. *Methods Enzymol.*, 1995. **246**: 34–71.

98. Smyth, E., C. D. Syme, E. W. Blanch, L. Hecht, M. Vasak, and L. D. Barron, Solution structure of native proteins with irregular folds from Raman optical activity. *Biopolymers*, 2001. **58**(2): 138–51.

99. Uversky, V.N., A multiparametric approach to studies of self-organization of globular proteins. *Biochemistry (Mosc)*, 1999. **64**(3): 250–66.

100. Receveur-Brechot, V., J. M. Bourhis, V. N. Uversky, B. Canard, and S. Longhi, Assessing protein disorder and induced folding. *Proteins*, 2006. **62**(1): 24–45.

101. Markus, G., Protein substrate conformation and proteolysis. *Proc Natl Acad Sci USA*, 1965. **54**: 253–8.

102. Mikhalyi, E., *Application of Proteolytic Enzymes to Protein Structure Studies.* 1978, Boca Raton, FL: CRC Press.

103. Hubbard, S. J., F. Eisenmenger, and J. M. Thornton, Modeling studies of the change in conformation required for cleavage of limited proteolytic sites. *Protein Sci*, 1994. **3**(5): 757–68.

104. Fontana, A., P. Polverino de Laureto, V. De Filippis, E. Scaramella, and M. Zambonin, Probing the partly folded states of proteins by limited proteolysis. *Fold Des*, 1997. **2**(2): R17–26.

105. Fontana, A., M. Zambonin, P. Polverino de Laureto, V. De Filippis, A. Clementi, and E. Scaramella, Probing the conformational state of apomyoglobin by limited proteolysis. *J Mol Biol*, 1997. **266**(2): 223–30.

106. Iakoucheva, L. M., A. L. Kimzey, C. D. Masselon, R. D. Smith, A. K. Dunker, and E. J. Ackerman, Aberrant mobility phenomena of the DNA repair protein XPA. *Protein Sci*, 2001. **10**(7): 1353–62.

107. Privalov, P. L., Stability of proteins: Small globular proteins. *Adv Protein Chem*, 1979. **33**: 167–241.
108. Ptitsyn, O., Molten globule and protein folding. *Adv Protein Chem*, 1995. **47**: 83–229.
109. Ptitsyn, O. B., Kinetic and equilibrium intermediates in protein folding. *Protein Eng*, 1994. **7**(5): 593–6.
110. Uversky, V. N. and O. B. Ptitsyn, Further evidence on the equilibrium "pre-molten globule state": Four-state guanidinium chloride-induced unfolding of carbonic anhydrase B at low temperature. *J Mol Biol*, 1996. **255**(1): 215–28.
111. Westhof, E., D. Altschuh, D. Moras, A. C. Bloomer, A. Mondragon, A. Klug, and M. H. Van Regenmortel, Correlation between segmental mobility and the location of antigenic determinants in proteins. *Nature*, 1984. **311**(5982): 123–6.
112. Berzofsky, J. A., Intrinsic and extrinsic factors in protein antigenic structure. *Science*, 1985. **229**(4717): 932–40.
113. Brown, H. G., J. C. Troncoso, and J. H. Hoh, Neurofilament-L homopolymers are less mechanically stable than native neurofilaments. *J Microsc*, 1998. **191**(Pt 3): 229–37.
114. Kaltashov, I. A. and A. Mohimen, Estimates of protein surface areas in solution by electrospray ionization mass spectrometry. *Anal Chem*, 2005. **77**(16): 5370--9.
115. Uversky, V. N. and O. B. Ptitsyn, "Partly folded" state, a new equilibrium state of protein molecules: Four-state guanidinium chloride-induced unfolding of beta-lactamase at low temperature. *Biochemistry*, 1994. **33**(10): 2782–91.
116. Uversky, V. N., Natively unfolded proteins, in *Unfolded Proteins: From Denatured to Intrinsically Disordered*. T. P. Creamer, Editor. 2008, Hauppauge, NY: Nova Science Publishers, pp. 237–94.
117. Tcherkasskaya, O. and V. N. Uversky, Denatured collapsed states in protein folding: Example of apomyoglobin. *Proteins*, 2001. **44**(3): 244–54.
118. Crick, S. L., M. Jayaraman, C. Frieden, R. Wetzel, and R. V. Pappu, Fluorescence correlation spectroscopy shows that monomeric polyglutamine molecules form collapsed structures in aqueous solutions. *Proc Natl Acad Sci USA*, 2006. **103**(45): 16764–9.
119. Vitalis, A., X. Wang, and R. V. Pappu, Quantitative characterization of intrinsic disorder in polyglutamine: Insights from analysis based on polymer theories. *Biophys J*, 2007. **93**(6): 1923–37.
120. Tran, H. T., A. Mao, and R. V. Pappu, Role of backbone-solvent interactions in determining conformational equilibria of intrinsically disordered proteins. *J Am Chem Soc*, 2008. **130**(23): 7380–92.
121. Dill, K. A. and D. Shortle, Denatured states of proteins. *Annu Rev Biochem*, 1991. **60**: 795–825.
122. Pappu, R. V., R. Srinivasan, and G. D. Rose, The Flory isolated-pair hypothesis is not valid for polypeptide chains: Implications for protein folding. *Proc Natl Acad Sci USA*, 2000. **97**(23): 12565–70.
123. Shortle, D., The denatured state (the other half of the folding equation) and its role in protein stability. *FASEB J*, 1996. **10**(1): 27–34.
124. Shortle, D. R., Structural analysis of non-native states of proteins by NMR methods. *Curr Opin Struct Biol*, 1996. **6**(1): 24–30.
125. Gillespie, J. R. and D. Shortle, Characterization of long-range structure in the denatured state of staphylococcal nuclease. II. Distance restraints from paramagnetic relaxation and calculation of an ensemble of structures. *J Mol Biol*, 1997. **268**(1): 170–84.
126. Gillespie, J. R. and D. Shortle, Characterization of long-range structure in the denatured state of staphylococcal nuclease. I. Paramagnetic relaxation enhancement by nitroxide spin labels. *J Mol Biol*, 1997. **268**(1): 158–69.
127. Wang, Y. and D. Shortle, Residual helical and turn structure in the denatured state of staphylococcal nuclease: Analysis of peptide fragments. *Fold Des*, 1997. **2**(2): 93–100.

128. Eliezer, D., E. Kutluay, R. Bussell, Jr., and G. Browne, Conformational properties of alpha-synuclein in its free and lipid-associated states. *J Mol Biol*, 2001. **307**(4): 1061–73.

129. Bussell, R., Jr. and D. Eliezer, Residual structure and dynamics in Parkinson's disease-associated mutants of alpha-synuclein. *J Biol Chem*, 2001. **276**(49): 45996–46003.

130. Sung, Y. H. and D. Eliezer, Residual structure, backbone dynamics, and interactions within the synuclein family. *J Mol Biol*, 2007. **372**(3): 689–707.

131. Fuxreiter, M., I. Simon, P. Friedrich, and P. Tompa, Preformed structural elements feature in partner recognition by intrinsically unstructured proteins. *J Mol Biol*, 2004. **338**(5): 1015–26.

132. Parker, D., M. Rivera, T. Zor, A. Henrion-Caude, I. Radhakrishnan, A. Kumar, L. H. Shapiro, P. E. Wright, M. Montminy, and P. K. Brindle, Role of secondary structure in discrimination between constitutive and inducible activators. *Mol Cell Biol*, 1999. **19**(8): 5601–7.

133. Radhakrishnan, I., G. C. Perez-Alvarado, H. J. Dyson, and P. E. Wright, Conformational preferences in the Ser133-phosphorylated and non-phosphorylated forms of the kinase inducible transactivation domain of CREB. *FEBS Lett*, 1998. **430**(3): 317–22.

134. Lee, H., K. H. Mok, R. Muhandiram, K. H. Park, J. E. Suk, D. H. Kim, J. Chang, Y. C. Sung, K. Y. Choi, and K. H. Han, Local structural elements in the mostly unstructured transcriptional activation domain of human p53. *J Biol Chem*, 2000. **275**(38): 29426–32.

135. Daughdrill, G. W., L. J. Hanely, and F. W. Dahlquist, The C-terminal half of the anti-sigma factor FlgM contains a dynamic equilibrium solution structure favoring helical conformations. *Biochemistry*, 1998. **37**(4): 1076–82.

136. Dedmon, M. M., C. N. Patel, G. B. Young, and G. J. Pielak, FlgM gains structure in living cells. *Proc Natl Acad Sci USA*, 2002. **99**(20): 12681–4.

137. Hauer, J. A., P. Barthe, S. S. Taylor, J. Parello, and A. Padilla, Two well-defined motifs in the cAMP-dependent protein kinase inhibitor (PKIalpha) correlate with inhibitory and nuclear export function. *Protein Sci*, 1999. **8**(3): 545–53.

138. Domanski, M., M. Hertzog, J. Coutant, I. Gutsche-Perelroizen, F. Bontems, M. F. Carlier, E. Guittet, and C. van Heijenoort, Coupling of folding and binding of thymosin beta4 upon interaction with monomeric actin monitored by nuclear magnetic resonance. *J Biol Chem*, 2004. **279**(22): 23637–45.

139. Longhi, S., V. Receveur-Brechot, D. Karlin, K. Johansson, H. Darbon, D. Bhella, R. Yeo, S. Finet, and B. Canard, The C-terminal domain of the measles virus nucleoprotein is intrinsically disordered and folds upon binding to the C-terminal moiety of the phosphoprotein. *J Biol Chem*, 2003. **278**(20): 18638–48.

140. Sivakolundu, S. G., D. Bashford, and R. W. Kriwacki, Disordered p27Kip1 exhibits intrinsic structure resembling the Cdk2/cyclin A-bound conformation. *J Mol Biol*, 2005. **353**(5): 1118–28.

141. Oldfield, C. J., Y. Cheng, M. S. Cortese, P. Romero, V. N. Uversky, and A. K. Dunker, Coupled folding and binding with alpha-helix-forming molecular recognition elements. *Biochemistry*, 2005. **44**(37): 12454–70.

142. Garner, E., P. Romero, A.K. Dunker, C. Brown, and Z. Obradovic, Predicting binding regions within disordered proteins. *Genome Inform Ser Workshop Genome Inform*, 1999. **10**: 41–50.

143. Mader, S., H. Lee, A. Pause, and N. Sonenberg, The translation initiation factor eIF-4E binds to a common motif shared by the translation factor eIF-4 gamma and the translational repressors 4E-binding proteins. *Mol Cell Biol*, 1995. **15**(9): 4990–7.

144. Fletcher, C. M. and G. Wagner, The interaction of eIF4E with 4E-BP1 is an induced fit to a completely disordered protein. *Protein Sci*, 1998. **7**(7): 1639–42.

145. Cheng, Y., C. J. Oldfield, J. Meng, P. Romero, V. N. Uversky, and A. K. Dunker, Mining alpha-helix-forming molecular recognition features with cross species sequence alignments. *Biochemistry*, 2007. **46**(47): 13468–77.

146. Mohan, A., C. J. Oldfield, P. Radivojac, V. Vacic, M. S. Cortese, A. K. Dunker, and V. N. Uversky, Analysis of molecular recognition features (MoRFs). *J Mol Biol*, 2006. **362**(5): 1043–59.

147. Pontius, B. W., Close encounters: Why unstructured, polymeric domains can increase rates of specific macromolecular association. *Trends Biochem Sci*, 1993. **18**(5): 181–6.

148. Csizmok, V., M. Bokor, P. Banki, É. Klement, K. F. Medzihradszky, P. Friedrich, K. Tompa, and P. Tompa, Primary contact sites in intrinsically unstructured proteins: The case of calpastatin and microtubule-associated protein 2. *Biochemistry*, 2005. **44**: 3955–64.

149. Kalthoff, C., J. Alves, C. Urbanke, R. Knorr, and E. J. Ungewickell, Unusual structural organization of the endocytic proteins AP180 and epsin 1. *J Biol Chem*, 2002. **277**(10): 8209–16.

150. Scheele, U., J. Alves, R. Frank, M. Duwel, C. Kalthoff, and E. Ungewickell, Molecular and functional characterization of clathrin- and AP-2-binding determinants within a disordered domain of auxilin. *J Biol Chem*, 2003. **278**(28): 25357–68.

151. Dafforn, T. R. and C. J. Smith, Natively unfolded domains in endocytosis: Hooks, lines and linkers. *EMBO Rep*, 2004. **5**(11): 1046–52.

152. Vise, P., B. Baral, A. Stancik, D. F. Lowry, and G. W. Daughdrill, Identifying long-range structure in the intrinsically unstructured transactivation domain of p53. *Proteins*, 2007. **67**(3): 526–30.

153. Neduva, V. and R. B. Russell, Linear motifs: Evolutionary interaction switches. *FEBS Lett*, 2005. **579**(15): 3342–5.

154. Neduva, V. and R. B. Russell, DILIMOT: Discovery of linear motifs in proteins. *Nucleic Acids Res*, 2006. **34**(Web Server issue): W350–5.

155. Davey, N. E., D. C. Shields, and R. J. Edwards, SLiMDisc: Short, linear motif discovery, correcting for common evolutionary descent. *Nucleic Acids Res*, 2006. **34**(12): 3546–54.

156. Dyson, H. J. and P. E. Wright, Coupling of folding and binding for unstructured proteins. *Curr Opin Struct Biol*, 2002. **12**(1): 54–60.

157. Bienkiewicz, E. A., J. N. Adkins, and K. J. Lumb, Functional consequences of preorganized helical structure in the intrinsically disordered cell-cycle inhibitor p27(Kip1). *Biochemistry*, 2002. **41**(3): 752–9.

158. Uversky, V. N., Neuropathology, biochemistry, and biophysics of alpha-synuclein aggregation. *J Neurochem*, 2007. **103**(1): 17–37.

159. Uversky, V. N., alpha-Synuclein misfolding and neurodegenerative diseases. *Curr Protein Peptide Sci*, 2008. **9**(5): 507–40.

160. Uversky, V. N., A protein-chameleon: Conformational plasticity of alpha-synuclein, a disordered protein involved in neurodegenerative disorders. *J Biomol Struct Dyn*, 2003. **21**(2): 211–34.

161. Dill, K. A. and H. S. Chan, From Levinthal to pathways to funnels. *Nat Struct Biol*, 1997. **4**(1): 10–19.

162. Nymeyer, H., N. D. Socci, and J. N. Onuchic, Landscape approaches for determining the ensemble of folding transition states: Success and failure hinge on the degree of frustration. *Proc Natl Acad Sci USA*, 2000. **97**(2): 634–9.

163. Onuchic, J. N., H. Nymeyer, A. E. Garcia, J. Chahine, and N. D. Socci, The energy landscape theory of protein folding: Insights into folding mechanisms and scenarios. *Adv Protein Chem*, 2000. **53**: 87–152.

164. Gunasekaran, K., C. J. Tsai, and R. Nussinov, Analysis of ordered and disordered protein complexes reveals structural features discriminating between stable and unstable monomers. *J Mol Biol*, 2004. **341**(5): 1327–41.

165. Lowe, E. D., I. Tews, K. Y. Cheng, N. R. Brown, S. Gul, M. E. Noble, S. J. Gamblin, and L. N. Johnson, Specificity determinants of recruitment peptides bound to phospho-CDK2/cyclin A. *Biochemistry*, 2002. **41**(52): 15625–34.

166. Avalos, J. L., I. Celic, S. Muhammad, M. S. Cosgrove, J. D. Boeke, and C. Wolberger, Structure of a Sir2 enzyme bound to an acetylated p53 peptide. *Mol Cell*, 2002. **10**(3): 523–35.

167. Mujtaba, S., Y. He, L. Zeng, S. Yan, O. Plotnikova, Sachchidanand, R. Sanchez, N. J. Zeleznik-Le, Z. Ronai, and M. M. Zhou, Structural mechanism of the bromodomain of the coactivator CBP in p53 transcriptional activation. *Mol Cell*, 2004. **13**(2): 251–63.

168. Wu, H., M. W. Maciejewski, A. Marintchev, S. E. Benashski, G. P. Mullen, and S. M. King, Solution structure of a dynein motor domain associated light chain. *Nat Struct Biol*, 2000. **7**(7): 575–9.

169. Chuikov, S., J. K. Kurash, J. R. Wilson, B. Xiao, N. Justin, G. S. Ivanov, K. McKinney, P. Tempst, C. Prives, S. J. Gamblin, N. A. Barlev, and D. Reinberg, Regulation of p53 activity through lysine methylation. *Nature*, 2004. **432**(7015): 353–60.

170. Poux, A. N. and R. Marmorstein, Molecular basis for Gcn5/PCAF histone acetyltransferase selectivity for histone and nonhistone substrates. *Biochemistry*, 2003. **42**(49): 14366–74.

171. Bochkareva, E., L. Kaustov, A. Ayed, G. S. Yi, Y. Lu, A. Pineda-Lucena, J. C. Liao, A. L. Okorokov, J. Milner, C. H. Arrowsmith, and A. Bochkarev, Single-stranded DNA mimicry in the p53 transactivation domain interaction with replication protein A. *Proc Natl Acad Sci USA*, 2005. **102**(43): 15412–7.

172. Kussie, P. H., S. Gorina, V. Marechal, B. Elenbaas, J. Moreau, A. J. Levine, and N. P. Pavletich, Structure of the MDM2 oncoprotein bound to the p53 tumor suppressor transactivation domain. *Science*, 1996. **274**(5289): 948–53.

173. Di Lello, P., L. M. Jenkins, T. N. Jones, B. D. Nguyen, T. Hara, H. Yamaguchi, J. D. Dikeakos, E. Appella, P. Legault, and J. G. Omichinski, Structure of the Tfb1/p53 complex: Insights into the interaction between the p62/Tfb1 subunit of TFIIH and the activation domain of p53. *Mol Cell*, 2006. **22**(6): 731–40.

174. Kuszewski, J., A. M. Gronenborn, and G. M. Clore, Improving the packing and accuracy of NMR structures with a pseudopotential for the radius of gyration. *J Am Chem Soc*, 1999. **121**(10): 2337–8.

175. Cho, Y., S. Gorina, P. D. Jeffrey, and N. P. Pavletich, Crystal structure of a p53 tumor suppressor-DNA complex: Understanding tumorigenic mutations. *Science*, 1994. **265**(5170): 346–55.

176. Joo, W. S., P. D. Jeffrey, S. B. Cantor, M. S. Finnin, D. M. Livingston, and N. P. Pavletich, Structure of the 53BP1 BRCT region bound to p53 and its comparison to the Brca1 BRCT structure. *Genes Dev*, 2002. **16**(5): 583–93.

177. Gorina, S. and N. P. Pavletich, Structure of the p53 tumor suppressor bound to the ankyrin and SH3 domains of 53BP2. *Science*, 1996. **274**(5289): 1001–5.

178. Meszaros, B., P. Tompa, I. Simon, and Z. Dosztanyi, Molecular principles of the interactions of disordered proteins. *J Mol Biol*, 2007. **372**(2): 549–61.

13 Identification of Druggable Hot Spots on Proteins and in Protein–Protein Interfaces

Dmitri Beglov, Ryan Brenke, Gwo-Yu Chuang, David Hall, Melissa Landon, Chi Ho Ngan, Yang Shen, Spencer Thiel, Brandon Zerbe, Dima Kozakov, and Sandor Vajda

CONTENTS

OVERVIEW

The interactions of proteins with each other and other biochemical compounds play a central role in various aspects of the structural and functional organization of the cell. Elucidation of such interactions is a major step toward understanding cellular pathways and processes and also suggests avenues for drug design. One observation that emerges from these studies is that the various residues in the binding region do not equally contribute to the binding free energy. By replacing individual interface residues with alanine (known as *alanine scanning mutagenesis*), Clackson and Wells found that a central hydrophobic region of human growth hormone receptor accounts for more than three-quarters of the binding free energy.[1] This led the authors to introduce the notion of hot spots.

253

Subsequent studies have shown that hot spots can be identified in most if not all protein–protein and protein–small molecule interfaces.[2] A database of hot spots in proteins, ASEdb,[3] collected single alanine mutations and associated binding affinity in protein–protein, protein–nucleic acids, and protein–small molecule interactions. Efforts were made to interpret these data so that hot spot residues could be characterized. For instance, after analyzing 2325 alanine mutants, Bogan and Thorn[2] found that (1) hot spots tend to cluster at the center of the interface where they are protected from the bulk solvent and (2) some amino acids appear more often in hot spots. In particular, hot spots are often enriched in tryptophan, tyrosine, and arginine and are surrounded by residues that most likely serve to occlude the solvent but which do not contribute significantly to binding free energy. Work by Hu and co-workers[4] also showed that polar residues (arginine, glutamine, histidine, aspartic acid, and asparagine) were generally conserved in interfaces and further that these conserved polar residues are frequently in hot spots; however, as these authors and other investigators have observed, a residue's solvent accessibility, polarity, and charges are necessary but probably not sufficient conditions for the identification of hot spots.[5,6] In addition, some residues that are not within the interface can also contribute substantially to the free energy of binding when assayed by alanine scanning mutagenesis. This is most often explained by changes in three-dimensional conformation of the binding site.[5]

Another powerful approach to the identification of hot spots, which is frequently utilized in drug design, involves screening libraries of fragment-sized compounds for binding to the target protein.[7] Both NMR and x-ray crystallography techniques have been used for such fragment-based screening. Using NMR, the ^{15}N-labeled protein is screened against a library of fragment compounds.[8,9] Applications to a variety of proteins demonstrate that small organic compounds have a strong preference for binding at the energetic hot spots on protein surfaces although not necessarily with large binding free energies. A high hit rate is a good predictor of druggability: Indeed, high correlation was observed between the number of different probes binding to a site and the ability to identify high-affinity ($K_d < 300$ nM), nonpeptide, noncovalent inhibitors that bind there.[9] The x-ray technique application of this idea, known as multiple solvent crystal structures (MSCS) method, is based on solving the structure of the protein in aqueous solutions of various compounds, primarily organic solvents.[10,11] Each structure shows a few organic molecules associated with the protein surface in the first shell of water molecules. The power of the method arises from superimposing a number of structures solved in different solvents. Most organic compounds generally cluster in the binding site, and the overlapping probe clusters form "consensus" sites that highlight the functionally most important subsites. As demonstrated by applications to porcine elastase[10–12] and thermolysin,[13,14] some organic molecules may also bind at crystal contacts or in small buried pockets, but the large consensus sites generally occur in the hot spots of the binding site.

Since the identification of hot spots by alanine scanning mutagenesis, nuclear magnetic resonance (NMR), or x-ray crystallography is expensive, it is important to explore whether similar information can be obtained via a computational approach. In fact, the first report that only a small number of amino acids contribute actively to binding energetics was based on simple free energy calculations by Novotny and co-workers.[15] More recent approaches have simulated computational alanine scanning mutagenesis employing free energy perturbation and thermodynamic cycles with the expectation that a full description of the structural and energetic consequences of mutagenesis can be predicted. Massova and Kollman[16] estimated the binding free energy from energies in the gas phase (electrostatic, van der Waals, and internal energy terms); solvation free energies (Poisson–Boltzmann and nonpolar contributions); and vibrational, rotational, and translational entropies for the complex and the component molecules. They used these energy terms to study the interactions between the tumor suppressor protein p53 and oncoprotein Mdm2 demonstrating excellent agreement with experimental binding data. The accuracy of this approach relies on the accuracy of the free energy estimates, and the adequate calculation of the entropic terms requires extensive conformational sampling at considerable computational costs. Kortemme and Baker proposed a simplified free energy model[6] with the following physical considerations of dominant molecular interactions: (1) shape complementarity, (2) polar interactions involving ion pairs and hydrogen bonds, (3) the interactions of protein atoms with the solvent including a penalty for the desolvation of buried polar groups, and (4) the effects of mutations on both the protein–protein complex and the unbound partners. The relative weight of the energy terms and amino acid dependent reference energies were parameterized using a training set of stability changes measured in 743 single alanine mutations in monomeric proteins. The parameterized free energy function was tested on a database of 19 complexes with known crystal structures and with measured changes in binding energy on alanine mutagenesis. Remarkably, this methodology predicted 79% of the interface hot spot residues. Furthermore, the authors were able to make more precise explanations of some earlier observations on hot spots attributing much of their success on the inclusion of an environment-dependent hydrogen bonding term.[6]

In this chapter we focus on computational fragment mapping, a method that can be considered a direct computational analogue of the fragment-based approaches based on the use of NMR or x-ray crystallography. After a brief description of the method, we present mapping results for the zinc endopeptidase thermolysin, which has its x-ray structure determined in four aqueous solutions of different organic solvents.[13,14] This validation study shows that the same or even more complete information can be obtained by computational mapping. The second application is to renin aspartic protease, a long-standing pharmaceutical target for the treatment of hypertension. We do not have experimental mapping data for renin, so our goal is to instead show that mapping can reliably identify the hot spots that substantially contribute to the free energy of ligand binding and hence should be the primary targets of drug design efforts. Indeed, the major consensus sites found trace the shape of the first approved renin inhibitor, aliskiren,[17,18] rather than that of peptidomimetic inhibitors that have been studied for several decades but which have not provided any successful drug candidates. In the third application we explore the binding site structure of PPARγ, a ligand activated transcription factor, and show that mapping identifies the hot spots both in the ligand binding site and in the surface regions of PPARγ that interact with other proteins. Finally, we provide a number of examples of hot spots in protein–protein interfaces.

COMPUTATIONAL PROTEIN MAPPING

Computational mapping methods place molecular probes on the protein surface to explore the protein's binding properties. A number of methods identify potential binding sites.[19–21] Some early methods such as GRID[22] and multiple copy simultaneous search (MCSS)[23,24] have been developed to find favorable binding positions for specific molecules or functional groups rather than to identify hot spots. Both methods result in many energy minima, and, consequently, it is difficult to determine which of the minima are actually relevant.[10] For example, English et al.[14] compared the GRID and MCSS results with x-ray structures of thermolysin determined in four different organic solvents and showed that both methods found minima close to the experimentally observed positions; however, the closest minima were not among those with the lowest free energies. Thus, these methodologies resulted in false positives (i.e., conformations with favorable energy that are not located near any experimentally observed binding site). The mapping algorithms CS-Map[71] and FTMAP[70] were developed to alleviate these problems by using an improved energy function and thermodynamic inspired sampling. We first developed CS-Map to reproduce NMR and x-ray screening results using a number of organic molecules as virtual probes.[25,26] For each probe, the algorithm generates more than 6000 bound positions by a multistart rigid-body docking based on the nonlinear simplex algorithm, further refines the positions by energy minimization, clusters the resulting conformations, and ranks the clusters on the basis of this average free energy. Results are better than for GRID or MCSS because the simplex algorithm provides better sampling, the scoring potential includes a solvation term, and the final ranking is based on the average cluster free energy rather than the energy of individual docked conformations. Another difference is that the mapping algorithm's goal is to find consensus sites where several low energy probe clusters overlap instead of the goal of finding favorable binding positions. Each of these factors helps the method to avoid irrelevant local minima, which provides a strong safeguard against false positives. We have shown that CS-Map was able to identify the most important subsite in a number of proteins;[25–31] nevertheless, there was clear need for improvements. In some cases the sampling turned out to be inadequate, particularly if the protein had deep pockets.[28] Although the largest consensus site was generally reliable, results became noisy for the smaller sites, and some of the subsites were not identified. In addition, the algorithm was slow; it required more than 1000 CPU hours on a 1 GHz PIII processor to map a single small protein with 16 probes. Running the program on multiple processors makes the computations feasible, but it is still difficult to perform large-scale studies, particularly if the goal is to map several structures or models of a protein to account for its flexibility.[26]

More recently we have developed FTMAP,[70] an improved mapping algorithm based on the fast Fourier transform (FFT) correlation approach. FTMAP starts with sampling billions of probe positions on a dense translational and rotational grid. The positions are scored using an energy function that includes attractive and repulsive van der Waals terms, electrostatic interaction energy based on Poisson–Boltzmann calculations, a cavity term to represent the effect of nonpolar enclosures, and a structure-based pairwise interaction potential. In spite of its relative complexity, the energy expression is written as a sum of correlation functions with components defined on grids. This enables the use of the extremely efficient FFT correlation method for function evaluation.[32] The FTMAP algorithm consists of five steps as follows:

Step 1: Rigid-body docking of probe molecules. Protein structures are downloaded from the Protein Data Bank (PDB)[33] and all bound ligands and water molecules are removed. For each structure, the 16 small molecules shown in Figure 13.1 are used as probes. For each probe, billions of docked conformations are sampled by a special-purpose rigid-body docking algorithm based on the FFT correlation approach. Mapping requires only the atomic coordinates of the two molecules, that is, no a priori information on the binding site is used. The 2000 best poses for each probe are retained for further processing.

Step 2: Minimization and rescoring. The free energy of each of the 2000 complexes, generated in Step 1, is minimized using the CHARMM (Chemistry at HARvard Macromolecular Mechanics) potential with the analytic continuum electrostatic (ACE) model representing the electrostatics and solvation terms as implemented in version 27 of CHARMM[34] using the parameter set from version 19 of the program. The ACE model includes a surface area dependent term to account for the solute–solvent van der Waals interactions. The minimization is performed using an adopted basis Newton–Raphson method. During the minimization, the protein atoms are held fixed, while the atoms of the probe molecules are free to move.

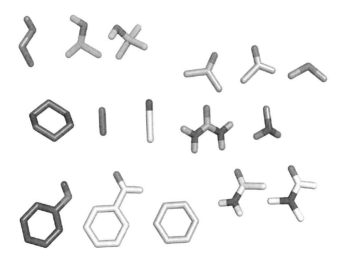

FIGURE 13.1 (SEE COLOR INSERT FOLLOWING PAGE 174.) Small organic molecules used as probes in protein mapping. The probes are colored as follows: ethanol, sky blue; isopropanol, green; isobutanol, cyan; acetone, violet; acetaldehyde, olive; dimethyl ether, sand; cyclohexane, chocolate; ethane, purple; acetonitrile, pale green; urea, orange; methylamine, zinc; phenol, deep blue; benzaldehyde, gold; benzene, silver; acetamide, lemon; N,N-dimethylformamide, yellow. Oxygen atoms are colored red; nitrogen atoms are colored blue. (From Brenke, R. et al., 2009. Reprinted with permission from *Bioinformatics*.)

Step 3: Clustering and ranking. The minimized probe conformations from Step 2 are grouped into clusters using a simple greedy algorithm. The lowest energy structure is selected and the structures within 3 Å RMSD are joined in the first cluster. The members of this cluster are removed, and the next lowest energy structure is selected to start the second cluster. This step is repeated until the entire set is exhausted. Clusters with less than 10 members are excluded from consideration thereby avoiding narrow energy minima with low entropy.[35] The retained clusters are ranked on the basis of their Boltzmann averaged energies.

Step 4: Determination of consensus sites. To determine the hot spots, FTMAP uses the consensus clustering idea based on the original MSCS experiments[10,36] and finds the regions on the protein where clusters of different probes overlap. Six clusters with the lowest average free energies are retained for each probe. The clusters of different probes are clustered using the distance between the centers of mass of the cluster centers as the distance measure. FTMAP again uses a simple greedy algorithm to find the cluster with the maximum number of neighbors (defined as cluster centers within 4 Å from each other), forming consensus site 1 (CS1). Members of CS1 are then removed from consideration, and the procedure is repeated until all clusters are exhausted. The structures are then redistributed among the consensus sites such that each structure is closest to the center of its own consensus site, and finally the consensus sites are ranked based on the number of their clusters. Duplicate clusters of the same type are considered in the count.

Step 5: Characterization of the binding site. First, FTMAP selects the largest consensus site (CS1) that generally identifies the most important subsite (or hot spot). CS1 forms the kernel of the binding site. Since additional clustering of probes close to the main consensus site is likely to indicate other subsites of the binding site, the binding site is expanded by adding any consensus site (irrespective of its size) within 7 Å from any consensus site already in the binding site. This procedure continues until no further expansion is possible. The resulting set of consensus sites is used to describe the binding site.

THE MAPPING OF THERMOLYSIN

Thermolysin is a thermostable extracellular bacterial zinc endopeptidase with a large active-site cleft consisting of at least four subsites (S_2, S_1, S_1', and S_2'). The S_1' subsite is the main specificity pocket.[13,37] It forms a distinct cavity that is lined with hydrophobic residues (F130, L33, V139, and L202), whereas polar residues (N112, E143, R203, and H231) are toward its edge.[38,39]

English et al.[13,14] determined high-resolution x-ray structures of thermolysin from crystals soaked in aqueous solutions of isopropanol, acetone, acetonitrile, and phenol. An increasing number of solvent interaction sites was identified as the organic solvent concentrations increased; however, the S_1' subsite is exceptional on two accounts. First, S_1' is the only probe binding site at low solvent concentration levels. Furthermore, the concentrations must be substantially increased (up to 80% in the case of isopropanol) before binding occurs at any other location. Second, S_1' is the only site where all four solvent molecules bind. The superimposition of the four mapped thermolysin structures solved in 5% isopropanol, 50% acetone, 80% acetonitrile, and 50 mM phenol (Figure 13.2A) shows the four organic solvent molecules binding within the S_1' pocket.

Superimposing all structures obtained in the four solvents at different concentrations shows isopropanol, acetone, and phenol binding at 12, 6, and 2 positions, respectively, whereas acetonitrile binds only at the S_1' site. Figure 13.2B shows the overlay of all probe molecules that bind in the active site. As described, the S_1' pocket binds all four probes. Two more isopropanol molecules from the structure of thermolysin in 90% isopropanol are shown below the main specificity pocket occupying the S_1 and S_2 subsites. Finally, the upper left side of Figure 13.2B along the active site cleft shows an isopropanol and an acetonitrile. This site binds a crystallographic water molecule in the original structure, and it starts to bind isopropanol and acetone at 80% and 70% concentrations, respectively.

As discussed by Mattos and Ringe,[10] the binding of the maximum number of probe molecules generally identifies the most important subsite in the active site of an enzyme. The S_1' pocket of thermolysin clearly satisfies this condition. The nearby positions that also bind clusters of probe molecules are likely to be further subsites of the active site (Figure 13.2B). Although probes binding far from the main consensus site often identify further important sites (e.g., allosteric sites, residues involved in dimerization, etc.), such secondary positions for thermolysin include only a buried pocket in the interior of the protein, several crystal contacts, and an interaction site created by a protein-bound molecule of dimethyl sulfoxide (DMSO).[13] Thus, the focus of this section will be on the identification of hot spots in the active site, and probes binding at secondary locations will be ignored.

Since no thermolysin structure without any bound ligand was available, a structure co-crystallized with the dipeptide Val-Lys, a cleavage product (PDB code 2tlx), was mapped. The peptide, the active site Zn^{2+} ion, and all crystallographic water molecules were removed before mapping. To explore the information that can be obtained by generic mapping methodology, all 16 probes shown in Figure 13.1 were used rather than only the four probes with experimental mapping results available. For each probe, the six lowest free energy clusters were superimposed to identify the consensus sites (CSs) defined by overlapping probe clusters. The largest consensus site (CS1) contains clusters of all 16 probes, plus second clusters of isopropanol, isobutanol, and N,N-dimethylformamide. Figure 13.2C shows that these centers (i.e., lowest energy structures of the 19 clusters in CS1) form a very tight supercluster in the S_1' pocket. This outcome is in excellent agreement with the results of the MSCS experiments that show S_1' being the only subsite binding all four solvent molecules.[14]

As described earlier, the algorithm first identifies the largest consensus site (always denoted as CS1), expands it if there is any further (even relatively small) consensus site within 7 Å, and continues this expansion until no further consensus sites can be reached. Figure 13.2D shows the consensus site at S_1', and the three adjacent sites. CS2 is a fairly elongated supercluster of 16 clusters, covering both the S_1 and

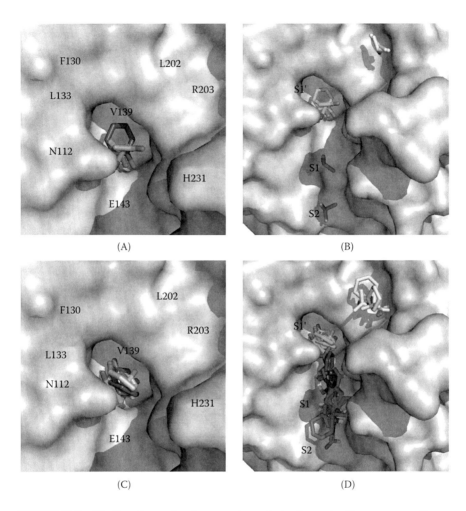

FIGURE 13.2 Binding of organic solvents to thermolysin, determined by x-ray crystallography and computational mapping. (A) Probe molecule binding in the S1' pocket of thermolysin, based on superimposing thermolysin structures solved in isopropanol, acetone, acetonitrile, and phenol. (B) Probe molecules in the active site of thermolysin, based on superimposing thermolysin structures solved in isopropanol, acetone, acetonitrile, and phenol. (C) Centers of probe clusters in the largest consensus site of thermolysin, located in the S1' pocket, from mapping the protein using the 16 probes shown in Figure 13.1. (D) Centers of probe clusters in the four consensus sites located in the active site of thermolysin, determined by mapping the thermolysin protein using the 16 probes shown in Figure 13.1. The probes are shaded to distinguish between the different consensus sites as follows: CS1, gray (at S1'); CS2, dark gray (near S2); CS6, black (near S1); and CS7, light gray (top right). (From Brenke, R. et al., 2009. Reprined with permission from *Bioinformatics*.)

S_2 subsites. The next consensus site within 7 Å is CS6, which includes eight probe clusters and expands from subsite S_1' to subsite S_1. Finally, CS7 includes seven probe clusters that overlap a methanol and an acetone molecule. Thus, the main consensus site CS1 and the three adjacent sites, CS2, CS6, and CS7, identify all subsites of the active site that bind any organic solvent in the MSCS experiments.[13,14]

The residues that interact with a large number of probes in the active site also tend to favorably interact with the specific ligands of the protein.[7,8] To show that computational mapping can provide such information, the nonbonded interactions and hydrogen bonds between each thermolysin residue and the probes, which are taken both from the experimental[13,14] and the computational mapping results, are shown in Figures 13.2B and 13.2D, respectively. FTMAP finds all residues that are important for substrate binding and most residues participating in the catalytic mechanism (Figures 13.3A and 13.3B): H142, H146, and E166 coordinate the Zn^{2+} ion in the active site; E143 serves as the general base; Y157 and H231 provide stabilization of the transition state; the backbone of W115 forms two hydrogen bonds with the substrate in the S_1 pocket; and N112 and the backbone of A113 form hydrogen bonds with the leaving group on one side and R203 on the opposite side of the S_1' pocket.[38,39] While these results are essentially complete, note that the experimental mapping misses residues E166 and H231 as well as some of the hydrogen bonds.

Note that the mapping results reflect the importance of each residue for substrate binding rather than for catalytic activity. For example, computational mapping finds a large number of hydrogen bonds for R203 (Figure 13.3B). Although R203 does not directly participate in hydrolysis, it forms hydrogen bonds with the carbonyl group of the residue at the P_1' position[38,39] and is known to be crucial for substrate binding.[40] In contrast, the mapping does not find D226, which is part of the catalytic mechanism,[38,39] although this residue is not particularly important (the D226A mutation introduces only a minor perturbation in the activity[41]). In spite of their completeness, the computational results do not seem to overpredict the importance of any residue. The experimental mapping is slightly less specific and finds a few residues (N116, G117, D150, Y211, and G212) that do not seem to play major roles.[38,39] Note that in computational mapping very few or no probes interact with these residues (Figure 13.3A).

IDENTIFICATION OF THE DRUGGABLE HOT SPOTS OF RENIN

Renin is an enzyme secreted by the kidneys that is involved in the first step of the rennin–angiotensin system (RAS), a system that regulates blood pressure and volume balance and is a common target for the treatment of hypertension. Renin hydrolyzes angiotensinogen into angiotensin I, which is in turn further hydrolyzed by angiotensin converting enzyme (ACE) into angiotensin II. Angiotensin II has numerous physiological effects including vasoconstriction and the stimulation of thirst. Inhibition of either step in RAS is possible, and ACE inhibitors have been Food and Drug Administration (FDA) approved since the early 1980s. Aliskiren, approved by the FDA in 2007, is the first non-peptidomimetic drug that inhibits renin. Following the completion of computational mapping of renin detailed in this section, the structure of aliskiren-bound renin was deposited in the PDB with code 2v0z; however, at the time of our mapping studies,[30] no such structure was available.

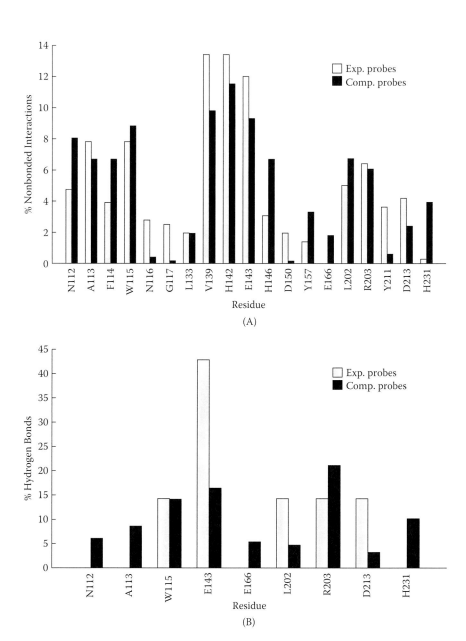

FIGURE 13.3 (A) Intermolecular nonbonded interactions between probes and thermolysin residues, determined by x-ray crystallography[13,14] and computational mapping. The experimental and computational results are based on the interactions found between various thermolysin residues and the probes in the clusters shown in Figures 13.2B and 13.2D, respectively. (B) The same as panel A, but for hydrogen bonds rather than nonbonded interactions. (From Brenke, R. et al., 2009. Reprinted with permission from *Bioinformatics*.)

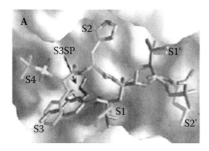 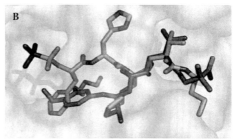

FIGURE 13.4 (SEE COLOR INSERT FOLLOWING PAGE 174.) (A) Published figure from Novartis depicting the conformation of aliskiren with respect to a peptidomimetic. (From J. M. Wood et al., Structure-based design of aliskiren, a novel orally effective renin inhibitor, *Biochem Biophys Res Commun*, 308, 698–705, 2003. Reprinted with permission from *Biochemical and Biophysical Research Communications*). (B) Resulting docked conformation of aliskiren, shown in purple, in the peptide-binding pocket of renin. The relative conformation of the same peptidomimetic from panel A is shown in green. The docked conformation of aliskiren is qualitatively consistent with the published figure. Key structural features, such as the occupation of the S_3^{SP} pocket, conjectured to be responsible for the affinity of aliskiren, are preserved in the docked conformation. (From M.R. Landon et al., Identification of hot spots within druggable binding sites of proteins by computational solvent mapping, *J Med Chem*, 50, 1231–1240, 2007. Reprinted with permission from the *Journal of Medicinal Chemistry*.)

For comparison with the mapping results, aliskiren was docked to a renin structure (PDB code 1RNE) using the GOLD program. Details regarding key contacts and hydrogen-bonding positions from the literature were used in conjunction with qualitative information, specifically Figure 13.4, to determine the most appropriate docked conformation for use in the analyses. The resulting conformation exhibited high similarity to the aliskiren-bound structure that has been later deposited in the PDB. Figure 13.4B shows the most similar docked conformation of aliskiren, colored in purple, in the binding pocket of renin. The conformation of the peptidomimetic inhibitor in 1RNE is shown in green. The resulting contact data between aliskiren and renin were used to compare the mapping results to residue interactions made by aliskiren and the peptidomimetic inhibitors of renin.

Five renin structures were mapped to predict favorable binding locations within the peptide binding pocket. As shown in Figure 13.5B, all consensus sites that were found to be located in the binding region are shown superimposed onto the structure of PDB code 1rne, renin with inhibitor. Consensus sites are shown in Figure 13.5A according to the structure from which they were derived; structures utilized from the PDB for the study were 1bil, 1bim, 1hrn, 1rne, and 2ren, where 2ren is an unliganded structure of renin and the other four structures are bound by peptidomimetic inhibitors. Given the confinement of all consensus sites to the same region of the binding pocket, it was concluded that small changes in the conformation of residues in the binding region do not affect the mapping results significantly.

Figure 13.5B is a close-up view of the consensus sites located in the binding pocket. In this depiction, all consensus sites resulting from the five structures are colored uniformly in light blue. For clarity, only the probe cluster representatives

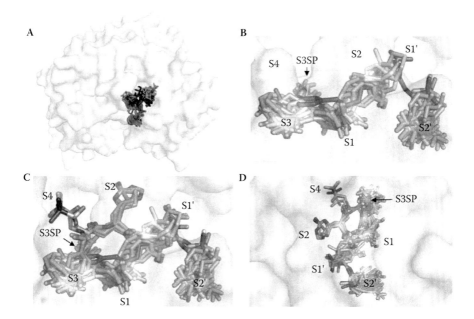

FIGURE 13.5   (SEE COLOR INSERT FOLLOWING PAGE 174.) Mapping results for five structures of renin. (A) The first and second ranked consensus sites resulting from the mapping of the different structures are superimposed in the peptide binding pocket of renin, demonstrating the reproducibility of the results. Each color represents the results of a distinctive protein. (B) Closer examination of the consensus sites depicted in panel A, now all colored light blue and shown in relation to the docked conformation of aliskiren, supports the importance of the S_1, S_2, S_3, and S_3^{SP} subsites for ligand affinity. (C and D) The preferred binding mode of aliskiren as compared to the peptidomimetics, shown in green, is confirmed by the mapping results. The S_2 and S_4 subsites are bound preferentially by the peptidomimetics, but not by aliskiren or the mapping probes. (From M.R. Landon et al., Identification of hot spots within druggable binding sites of proteins by computational solvent mapping, *J Med Chem*, 50, 1231–1240, 2007. Reprinted with permission from the *Journal of Medicinal Chemistry*.)

that comprise the consensus sites are displayed. The conformation of GOLD docked aliskiren is shown in purple in Figure 13.4B to allow for comparison to the mapping results. It is readily apparent from this view that the consensus sites in the binding region overlap significantly with the region occupied by aliskiren, making contacts primarily in the S_1, S_3, and S_2' subsites. The absence of probe clusters in the S_2 and S_4 subsites is illustrated in Figures 13.5C and 13.5D, where the four peptidomimetic inhibitors taken from the bound PDB structures that were used for mapping are added in green. Figure 13.5D is a side view of the binding region. Figures 13.5C and 13.5D emphasize that while each of the peptidomimetic inhibitors makes significant contacts in the S_2 and S_4 regions, neither aliskiren nor the mapping probes do so to a visible extent. Interestingly, one of the densest regions of probe molecules is found in the S_3^{SP} subsite, a region of the binding pocket that was described in the Novartis publication as being unique to the binding modality of aliskiren versus other renin inhibitors.

TABLE 13.1

Rankings of the Consensus Sites Present in the Subsites of the Binding Pocket of Renin for the Five Structures Mapped

PDB/Subsite	S4	S3	S2	S1	S1'	S2'
1RNE	NP	1 (28)	NP	1 (28)	6 (7)	2 (15)
1HRN	NP	1 (24)	4 (12)[a]	1 (24)	4 (12)[a]	2 (15)
1BIL	NP	1 (19)	NP	1 (19)	NP	2 (15)
1BIM	NP	1 (24)	NP	1 (24)	2 (14)	2 (14)
2REN	NP	1 (11)	NP	2 (8)	NP	NP

Source: From M.R. Landon et al., Identification of hot spots within druggable binding sites of proteins by computational solvent mapping, *J Med Chem, 50*, 1231–1240, 2007. Reprinted with permission from the *Journal of Medicinal Chemistry*.

Note: The number in parentheses indicates the number of probe clusters used to create the consensus site. NP = not present.

[a] Both consensus sites have the same number of clusters.

The ranks of the consensus sites occupying the different subsites of the binding pocket are summarized in Table 13.1, where the assignment of residues to a subsite was utilized from a previous publication.[42] With the exception of the unbound structure, the top ranked consensus sites occupy both the S_1 and S_3 subsites of the binding pocket for each structure; in the case of the unbound structure, the consensus site in the S_3 pocket is first and the consensus site in the S_1 pocket is second. Conformational changes undergone by aspartyl proteases upon ligand binding may account for the difference in mapping results existing between the unbound and the bound structures, in particular the change in shape of the S_3 region of the active site.[43] The important result is that no significantly populated consensus site is present in the S_4 pocket, and only a single, low-ranked consensus site is found in the S_2 subsite. This analysis suggests that the S_2 and S_4 subsites bind druglike functional groups with lower affinity than the other subsites of the peptide binding pocket. Based on these results, we can conclude that the S_1, S_3, and S_3^{SP} subsites of the binding pocket are hot spots for fragment binding and, within the S_3 subsite, the S_3^{SP} region displays particularly high affinity for small molecules.

Subsequent to the characterization of hot spots within the binding pocket of renin, the mapping algorithm was applied to the identification of specific residues that are crucial for ligand affinity within the hot spot regions. Residues were defined as part of the binding pocket if, based on the docking, any of their atoms were within 6 Å of an atom of the bound aliskiren. In addition, calculating interactions for the probe molecules, we determined interactions for both aliskiren and the four peptidomimetic inhibitors shown in Figures 13.5C and 13.5D for comparison. The resulting residue-based interaction distributions are shown in Figure 13.6, with the residues composing the binding pocket placed in sequence order on the horizontal axis. Residue interactions were calculated separately for each peptidomimetic inhibitor and then averaged to create one value. A high level of agreement exists between the

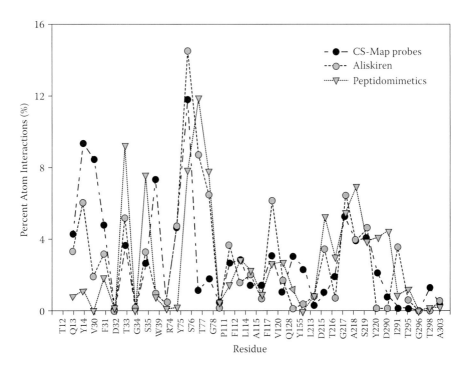

FIGURE 13.6 Distribution of atom-based residue interactions in the binding pocket of renin for the mapping probes, aliskiren, and three different peptidomimetics. In the case of the peptidomimetics, the average number of interactions at each residue was utilized. The Pearson correlation coefficient between the mapping probes and aliskiren is 0.72, while that between the mapping probes and the peptidomimetics is 0.19. (From M.R. Landon et al., Identification of hot spots within druggable binding sites of proteins by computational solvent mapping, *J Med Chem*, 50, 1231–1240, 2007. Reprinted with permission from the *Journal of Medicinal Chemistry*.)

distributions for aliskiren, shown in dark gray, and the mapping probes, shown in black (Figure 13.6); in particular, atom interactions are enriched in both distributions for residues G13 and Y75, located in the S_3^{SP} and S_1 subsites, respectively. However, while both aliskiren and the peptidomimetic inhibitors interact significantly with the catalytic D32, located in the S_1 subsite, the residue was not highly interactive with mapping probes; because both aliskiren and the mapping probes make the highest level of interactions in the S_1 subsite with Y75, this may suggest that while D32 is necessary for catalysis, it may contribute less to the binding affinity of ligands than the surrounding residues.

As compared to the high level of agreement existing between the GOLD docked aliskiren's and the mapping probes' distribution of residue interactions, the comparison of interaction distributions between the mapping probes and the peptidomimetics yielded a very low level of correlation. Residues predicted to be hot spots based on the analysis of interactions made with the peptidomimetics, shown in Figure 13.6 as the light gray distribution, were located primarily in the S2 subsite, such as residues S76 and A218. Agreement between the peptidomimetics and aliskiren was only seen in the

S1 region with the two catalytic aspartic acids, D32 and D215. As a quantitative measure of similarity, the Pearson correlation coefficients (R) in a pairwise fashion for the three distributions were calculated. Assuming that there are 20 residues in the entire binding site, the threshold for a correlation coefficient to be significant with a p-value of less than 0.01 is R = 0.52. The calculated R-value between the mapping probes and the GOLD bound aliskiren was 0.72, significantly higher than the R-value of 0.19 existing between the mapping probes and the peptidomimetic inhibitors. The correlation between aliskiren and the peptidomimetics was an intermediate value of 0.53; the main reason for the differences in correlation existing between the two different inhibitor types when compared to the mapping results is due to proximity of the mapping probes and aliskiren to S3SP compared to the peptidomimetic inhibitor's absence in this subregion. Strong hydrophobic interactions in this region allow aliskiren to exhibit high affinity despite its decrease in peptide-like character. This suggests that residues predicted by mapping as being highly interacting with the probes can serve as starting points for the development of high-affinity, druglike inhibitors.

HOT SPOTS IN THE LIGAND BINDING DOMAIN OF PPARγ

The peroxisome proliferator activated receptor-γ (PPARγ) is a ligand-activated transcription factor and a member of the nuclear receptor superfamily that plays an important role in adipogenesis and glucose homeostasis. PPARγ and the closely related receptors PPARα and PPARδ bind a variety of fatty acids and their metabolites.[44–46] Synthetic PPARγ agonists, including thiazolidinediones (TZDs), have been shown to be effective as insulin sensitizing agents, in reducing insulin resistance, and in lowering plasma glucose levels in patients with type 2 diabetes.[47] The effects of ligands on PPARγ are mediated through the ligand-binding domain (LBD), a region of 270 amino acid residues in the C-terminal half of the receptor.[45] In addition to its role in ligand binding, the LBD also contains dimerization and transactivation regions, which includes the transcriptional activation function 2 (AF-2) associated with helix 12 (H12). Structural[48–55] and biochemical[56,57] studies have helped to elucidate the mechanism of ligand-induced transcription activation by PPARγ. Upon binding of an agonist, the PPARγ LBD undergoes conformational changes, most notably in the AF-2 region. These changes result in the displacement of corepressor proteins that inhibit transcription and recruitment of coactivator proteins, which are required for transcriptional activation.

The LBD is comprised of a three-layer antiparallel α-helical sandwich of 13 helices and a small four-stranded β-sheet (Figure 13.7A). This architecture is very similar to that of other nuclear receptors with the exception of an extra helix, designated H2′, between the first β-strand and H3. The three long helices (H3, H7, and H10/H11) form the two outer layers of the sandwich. The middle layer of helices (H4, H5, H8, and H9) occupies the top half of the domain and is absent from the bottom half, thereby creating a very large cavity (1400 Å^3) for ligand binding. This large ligand-binding cavity has a distinct, three-arm Y-shape, which allows PPARγ to bind ligands with multiple branches or singly branched ligands in multiple conformations. On its lower half, the right-hand side of the LBD is sealed by a two-stranded β-sheet, while the left side is sealed by the

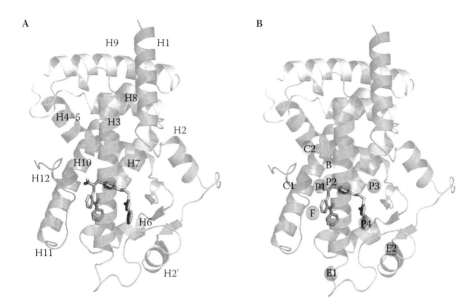

FIGURE 13.7 Structure of the PPARγ LBD. (A) The polypeptide backbone is shown as
a cartoon, indicating the 12 α-helices that comprise the domain. The agonist farglitazar
(GI262570, from PDB structure 1fm9) is shown as a stick representation. (B) Approximate
location of the hot spots or sites identified by the uniform mapping. Sites P1 through P4 are
subsites of the ligand binding site of PPARγ, with P1 located at the TZD or carboxyl head
group of the bound agonist, P2 slightly to the right, and P3 and P4 at the upper and lower
distal ends, respectively, of the ligand binding site. Sites F and B are, respectively, in the
front and back of the LBD, the latter in the dimerization domain. Sites C1 and C2 are located
in the region of coactivator binding, with C2 overlapping with the SRC-1 peptide. Finally,
sites E1 and E2 are in channels leading to the binding site, with E2 overlapping the putative ligand entrance. (From S.-H. Sheu et al., PPAR-γ ligand binding domain by computational solvent mapping, *Biochemistry* 44, 1193–1209, 2005. Reprinted with permission from
Biochemistry.)

short C-terminal α-helix (H12) of the receptor, which constitutes the receptor's
ligand dependent activating function 2 region (AF-2). A critical step during the
activation process involves ligand-induced alteration of the conformation of H12
to an active or "on" position, shown in Figure 13.7A, which acts as a molecular
switch and creates a binding cleft on the receptor for the coactivator.[45,46] A cleft
for the corepressor is formed in the same surface region when H12 is in the inactive or "off" position. In the unliganded LBD, the flexible pocket is believed to
be in an equilibrium of conformational states and can adopt the active state even
in absence of an agonist.[45,56,57]

To identify binding hot spots on the surface of PPARγ LBD, 12 LBD structures[28]
listed in Table 13.2 were mapped. The mapping used seven solvent molecules as
probes: acetone, acetonitrile, urea, methanol, isopropanol, tert-butanol, and phenol. The results show 10 binding consensus sites (Figure 13.7B). For reference,
Figure 13.7b also shows the bound agonist farglitazar (GI262570) from the structure
1fm9 (see Table 13.2).

TABLE 13.2
PPARγ Structures in the Protein Data Bank (PDB), Studied by Computational Solvent Mapping

Chain[a]	Structure	H12[b]	Ligand	Ref.
1prg(a)	Homodimer	On	none	48, 49
1prg(b)	Homodimer	Off	none	48, 49
4prg(a)	Homodimer	On	GW0072 (partial agonist)	50
4prg(b)	Homodimer	Off	GW0072 (partial agonist)	50
2prg(a)	Homodimer with SRC-1 peptide	On	Rosiglitazone (TZD)	48
1fm6(d)	PPARγ/RXRα heterodimer with SRC-1 peptide	On	Rosiglitazone (TZD)	51
1fm9(d)	PPARγ/RXRα heterodimer with SRC-1 peptide	On	Farglitazar (GI262570)	51
1k74(d)	PPARγ/RXRα heterodimer with SRC-1 peptide	On	GW409544	52
1i7i(a)	Homodimer	On	Tesaglitazar (AZ242)	53
1i7i(b)	Homodimer	Off	Tesaglitazar (AZ242)	53
1nyx(a)	Homodimer	On	Ragaglitazar (DRF2725)	54
1knu(a)	Homodimer	On	3q (carbazole analog of ragaglitazar)	55

Source: Sheu, S.-H., Kaya, T., Waxman, D. J., and Vajda, S. PPAR-γ ligand binding domain by computational solvent mapping. *Biochemistry 44*:1193–1209, 2005. With permission.

[a] PDB code. The number in parentheses specifies the chain studied.

[b] Position of the H12 helix. On and off denote active (or activelike) and inactive positions of the helix, respectively.

Four of the sites identified (P1 through P4) are within the ligand binding site (Figure 13.7B). Site P1 is on the left arm of the Y-shaped cavity adjacent to the carboxyl group of the PPARγ ligand GI262570, which makes hydrogen bonds with residues S289, H323, H449, and Y473. This largely hydrophilic pocket is important for the binding of all strong agonists, and accommodates the polar TZD or carboxyl headgroup that interacts with H12. The cavity in the structure with GI262570 extends downward to accommodate the benzophenone tail of the agonist. The P2 site is buried slightly to the right of P1 between helices H3 and H5. Sites P1 and P2 are adjacent to the AF-2 motif. Site P1 is present only in structures co-crystallized with a strong agonist. In contrast, site P2 is already formed in the ligand-free PPARγ. In structure 4prg, which has the bound partial agonist GW0073, one finds site P2 rather than site P1. Sites P1 and P2 are very close to each other and share a number of residues. However, a number of residues are absent from site P2 but become accessible to the solvent upon agonist binding, thereby opening up the new binding site P1. These newly accessible residues include Q286, H323, and Y473. Y473 is located on helix H12, and access to it contributes to the stabilization of H12 in the active conformation. It appears that the strong nonbonded interactions and hydrogen bonds between the head group of TZD, a carboxyl group, and the protein are necessary for creating a pocket at site P1. Indeed, as shown in Table 13.3, neither the ligand-free

TABLE 13.3

Number of Clusters and Ranking of the Binding Sites, Determined by Targeted Mapping

	PPARγ (Chain)											
Site	1prg (a)	1prg (b)	4prg (a)	4prg (b)	2prg (a)	1fm6 (d)	1fm9 (d)	1k74 (d)	1i7l (a)	1i7l (b)	1knu (a)	1nyx (a)
P1					7(1)	9(1)	6(2)	9(1)	8(2)	7(2)	6(2)	9(1)
P2	9(2)	11(1)	3(5)	5(4)								
P3	9(1)	9(2)	8(1)	6(2)	5(4)	7(2)	3(6)	3(5)	5(3)	6(3)	4(5)	4(5)
P4	8(3)	9(3)	5(3)	7(1)	5(5)	5(4)	8(1)	5(4)	10(1)	8(1)	9(1)	2(7)
B		2(5)	3(6)	4(5)					3(6)	5(4)		4(4)
F			7(2)	5(3)	6(2)	2(7)				5(5)	2(7)	
C1	7(4)		4(4)		2(7)	3(5)	5(4)	5(3)	5(4)	4(6)	6(3)	6(3)
C2	2(5)	2(4)	3(7)		6(3)	6(3)	3(7)	3(6)	4(5)		5(4)	6(2)
E1							4(5)	6(2)				
E2					4(6)	2(6)	5(3)	2(7)			3(6)	3(6)

Source: Sheu, S.-H., Kaya, T., Waxman, D. J., and Vajda, S. PPAR-γ ligand binding domain by computational solvent mapping. *Biochemistry* 44:1193–1209, 2005. With permission.

Note: For each PPARγ structure, the table shows the number of clusters found at each of the 10 sites, P1 through E2. The numbers in parentheses indicate the rank of the corresponding consensus site among all consensus sites for that structure, ranked on the basis of the number of clusters.

structure 1prg nor the partial agonist-bound structure 4prg has any consensus site in the vicinity of site P1. However, once the hydrophilic, high affinity pocket at P1 becomes accessible, it siphons away the probes from P2 resulting in the loss of P2.

Table 13.3 indicates that site P1 binds about the same number of probes in each of the structures with strong agonists bound. This suggests that strong agonists are likely to induce similar conformational changes. Furthermore, it appears that pocket P1 is the result of a complex set of cooperative conformational changes in a number of residues (primarily Q286, F363, H449, Y473, H323, S289, and I341), which are somewhat coordinated by the large movement of R288.

Site P3 is important for partial agonist binding and domain stabilization. P3 is at the end of the right arm of the Y-shaped cavity, which is defined by helix H3 from the left, helix H2 from the right, the loop connecting helices H1 and H2 from the top, and helix H5 from the back. The only ligand known to reach into this site is the partial agonist GW0072, which places one of its two benzylamide groups in P3. The P3 site is a large pocket, which is at least partially open in all PPARγ structures. In the apo structure, the P3 site is open and available to the probes as suggested by the data in Table 13.3. Based on the number of probe clusters bound, the binding of the partial agonist yields a smaller P3 site, and the pocket is even smaller in structures with strong agonists. In all such structures, the R288 side chain moves upward into the P3 site and closes down part of the pocket. A less open pocket at site P3 in all agonist-bound structures agrees with the observations that ligand binding globally stabilizes the LBD domain.[45,53,56,57]

Site P4 is the most hydrophobic part of the ligand binding pocket. Site P4 is close to the lower end of the ligand binding site, which is defined by H3 from the left and by the β-sheet and helix H2′ from the right. As seen in Table 13.3, site P4 is accessible to probes in the ligand-free PPARγ structures. The pocket at site P4 is slightly smaller after binding the partial agonist GW00720 (4prg), which places its carboxyl group in this region. P4 is also reduced in structures with short agonists such as rosiglitazone (2prg and 1fm6) in which the distal end does not reach site P4. P4 generally remains large for longer agonists that have their distal end group bound at P4. Ragaglitazar in 1nyx has a bulky phenoxazine end group, and it is too short to reach site P4, which is small and binds only two probe clusters. A carbazole ring of 3q in 1knu is rotated downward into the P4 pocket resulting in its larger size (Table 13.3).

Site B is a potential hot spot in the dimerization region. Site B is located in the back of the LBD between helices H7 and H10/H11, and close to residue Q444 on the latter. This site is part of the protein surface involved in forming both the PPARγ–RXRα heterodimer and the PPARγ homodimer. Nettles et al.[58] demonstrated that, due to an allosteric effect, H11 conveys structural information between the ligand and H12, thereby affecting receptor activation. Since the size of site B depends on both the ligand and dimerization partner, it may play a role in this communication.

Site F is a surface pocket between H3 and H12. This site is created either almost exclusively by the inward motion of the Q286 side chain upon binding of the partial agonist in 4prg or by the binding of agonists that do not have a bulky group protruding downward from the site P1 pocket. The significance of this pocket for coactivator recognition was considered unknown; however, more recently it was demonstrated that this region is likely to be the binding site for a secondary coactivator.[59]

C1 and C2 are sites in the coactivator binding region around helix H12 and close to the putative region of the coactivator/corepressor binding. Site C2 is located between helices H12 and the H4/H5 boundary. The pocket overlaps with the binding site of the SRC-1 coactivator peptide, which is present in several x-ray structures. As shown in Table 13.3, site C2 is present in the ligand-free and partial-agonist-bound structures but only when H12 is in the active state. Agonist binding stabilizes site C2, which accommodates about the same number (five or six) clusters in all structures with agonists. Site C1 is located on the opposite side of H12 close to the center of a triangle formed by H10, H12, and the loop between the two helices. Although there is no direct evidence that this site is involved in coactivator binding, the coactivators of PPARγ are large proteins and hence are likely to extend beyond the known SRC-1 peptide binding site, thus also covering the site C1. The C1 site is very weak or absent in structures in which H12 is in the inactive position.

The numbers of probe clusters in C1 and C2 appear to provide information on the coactivator binding specificity of a particular PPARγ structure. Without a ligand, site C2 is very small in both chains of 1prg (Table 13.3), whereas site C1 discriminates between the activelike and inactive forms. The binding of the partial agonist creates both C1 and C2 in the active chain of 4prg, but both sites are of modest size. In addition, none of the sites are present in chain b, and thus the binding of the partial agonist GW0072 is unable to stabilize the coactivator recognition pockets. All active agonist-bound structures exhibit stable C1 and C2 binding sites, suggesting that they may be prerequisites for coactivator binding. The relative strengths of the two pockets seem to depend on the properties of the agonist. Both mapped rosiglitazone-bound structures 2prg and 1fm6 have a modest C1 and very strong C2, the latter containing six clusters (Table 13.3). The binding of the L-tyrosine derivatives in 1fm9 and 1k74 creates a relatively large C1 site (five clusters) and a smaller C2 site (three clusters).

E1 and E2 are putative entrance sites, surface pockets at the openings on the two ends of the PPARγ LBD binding site. Site E2 is located between helix H2 and the β-sheet. This position is frequently mentioned as the putative entrance to the LBD binding site.[53] In view of the lack of any alternative site at the distal end of the binding site, our results support the hypothesis that the ligands are likely to enter through the pocket at site E2. Indeed, the loop connecting H2' and H3 is very flexible and does not prevent the entrance of large ligands into the binding site. Site E1 is most likely produced specifically by L-tyrosine agonists, primarily through their effects on the orientation of the F182 side chain.

IDENTIFICATION OF DRUGGABLE HOT SPOTS IN PROTEIN–PROTEIN INTERFACES

Finding small-molecule inhibitors against protein–protein interaction (PPI) targets has always been a challenge. It is very hard to obtain a low-nanomolar inhibitor that does not violate Lipinski's rule of five[60] due to the fact that PPI interfaces are typically large and flat, resulting in the necessary noncovalent contacts needed for affinity to not converge to a small volume.[61] Nevertheless, small-molecule PPI inhibitors are often more desirable than their protein counterparts due to their oral availability and their ability to pass through cell membranes. Current examples of PPI inhibition targets include IL-2, Bcl-X_L, HDM2, HPV E2, ZipA binders, TNF disruptors, and others.[62] Binding hot spot determination is important because identifying a small number of residues that provide a relatively large portion of the binding free energy helps to select the regions to target for inhibitor design. This can be done by alanine scanning mutagenesis or computational methods such as protein mapping. The plasticity of the PPI interfaces should also be considered because a number of findings showed that binding of small molecules on flat PPI interfaces creates a substantial cavity at the binding site. Fragment-based screening, using chemical fragments with an average molecular weight between 100 to 300 Da, can be more efficient than traditional high throughput screening (HTS) because it can cover more chemical space. Moreover, the HTS libraries are often biased toward past drug discovery research on traditional druggable targets, which is also overcome by fragment-based screening.

Computational analyses of PPI interfaces provide some insights on the physical nature of these binding surfaces. Eyrisch and Helms[63] ran molecular dynamics (MD) simulations on PPI interfaces in Bcl-X_L, MDM2, and IL-2 and used the cavity identifier program PASS to scan for cavities on the trajectory snapshots. As a result, 20–36 transient pockets were observed during MD simulations as compared to the 2–5 pockets identified from the ligand-free crystal structures. The pockets were all found to be open up to 440 ps, vanished and reappeared again several times. The pocket polarity ratio (ratio of the sum of N, O, and S atoms to the sum of N, O, S, and C atoms) is lower in the larger pockets than overall on the protein surface. This suggests that the hydrophobic interiors open up to form these pockets and the hydrophobicity increases the likelihood of ligand binding. Notably in all three cases, full openings of cavities are identified at the known inhibitor binding site despite the fact that PASS did not identify the inhibitor binding site in the static ligand-free IL-2 structure. The corresponding inhibitors that docked computationally in these simulated transient pockets had very low root mean square deviation (RMSD) from the crystallized structures. The protocol demonstrated that the flat PPI interfaces shown in the unbound structures are flexible and can yield favorable binding cavities. As a result, this plasticity has to be taken into account when designing PPI inhibitors.

Protein mapping can be applied to identify the binding hot spots on the PPI interface. For example, a PPI inhibitor of IL-2 shown in Figures 13.8A and 13.8B is composed of two components, a hydrophilic fragment containing a piperidyl guanidine forming a hydrogen bond with the side-chain carboxylate of E62, and a hydrophobic fragment containing a biaryl alkyne located in a narrow channel created by the hydrophobic side chains of M39, R38, F42, L72, and K76. As shown in Figure 13.8A and 13.8B, the binding of the inhibitor creates a long cavity not seen in the unbound structure. For the protein mapping of the unbound IL-2 shown in Figure 13.8C the largest consensus site, containing 20 probe clusters, is located near E62 where the hydrophilic part of the compound interacts. The fourth largest consensus site, containing 10 probe clusters, is located near R38, L72, and F42 where the hydrophobic end of the compound resides. This suggests that these two regions are the binding hot spots of this interface for small molecules. Similar binding hot spots are observed in the mapping of the compound bound IL-2 (Figure 13.8D).

Another example of a PPI target is the bacterial protein RecA. RecA is a multifunctional DNA-dependent ATPase that catalyzes recombination and acts as the signal protein to start the SOS repair response. The SOS response results in the synthesis of approximately 30 proteins, many of which work to repair damaged DNA. Free RecA binds to single-stranded DNA (ssDNA), which then acts as a substrate for additional RecA molecules to bind. This results in long nucleoprotein filaments consisting of RecA, ssDNA, and either ATP or ADP depending on the state of the protein. Electron micrograph studies show that these filaments exist in two conformational states.[64] The active state with bound ATP is represented by an extended filament with a pitch of ~95 Å, while the first crystal structures of RecA, solved in the absence of DNA and ATP, show an inactive form with a pitch of 83 Å.[65]

Because the activity of RecA lies in its ability to form long filaments on ssDNA, the protein–protein interface between neighboring RecA subunits in these filaments is of particular interest. Roca and Cox defined nine residues (A214–R222) in the protein–protein interface as the RecA "signature series."[66] Five of the nine residues are identical in 64 RecA sequences. Three residues that make specific contacts in a region of the neighboring subunit, K216, F217, and R222, have been shown to be intolerant to most mutations.[67] Least tolerant is F217, which, when mutated to 14 other residues, showed little or no RecA activity; however, when mutated to tyrosine, the protein retained full functionality and also increased the cooperative interaction between RecA subunits 250-fold.[68] This suggests that the hot spot to which F217 binds contains the capacity to bind tightly to the additional hydroxyl group of tyrosine.

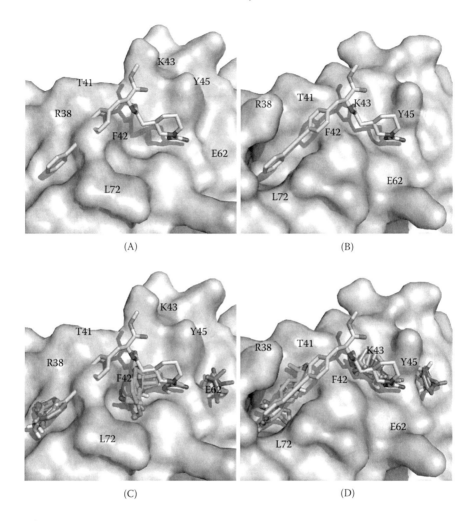

FIGURE 13.8 (A) Overlap of the compound on unbound IL-2 crystal structure. (B) The compound bound IL-2 crystal structure. (C) Protein mapping result of unbound IL-2. Consensus sites shown in the figure from left to right are fourth, first, and sixth largest consensus sites, respectively. (D) Protein mapping result of compound bound IL-2. Consensus sites shown in the figure from left to right are first, fourth, and sixth largest consensus sites, respectively.

An inactive RecA structure bound to the ATP analog MnANP-PNP was mapped (PDB code 1xms).[69] All ligands and waters were removed during mapping, and mapping was performed only on single monomers. Thirteen probes were used in this study. Seven high ranking probes were found in the ATP binding site, while all 13 of the probes were found in the largest consensus site (CS1). CS1 was identified to be in a hydrophobic pocket located in the interface between two RecA monomers, consisting of the residues I93-A95, Q118-G122, A125, A148-T150, and I155. This was identified to be the location where the F217 residue from the neighboring RecA monomer interacts. Figure 13.9 shows detail of the mapping results in CS1 with the

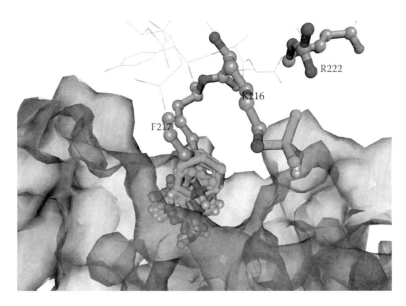

FIGURE 13.9   (SEE COLOR INSERT FOLLOWING PAGE 174.) Detail of the *E. coli* RecA CS1 pocket with the consensus cluster shown in green and neighboring subunit shown in orange. The probes are shown in atomic colors with carbons colored green, oxygens colored red, and nitrogen colored blue.

neighboring monomer overlapping. What is noteworthy is that many of the probes that contain hydroxyl groups or other hydrogen donors have these groups oriented toward the bottom of this pocket. Together with the mutation results suggesting that tyrosine may bind to this pocket tighter than phenylalanine, this shows that mapping not only reveals hot spots that are important for protein–protein interactions but may also elucidate details of the binding properties of these hot spots.

CONCLUSIONS

As demonstrated by alanine scanning mutagenesis, the interfaces in protein–protein complexes and the binding sites of proteins generally contain smaller regions that provide major contributions to the binding free energy and hence are the prime targets in drug design. Screening of compound libraries by NMR or x-ray crystallography shows that such hot spot regions bind a large variety of small organic molecules of various sizes and polarities. Although the binding of most small compounds is weak, a relatively high hit rate is predictive of target sites that are likely to bind druglike ligands with high affinity. The goal of this review was to show that mapping algorithms can provide similar information computationally. By selecting 16 probe molecules with different properties, such as hydrogen bond donors and acceptors or aliphatic and aromatic hydrocarbons, mapping provides an excellent characterization of protein binding sites. The applications presented demonstrate the type of information that can be obtained by mapping. We have mapped thermolysin because the structure of this protein has been determined in aqueous solutions of four organic solvents. Mapping correctly identifies all-important subsites of the binding site and finds all-important residues. As we have shown, the mapping results are even more complete than the ones provided by x-ray crystallography. In the second example, we considered renin, a long standing pharmaceutical target for the treatment of hypertension, and showed that the few largest consensus sites trace out the shape of the first approved renin inhibitor, aliskiren, rather than that of peptidomimetic inhibitors that have been studied for several decades but did not provide any successful drug candidate. It is important that the mapping reveals the better fit of aliskiren into the hot spots even when applied to a renin structure without any bound ligand or to structures co-crystallized with peptidomimetic inhibitors. The next application was to the peroxisome proliferator activated receptor (PPARγ) ligand binding domain (LBD). The mapping identified 10 sites as hot spots for the recognition of interaction partners by the LBD. Four of these sites (P1 through P4) are in the ligand binding site, two (C1 and C2) in the coactivator binding area, one in the dimerization region (B), two (E1 and E2) in the lower half of LBD where E2 is at the putative ligand entrance site, and an additional site of unknown function is on the surface (F). We note that this last pocket has been recently identified as a secondary coactivator binding site.[59] Finally, we showed that mapping can find hot spot regions in protein–protein interface even before the pockets are formed due to the interactions with small molecular ligands. Thus, we concluded that mapping provides potentially very useful information for drug design.

We note that the method of screening libraries of small compounds by NMR to explore the binding properties of proteins (known as SAR by NMR) has been developed by the Fesik group for Abbott Laboratories, and it has been applied to many drug targets during the last decade.[9] At present, several other companies also screen libraries of fragment-sized compounds, either by NMR or by x-ray crystallography, as part of their core technology. This includes small companies such as Astex Therapeutics, Locus Pharmaceuticals, and De Novo Pharmaceuticals as well as many larger companies such as Vertex, Astra-Zeneca, Lilly, Novartis, and Roche; however, there are almost no detailed results from this work in the public domain. Since the methodology is costly and labor intensive, academic labs rarely report new studies. In fact, the available data are scarce enough that the interplay between experimental and computational solvent mapping is truly symbiotic; that is, experiments validate the computational method, which in turn supports the generality and meaning of the small organic molecules clustering in the hot spot regions of protein binding sites and potential protein–protein interfaces. We note that the FTMAP program is freely available for academic use by contacting Sandor Vajda (vajda@bu.edu) or Dima Kozakov (midas@bu.edu).

ACKNOWLEDGMENTS

This work has been supported by grants GM061867 and GM064700 from the National Institutes of Health, and P42ES07381 from the National Institute of Environmental Health Sciences.

REFERENCES

1. Clackson, T. and Wells, J. A hot spot of binding energy in a hormone-receptor interface. *Science 267*:383–386, Jan, 1995.
2. Bogan, A. and Thorn, K. Anatomy of hot spots in protein interfaces. *J. Mol. Biol. 280*:1–9, Jul, 1998.
3. Thorn, K. and Bogan, A. ASEdb: A database of alanine mutations and their effects on the free energy of binding in protein interactions. *Bioinformatics 17*:284–285, Mar, 2001.
4. Hu, Z., Ma, B., Wolfson, H., and Nussinov, R. Conservation of polar residues as hot spots at protein interfaces. *Proteins 39*:331–342, Jun, 2000.
5. DeLano, W. Unraveling hot spots in binding interfaces: Progress and challenges. *Curr Opin Struct Biol 12*:14–20, Feb, 2002.
6. Kortemme, T. and Baker, D. A simple physical model for binding energy hot spots in protein–protein complexes. *Proc Natl Acad Sci USA 99*(22):14116–14121, Oct, 2002.
7. Vajda, S. and Guarnieri, F. Characterization of protein-ligand interaction sites using experimental and computational methods. *Curr Opin Drug Discov Devel 9*(3):354–362, May, 2006.
8. Hajduk, P. J., Huth, J. R., and Fesik, S. W. Druggability indices for protein targets derived from NMR-based screening data. *J Med Chem 48*(7):2518–2525, Apr, 2005.
9. Hajduk, P., Huth, J., and Tse, C. Predicting protein druggability. *Drug Discov Today 10*:1675–1682, Dec, 2005.
10. Mattos, C. and Ringe, D. Locating and characterizing binding sites on proteins. *Nat Biotechnol 14*(5):595–599, May, 1996.
11. Mattos, C., Bellamacina, C. R., Peisach, E., Pereira, A., Vitkup, D., Petsko, G. A., and Ringe, D. Multiple solvent crystal structures: Probing binding sites, plasticity and hydration. *J Mol Biol 357*(5):1471–1482, Apr, 2006.
12. Allen, K., Bellamacina, C. R., Ding, X., Jeffery, C. J., Mattos, C., Petsko, G. A., and Ringe, D. An experimental approach to mapping the binding surfaces of crystalline proteins. *J Phys Chem 100*:2605–2611, 1996.
13. English, A. C., Done, S. H., Caves, L. S., Groom, C. R., and Hubbard, R. E. Locating interaction sites on proteins: The crystal structure of thermolysin soaked in 2% to 100% isopropanol. *Proteins 37*(4):628–640, Dec, 1999.
14. English, A. C., Groom, C. R., and Hubbard, R. E. Experimental and computational mapping of the binding surface of a crystalline protein. *Protein Eng 14*(1):47–59, Jan, 2001.
15. Novotny, J., Bruccoleri, R., and Saul, F. On the attribution of binding energy in antigen–antibody complexes McPC 603, D1.3, and HyHEL-5. *Biochemistry 28*:4735–4749, May, 1989.
16. Massova, I. and Kollman, P. A. Computational alanine scanning to probe protein–protein interactions: A novel approach to evaluate binding free energies. *J Amer Chem Soc 121*(36):8133–8143, Aug, 1999.
17. Rahuel, J., Rasetti, V., Maibaum, J., Rueger, H., Goschke, R., Cohen, N. C., Stutz, S., Cumin, F., Fuhrer, W., Wood, J. M., and Grutter, M. G. Structure-based drug design: The discovery of novel nonpeptide orally active inhibitors of human renin. *Chem Biol 7*(7):493–504, Jul, 2000.
18. Wood, J. M., Maibaum, J., Rahuel, J., Grutter, M. G., Cohen, N.-C., Rasetti, V., Ruger, H., Goschke, R., Stutz, S., Fuhrer, W., Schilling, W., Rigollier, P., Yamaguchi, Y., Cumin, F., Baum, H.-P., Schnell, C. R., Herold, P., Mah, R., Jensen, C., O'Brien, E., Stanton, A., and Bedigian, M. P. Structure-based design of aliskiren, a novel orally effective renin inhibitor. *Biochem Biophys Res Commun 308*(4):698–705, Sep, 2003.

19. An, J. H., Totrov, M., and Abagyan, R. Pocketome via comprehensive identification and classification of ligand binding envelopes. *Mol Cell Proteomics 4*:752–761, 2005.

20. Laurie, A. T. R. and Jackson, R. M. Q-SiteFinder: An energy-based method for the prediction of protein-ligand binding sites. *Bioinformatics 21*:1908–1916, 2005.

21. Glaser, F., Morris, R. J., Najmanovich, R. J., Laskowski, R. A., and Thornton, J. M. A method for localizing ligand binding pockets in protein structures. *Proteins 62*:479–488, 2006.

22. Goodford, P. J. A computational procedure for determining energetically favorable binding sites on biologically important macromolecules. *J Med Chem 28*:849–875, 1985.

23. Miranker, A. and Karplus, M. Functionality maps of binding sites: A multiple copy simultaneous search method. *Proteins 11*:29–34, 1991.

24. Caflisch, A., Miranker, A., and Karplus, M. Multiple copy simultaneous search and construction of ligands in binding sites: Application to inhibitors of HIV-1 aspartic proteinase. *J. Med. Chem. 36*:2142–2167, 1993.

25. Dennis, S., Kortvelyesi, T., and Vajda, S. Computational mapping identifies the binding sites of organic solvents on proteins. *Proc Natl Acad Sci USA 99*:4290–4295, 2002.

26. Silberstein, M., Dennis, S., Brown, L. III, Kortvelyesi, T., Clodfelter, K., and Vajda, S. Identification of substrate binding sites in enzymes by computational solvent mapping. *J Mol Bio 332*:1095–1113, 2003.

27. Kortvelyesi, T., Dennis, S., Silberstein, M., Brown, L. III, and Vajda, S. Algorithms for computational solvent mapping of proteins. *Proteins 51*(3):340–351, May, 2003.

28. Sheu, S.-H., Kaya, T., Waxman, D. J., and Vajda, S. PPAR-γ ligand binding domain by computational solvent mapping. *Biochemistry 44*:1193–1209, 2005.

29. Silberstein, M., Damborsky, J., and Vajda, S. Exploring the binding sites of the haloalkane dehalogenase DhlA from Xanthobacter autotrophicus GJ10. *Biochemistry 46*:9239–9249, 2007.

30. Landon, M. R., Lancia, D. R. J., Yu, J., Thiel, S. C., and Vajda, S. Identification of hot spots within druggable binding sites of proteins by computational solvent mapping. *J Med Chem 50*:1231–1240, 2007.

31. Landon, M. R., Amaro, R. E., Baron, R., Ngan, C. H., Ozonoff, D., McCammon, J. A., and Vajda, S. Novel druggable hot spots in avian influenza neuraminidase H5N1 revealed by computational solvent mapping of a reduced and representative receptor ensemble. *Chem Biol Drug Des 71*:106–116, 2008.

32. Kozakov, D., Brenke, R., Comeau, S., and Vajda, S. PIPER: An FFT-based protein docking program with pairwise potentials. *Proteins 65*:392–406, Nov, 2006.

33. Berman, H., Westbrook, J., Feng, Z., Gilliland, G., Bhat, T., Weissig, H., Shindyalov, I., and Bourne, P. The Protein Data Bank. *Nucleic Acids Res 28*:235–242, 2000.

34. Brooks, B. R., Bruccoleri, R. E., Olafson, B. D., States, D. J., Swaminathan, S., and Karplus, M. CHARMM: A program for macromolecular energy, minimization, and dynamics calculations. *J Comput Chem 4*:187–217, 1983.

35. Ruvinsky, A. M. and Kozintsev, A. V. Novel statistical-thermodynamic methods to predict protein-ligand binding positions using probability distribution functions. *Proteins 62*:202–208, 2006.

36. Mattos, C. and Ringe, D. Proteins in organic solvents. *Curr Opin Struct Biol 11*(6):761–764, Dec, 2001.

37. Tronrud, D. E., Roderick, S. L., and Matthews, B. W. Structural basis for the action of thermolysin. *Matrix Suppl 1*:107–111, 1992.

38. Matthews, B. W. Structural basis of the action of thermolysin and related zinc peptidases. *Acc Chem Res 21*:333–340, 1988.

39. Lipscomb, W. N. and Strater, N. Recent advances in zinc enzymology. *Chem Rev 96*:2375–2433, 1996.

40. Marie-Claire, C., Ruffet, E., Antonczak, S., Beaumont, A., O'Donohue, M., Roques, B. P., and Fournie-Zaluski, M. C. Evidence by site-directed mutagenesis that arginine 203 of thermolysin and arginine 717 of neprilysin (neutral endopeptidase) play equivalent critical roles in substrate hydrolysis and inhibitor binding. *Biochemistry 36*:13938–13945, 1988.

41. Marie-Claire, C., Ruffet, E., Tiraboschi, G., and Fournie-Zaluski, M. C. Differences in transition state stabilization between thermolysin (ec 3.4.24.27) and neprilysin (ec 3.4.24.11). *FEBS Lett 438*:215–219, 1998.

42. Rahuel, J., Priestle, J., and Grtter, M. The crystal structures of recombinant glycosylated human renin alone and in complex with a transition state analog inhibitor. *J Struct Biol 107*:227–236, Dec, 1991.

43. Sali, A., Veerapandian, B., Cooper, J., Moss, D., Hofmann, T., and Blundell, T. Domain flexibility in aspartic proteinases. *Proteins 12*:158–170, Feb, 1992.

44. Berger, J. and Moller, D. The mechanism of action of PPARs. *Annu Rev Med 53*:409–435, 2002.

45. Nagy, L. and Schwabe, J. Mechanism of the nuclear receptor molecular switch. *Trends Biochem Sci 29*:317–324, Jun, 2004.

46. Li, Y., Lambert, M., and Xu, H. Activation of nuclear receptors: A perspective from structural genomics. *Structure 11*:741–746, Jul, 2003.

47. Willson, T., Brown, P., Sternbach, D., and Henke, B. The PPARs: From orphan receptors to drug discovery. *J Med Chem 43*:527–550, Feb, 2000.

48. Nolte, R., Wisely, G., Westin, S., Cobb, J., Lambert, M., Kurokawa, R., Rosenfeld, M., Willson, T., Glass, C., and Milburn, M. Ligand binding and co-activator assembly of the peroxisome proliferator-activated receptor-gamma. *Nature 395*:137–143, Sep, 1998.

49. Uppenberg, J., Svensson, C., Jaki, M., Bertilsson, G., Jendeberg, L., and Berkenstam, A. Crystal structure of the ligand binding domain of the human nuclear receptor PPARgamma. *J Biol Chem 273*:31108–31112, Nov, 1998.

50. Oberfield, J., Collins, J., Holmes, C., Goreham, D., Cooper, J., Cobb, J., Lenhard, J., Hull-Ryde, E., Mohr, C., Blanchard, S., Parks, D., Moore, L., Lehmann, J., Plunket, K., Miller, A., Milburn, M., Kliewer, S., and Willson, T. A peroxisome proliferator-activated receptor gamma ligand inhibits adipocyte differentiation. *Proc Natl Acad Sci USA 96*:6102–6106, May, 1999.

51. Gampe, R., Montana, V., Lambert, M., Miller, A., Bledsoe, R., Milburn, M., Kliewer, S., Willson, T., and Xu, H. Asymmetry in the PPARgamma/RXRalpha crystal structure reveals the molecular basis of heterodimerization among nuclear receptors. *Mol Cell 5*:545–555, Mar, 2000.

52. Xu, H., Lambert, M., Montana, V., Plunket, K., Moore, L., Collins, J., Oplinger, J., Kliewer, S., Gampe, R., McKee, D., Moore, J., and Willson, T. Structural determinants of ligand binding selectivity between the peroxisome proliferator-activated receptors. *Proc Natl Acad Sci USA 98*:13919–13924, Nov, 2001.

53. Cronet, P., Petersen, J., Folmer, R., Blomberg, N., Sjblom, K., Karlsson, U., Lindstedt, E., and Bamberg, K. Structure of the PPARalpha and -gamma ligand binding domain in complex with AZ 242; ligand selectivity and agonist activation in the PPAR family. *Structure 9*:699–706, Aug, 2001.

54. Ebdrup, S., Pettersson, I., Rasmussen, H., Deussen, H., Frost Jensen, A., Mortensen, S., Fleckner, J., Pridal, L., Nygaard, L., and Sauerberg, P. Synthesis and biological and structural characterization of the dual-acting peroxisome proliferator-activated receptor alpha/gamma agonist ragaglitazar. *J Med Chem 46*:1306–1317, Apr, 2003.

55. Sauerberg, P., Pettersson, I., Jeppesen, L., Bury, P., Mogensen, J., Wassermann, K., Brand, C., Sturis, J., Wldike, H., Fleckner, J., Andersen, A., Mortensen, S., Svensson, L., Rasmussen, H., Lehmann, S., Polivka, Z., Sindelar, K., Panajotova, V., Ynddal, L., and Wulff, E. Novel tricyclic-alpha-alkyloxyphenylpropionic acids: Dual PPARalpha/gamma agonists with hypolipidemic and antidiabetic activity. *J Med Chem 45*:789–804, Feb, 2002.

56. Kallenberger, B., Love, J., Chatterjee, V., and Schwabe, J. A dynamic mechanism of nuclear receptor activation and its perturbation in a human disease. *Nat Struct Biol 10*:136–140, Feb, 2003.

57. Pissios, P., Tzameli, I., Kushner, P., and Moore, D. Dynamic stabilization of nuclear receptor ligand binding domains by hormone or corepressor binding. *Mol Cell 6*:245–253, Aug, 2000.

58. Nettles, K., Sun, J., Radek, J., Sheng, S., Rodriguez, A., Katzenellenbogen, J., Katzenellenbogen, B., and Greene, G. Allosteric control of ligand selectivity between estrogen receptors alpha and beta: Implications for other nuclear receptors. *Mol Cell 13*:317–327, Feb, 2004.

59. Shiraki, T., Kodama, T., Jingami, H., and Kamiya, N. Rational discovery of a novel interface for a coactivator in the peroxisome proliferator-activated receptor gamma: Theoretical implications of impairment in type 2 diabetes mellitus. *Proteins 58*:418–425, Feb, 2005.

60. Lipinski, C. Drug-like properties and the causes of poor solubility and poor permeability. *J Pharmacol Toxicol Meth 44*:235–249, 2000.

61. Whitty, A. and Kumaravel, G. Between a rock and a hard place? *Nat Chem Biol 2*:112–118, Mar, 2006.

62. Wells, J. and McClendon, C. Reaching for high-hanging fruit in drug discovery at protein–protein interfaces. *Nature 450*:1001–1009, Dec, 2007.

63. Eyrisch, S. and Helms, V. Transient pockets on protein surfaces involved in protein–protein interaction. *J Med Chem 50*:3457–3464, Jul, 2007.

64. Egelman, E. and Stasiak, A. Electron microscopy of RecA-DNA complexes: Two different states, their functional significance and relation to the solved crystal structure. *Micron 24*(3):309–324, 1993.

65. Story, R., Weber, I., and Steitz, T. The structure of the E. coli recA protein monomer and polymer. *Nature 355*(6358):318–325, 1992.

66. Roca, A. and Cox, M. RecA protein: Structure, function, and role in recombinational DNA repair. *Prog Nucleic Acid Res Mol Biol 56*:129–223, 1997.

67. Skiba, M. and Knight, K. Functionally important residues at a subunit interface site in the RecA protein from Escherichia coli. *J Biol Chem 269*(5):3823–3828, Feb, 1994.

68. Kelley De Zutter, J., Forget, A., Logan, K., and Knight, K. Phe217 regulates the transfer of allosteric information across the subunit interface of the RecA protein filament. *Structure (Camb) 9*(1):47–55, 2001.

69. Xing, X. and Bell, C. Crystal structures of Escherichia coli RecA in complex with MgADP and MnAMP-PNP. *Biochemistry 43*(51):16142–16152, Dec, 2004.

70. Brenke, R., Kozakov, D., Chuang, G. Y., Beglov, D., Hall, D., Landon, M. R., Mattos, C., and Vajda, S. Fragment-based identification of druggable "hot spots" of proteins using Fourier domain correlation techniques. *Bioinformatics 25*:621–627, 2009.

71. Dennis, S., Kortvelyesi, T., and Vajda, S. Computational mapping identifies the binding sites of organic solvents on proteins. *Proc Natl Acad Sci USA 99*(7):4290–4295, 2002.

14 Designing Protein–Protein Interaction Inhibitors

Matthieu Montes

CONTENTS

Protein–protein interactions (PPIs) play a crucial role in many biological processes, such as cellular communication, immune defense, viral self-assembly, cell growth, proliferation, differentiation, and programmed cell death.[1,2] Since their disruption can lead to a disease state,[3] it is of great interest to consider them as potential drug targets.[4]

Whereas traditional active sites that are targeted in drug design projects are generally well defined and display limited size and complexity, PPIs are far more challenging.[5] There are several issues to consider such as a general lack of reference small molecules targeting these interfaces,[6] a relatively large buried surface area on each partner[7] with variable contact points,[5] and the difficulty of deriving short peptide inhibitors as the residues involved in the protein–protein interface are not contiguous in the sequence.[8]

Currently, most of the available molecules targeting PPIs are relatively large (peptides or small proteins,[9,10] aptamers, or antibodies). The lack of identified "druglike" small-molecule inhibitors is probably due to the traditional concept that PPIs can only be "slightly" modulated[11] and that structure-based design of small-molecule inhibitors targeting PPIs is still viewed as quite difficult[12] compared to "classical" drug targets that naturally bind small molecules.[6] Different methods can be used to identify small-molecule inhibitors of PPI including fragment-based screening and virtual screening of compound collections. In this chapter, a focus will be made on the virtual screening approaches that can be applied to the identification of novel and original PPI small-molecule inhibitors.

VIRTUAL SCREENING

In the early stage of research of a drug discovery program, high throughput screening (HTS) procedures are applied for hit identification into large small-molecule databases. These methods, which have been shown to be successful for many targets, cost from $100,000 to $1 million for 1 million compounds and can display low hit rates.[13–15] In silico screening allows a huge reduction of time and costs by rationally selecting, using a computational method, a subset of the compound collection on the basis of the structure of the target (structure-based virtual ligand screening, SBVLS) or on the basis of the structure of known actives (ligand-based virtual ligand screening, LBVLS) prior to their experimental testing. After a selection of a "clean and proper" compound collection and a retrieval of the data on the PPI targeted (known structure of the partners, known inhibitors, etc.), a virtual screening protocol can be defined and prepared. For SBVLS studies, it is mandatory to define a binding pocket on the protein–protein interface targeted and to assess its structural and physicochemical properties. For LBVLS studies, it is necessary to have at least the structure of one already known inhibitor (or a hit identified after SBVLS; see Figure 14.1). After a short introduction on protein–protein interfaces properties, the different steps and requirements of the procedure will be detailed in the following sections.

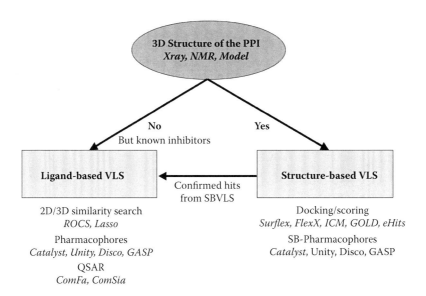

FIGURE 14.1   Example of the methods that can be used to identify PPI inhibitors depending on the starting material. If the 3D structure of the interface is known, it is recommended to use SBVLS methods. If inhibitors are already known or if hits have been identified with SBVLS, it can be interesting to use LBVLS methods to retrieve compounds that share a similar structure or pharmacophore and to derive QSAR for compound optimization. Italics indicate software that can be used.

PROTEIN–PROTEIN INTERFACES

Numerous investigations have described protein–protein interfaces.[7,16–23] Jones and Thornton proposed a classification of the PPIs into four groups, namely, homodimers, heterocomplexes, enzyme–inhibitor, and antibody–antigen[20] distinguishing obligatory interactions from transient interactions. Obligatory interactions are found between protomers, usually unstable on their own, that form permanent and optimized contacts with the other protomers involved in "obligate" complexes, for example, in the cytochrome c oxidase (Protein Data Bank [PDB] code 1ocr). Many heterocomplexes involve transient interactions between protomers that are not necessarily colocalized in the cell and exist independently as stable entities in solution.[20] The nature of the interface generally differs with the strength of the interaction as obligate complexes rely more on hydrophobic interactions compared to transient complexes that usually rely more on polar interactions like salt bridges and hydrogen bonds.[7,19,24] The shape complementarity of the interface is essential for the partner recognition and the strength of the interaction.[20,25] Antigen–antibody and heterocomplexes seem to have a lower shape complementarity compared to homodimeric and enzyme–inhibitor complexes.[26] In these cases, the level of packing can be as high as the interior of the protein fold.[7,20,27] The surface area of the interaction is also variable depending on the type of the complex. Approximately 750–1400 $Å^2$ of surface area is buried on each side of the interface.[7] Obligate complexes, in general, display a larger buried surface area (more than 1400 $Å^2$) compared to transient complexes and particularly to enzyme–inhibitor complexes that can be less than 1000 $Å^2$.[21,28,29]

It has been shown that a minority of residues, so-called hot spots, contributes to the majority of the binding energy.[30] The recognition site thus forms one or several "patches" composed by a cluster of surface atoms in the interface forming a signature of the protein interaction; the number of patches depends on the size of the interface.[31,32] Some of the residues composing the patch can be slightly more conserved than the rest of the surface residues, particularly for the residues composing an obligate interface.[33,34] Although identification of conserved residues alone is generally not sufficient to detect protein–protein interfaces[33] or hot spots,[35] it can still be helpful for characterization of functional regions of proteins like in the Web-based tools ConSurf [36–38] or Patchfinder.[39]

BINDING SITE IDENTIFICATION AND THE "DRUGGABLE POCKET"

A critical step for a structure-based in silico screening project is the definition of the binding pocket.[40] A "druggable pocket" is a cavity available at the surface of the protein displaying specific physicochemical properties compatible with the binding of a druglike small molecule.[41,42] These properties include a sufficient volume, a specific shape, the content in hydrophobic residues, or the roughness of the surface.[41,43–46] Protein interfaces are not flat and rigid but generally display pockets or voids when the partner is removed.[8,11,31,47] A large number of computational methods are available to identify binding pockets on protein surfaces based on geometric and/or energetic criterions, for example, LIGSITE,[48,49] PASS,[50] SiteMap,[45] PocketFinder,[51] or CAST.[52] Energetic algorithms like GRID,[53] VdW-FFT,[54] DrugSite,[41] or SCa[43] identify binding potential pockets by determining binding potentials or potential binding energies. Other methods such as computational solvent mapping,[55,56] SURFNET,[57] or Q-SiteFinder,[58] use molecular probes to screen the protein surface and evaluate binding potentials. Binding pockets and interfaces can also be identified and compared with programs using binding site databases like MED-SuMo,[59] CavBase,[60] SiteEngines,[61] SitesBase,[62] LigASite,[63] or sc-PDB.[64] When combined with hot spots or interface conserved residues prediction algorithms, these methods should be able to identify overlapping binding pockets.[11]

Once the pocket is identified, it is very important to assess its physicochemical properties (size, charges, solvent accessibility, hydrophobic character, roughness, etc.), as it will impact directly on the performance of the structure-based methods used in the drug discovery program.[65–67]

SCREENING LIBRARIES AND COMPOUND COLLECTIONS

To perform an in silico screening, it is necessary to use a compound collection. It is very difficult to evaluate the number of small molecules (especially because of the large number of compounds composing industrial proprietary compound collections), but several millions are available for purchase and have been used already in different virtual screening projects.[68–70] An interesting review of the different commercially available compound collections was written by Bradley in 2002 revising chemical substance databases, fine chemical databases, medicinal chemistry databases, and screening libraries containing purely synthetic or natural products.[71] Most of the screening libraries propose cherry-picking purchases, diversity subsets and subsets focused on a particular class of target (like ion channels or kinases). One can also use virtual small-molecule libraries generated in silico from chemical fragments, but it is mandatory to keep in mind that some of these virtual small molecules will not necessarily be easily synthesized.[72,73]

Small molecules are generally available in two dimensions in the MDL structure data format (SDF)[74] or in the simplified molecular input line entry system (SMILES).[75] Most of the algorithms need the three-dimensional (3D) structure of the small molecules to work correctly and use different 3D file formats like 3D SDF,[74] Tripos Mol2, or PDB.[76] This multiplicity of the file formats can lead to a large number of errors and a robust file conversion procedure, including 3D generation algorithms, are needed. Several programs exist like Omega,[77] Corina, Concord, or ICM[78] that allow the multiconformer generation needed by LBVLS or SBVLS packages. This step is crucial as the quality of the conformers will impact directly on the performance of the LBVLS or SBVLS procedures.[79] It is also very important to verify the protonation states and charges, and to sample, if possible, the different tautomeric forms of the compounds, as they can be considered as distinct molecules and impact on the performance of the screening.[80,81] If the final goal of the project is to design a drug candidate, it can also be interesting to filter the compound collection on the basis of the compounds predicted ADME-Tox (administration, distribution, metabolism, elimination, toxicity) to possibly reduce the attrition rate during the development.[82–91] These properties can be computed from simple molecular descriptors as in the Lipinski rule of five (RO5)[92,93] like molecular weight, logP, logS, number of hydrogen bond donors and acceptors, polar surface area, and so forth,[94–96] or using known undesirable fragments databases like unstable or reactive groups,[84,97–100] known frequent hitters,[84,101–103] and toxic groups.[84,91,97,104,105] Additional information about the metabolism of the compounds can be obtained via PGP-efflux[106] or cyto-chrome P450 binding[107] prediction methods. Recently, Neugebauer et al.[108] proposed a machine-learning method based on QSAR descriptors in order to predict the compounds' ability as a potential PPI inhibitor. Using the structural information of 25 PPI inhibitors combined with the information of 1137 non-PPI inhibitors, they calculated various descriptors and tried to identify the most significant combination of descriptors that allowed a good discrimination between PPI inhibitors and non-PPI inhibitors. This work is very promising in deriving RO5-like rules for the prediction of PPI inhibitor potential.

Of course all these methods are very simplified models, very far from the reality of the behavior of a small molecule in a whole organism, but they still can provide useful information for compound development.[109]

SBVLS: DOCKING AND SCORING

To perform SBVLS studies, it is necessary to use the structure of the target interface (obtained experimentally via nuclear magnetic resonance [NMR] or x-ray crystallo-graphy, or modeled using homology or threading methods). SBVLS consists gener-ally into predicting the orientation (molecular docking) and evaluating the interaction (scoring) of the small molecules composing the database into a binding pocket on the surface of the target interface (see earlier for the binding site prediction methods and structure preparation for SBVLS).

Different docking methods have been proposed for the prediction of ligand bind-ing modes. They are classified in two categories: rigid-body docking of multicon-former libraries (see earlier about conformer generation) and flexible docking.

During rigid-body docking, the receptor and ligand remain rigid while an ori-entation search of the ligand is performed. Different rigid-body docking programs exist and have been reported in the literature such as DOCK,[110] FRED, FLOG,[111] EUDOC,[112] or PLANTS[113] (which can also perform flexible docking). Despite being less accurate than flexible docking methods, rigid-body docking methods are inter-esting because they are very fast (e.g., FRED can dock several molecules per second on a single processor). They can be used as preliminary filters on a hierarchical pro-tocol to reduce the "noise" and remove from the compound collection the molecules that display a low surface complementarity with the binding pocket.[114–118]

Different algorithms exist to treat the problem of flexibility, as simulation methods, like molecular dynamics, are not applicable for high throughput prediction of ligand binding modes. The algorithms, as classified by Brooijmans and Kuntz in 2003, are systematic search algorithms, stochastic algorithms, and deterministic algorithms.[119]

Systematic search algorithms try to explore all the degrees of freedom of the molecules. Fragmentation/reconstruction algorithms reduce the complexity of the problem by dividing the molecules into rigid cores and flexible joints. After docking of the rigid cores, flexible joints are simulated until the full molecule is reconstructed incrementally. Different software use this method, including DOCK,[110] Surflex,[120] FlexX,[121] Glide,[122] or SLIDE.[123]

Stochastic search algorithms generate different ligand conformations by applying random modifications in the torsion angles of the molecule. Because they involve random, stochastic algorithms, they have to be rerun several times with different initial conditions to efficiently cover the conformational space. Different algorithms involve stochastic searches, like Monte Carlo (MC) and genetic algorithms (GAs), used in numerous programs. Monte Carlo uses the Metropolis criterion to accept or reject a new conformation generated at random. If the energy of the new conformation is preferred, it is accepted, otherwise it can be accepted with the Boltzmann probability. MC is used in ICM,[78] LigandFit,[124] QXP,[125] DockVision,[126] MCDOCK,[127] PRODOCK,[128] Glide,[122] or Autodock 2.x.[129] In evolutionary algorithms like GAs, a large number of conformations is generated as an initial population of solutions. The population evolves and the energetically preferred solutions survive (selection criteria) and are submitted to evolution criteria (crossing-overs and random mutations), thus constituting the next population. GAs are used in GOLD,[130] Autodock 3.x,[131] and DockVision.[126]

In deterministic search algorithms, the initial state determines the changes that will lead to the next steps of the simulation. Generally, the lowest-energy state is preferred at each step of the simulation. Unlike MC, molecular dynamics (MD) cannot cross high-energy barriers and are generally trapped in a local minimum[119] unless they are performed at high temperature. Few methods have been developed for SBVLS using molecular dynamics because they are, in general, very time consuming and thus not very applicable unless very few compounds remain in the hit list, such as at the very final steps of an SBVLS process, in order to accurately predict their binding energies.[117,132–134]

The evaluation of the proposed binding poses after docking and the ranking of the ligands constituting the database are crucial steps for the performance of an SBVLS protocol.[119,135] The final objective of an SBVLS protocol is to discriminate the actives from the inactives. Accurate evaluation of the free energy of binding can be performed by FEP/TI methods, but they are very time consuming and cannot be applied on large scale to hundred of thousands of small molecules. Alternate approximation methods have been developed to evaluate the relative binding energies of different compounds' so-called scoring functions. Scoring functions have two main purposes: they serve as fitness functions to optimize the positioning of the ligand into the binding pocket and they are used to rank the different ligands composing the database in order to select the most promising ones. Scoring functions can be classified into three groups: force-field-based scoring functions, knowledge-based scoring functions, and empirical scoring functions.[119,135]

Force-field-based scoring functions use the nonbonded terms of the classical molecular mechanics force fields. For example, DOCK,[110] Autodock,[129] and D-Score[136] scoring functions are based on the AMBER force field, whereas G-Score[136] is based on the TRIPOS[136] force field. Nonbonded terms used in force-field-based scoring functions are, in general, an electrostatic term described by a Coulombic potential and a van der Waals interaction term described by a "softened" 4-8, 6-9, or "classical" 6-12 Lennard-Jones potential as in DOCK

$$G_{bind} = \sum_{i=1}^{lig} \sum_{j=1}^{rec} \frac{q_i q_j}{\varepsilon r_{ij}} + \frac{A_{ij}}{r_{ij}^{12}} - \frac{B_{ij}}{r_{ij}^{6}}$$

where A_{ij} and B_{ij} are van der Waals repulsion and attraction parameters for atoms i and j, q_i and q_j their charge, r the distance between i and j, and ε the dielectric constant. "Soft" potentials are used to make the function more tolerant to small clashes between the ligand and the receptor. These functions have the drawback of evaluating only the enthalpic part of the interaction and neglect the entropic and solvation parts. Some advanced force-field-based scoring functions like the one in ICM[137] can treat the solvation part by resolving the Poisson equation using the boundary element method. The finite differences method has also been used by Wei et al. to treat the desolvation of ligands in receptor–ligand complexes.[138] Desolvation approached by the generalized Born method can also be used as an optional parameter in DOCK.[139] However, all these methods are computationally expensive and are only applicable on a reduced set of ligands (at the end of a hierarchical SBVLS protocol, for instance).

Knowledge-based scoring functions are issued from statistical mechanics, which define interaction potentials from atom pair distances observed in protein–ligand complexes.[140] Free energy of binding of an atom pair of atom types i and j in a receptor-ligand complex (or potential of mean force, PMF) would be of the form

$$A_{ij}(r) = -k_b T \ln \; f_{volcorr}^{j}(r) \frac{\rho_{seg}^{ij}(r)}{\rho_{bulk}^{ij}}$$

where k_b is the Boltzmann constant, r is the distance between the atoms involved in the pair, and the last part represents the interaction density between atom pairs when the interaction is occurring (ρ_{seg}) compared to when it is not occurring (ρ_{bulk}). These potentials of mean force have been used for protein-folding studies but can also be used as scoring functions to evaluate receptor–ligand interaction.

Contrary to the scoring functions described, empirical scoring functions are not based on physics but on experimental observations. Empirical scoring functions have the goal of evaluating the binding energy using noncorrelated terms that describe the different contribution to the interaction, each term weighted by a coefficient derived from statistical analysis of receptor–ligand complexes with experimentally measured free energy of binding. Different empirical scoring functions exist and differ with the terms they include and the way of simulating the different contributions to the binding energy. For example, LUDI [141] has two polar contact terms accounting for ionic and neutral hydrogen bonds, whereas they are included in the same term for Chemscore,[142] PLP,[143] and LigScore.[144] F-Score[136] and LUDI include a term accounting for aromatic contacts. LUDI, Chemscore, and Surflex include pseudo-entropic terms that take into account the number of rotatable bonds of the ligand to approximate the entropy loss penalty to the binding energy. Solvation terms can also be included, such as in LigScore or Fresno.[145] LigScore is an interesting scoring function containing both force field (a regular Lennard-Jones potential accounting for van der Waals interaction) and empirical terms (accounting for polar contacts and ligand desolvation).

Different groups have evaluated the relative performance of docking–scoring software for their ability to retrieve the experimental pose and to discriminate between actives and inactives in relatively large data sets,[114,115,146–153] and each software has its advantages and limits. The performance is generally dependent on the database used, the structural and physicochemical properties of the binding site, and the parameters used (thus the users' experience of the program[114,115,154]). It is thus of major importance to derive parameters for docking and scoring related to the binding pocket properties. Consensus scoring[148,155–158] and/or docking–scoring[114,115,159] can also be used to improve the performance of the enrichment.

Protein–protein interfaces sometimes display a relative high flexibility[160] and sometimes let transient pockets appear.[12] It is, thus, very important to assess the flexibility of the region targeted for SBVLS studies (backbone, side chains, or loops around the hot spots or interface conserved residues). Different docking methods accounting for partial flexibility of the receptor have been developed to face this phenomenon, in particular using softened 6-9 or 4-8 Lennard Jones potentials describing van des Waals interaction in the scoring function or by reducing the van der Waals radii of heavy atoms to let the scoring be more tolerant to small clashes after docking. Another option is to perform the SBVLS on multiple rigid experimental structures (x-ray, NMR)[161,162] or multiple simulated structures obtained by molecular dynamics or normal modes, for instance.[163–171] Some flexible docking algorithms can also include side chain flexibility,[113,164,172–174] but it seems that according for receptor flexibility during the docking process increases the "noise," and thus makes it more difficult to discriminate between true hits and false positives.[175,176]

LBVLS: SIMILARITY SEARCH, QSAR, AND PHARMACOPHORES

When one or several ligands have already been identified for a particular target, it is possible to develop new ligands based on this reference. Ligand-based virtual ligand screening (LBVLS) procedures consist of retrieving compounds with similar structure, shape, or physicochemical properties compared to the reference ligand(s).

Similarity search methods can be applied, based on the two-dimensional (2D) substructure of the reference structure[177] represented by molecular fingerprints derived from the fragments composing the molecule,[178] the connectivity between the fragments and sometimes even molecular descriptors computed from the structure.[179,180] Many molecular descriptors have already been described (see ADME-Tox earlier and Refs. 181–192) and can be used in such approaches. Shape[81,193] or surface descriptors[181,194] can also be used in similarity search methods, the choice of the descriptors used impacting directly on the screening performance.[195,196] Similarity search methods based on 3D descriptors and in particular shape descriptors like the ones used in ROCS[197] or LASSO[198] can be used for scaffold hopping, that is, change the molecular scaffold but keep the activity. Other scaffold hopping methods exist[199–204] and have been recently reviewed in Refs. 205–207.

Quantitative structure–activity relationship (QSAR) models can also be used against PPIs. A QSAR model is a statistical model that relates the properties of a molecule (such as its biological activity) to its structure. 3D QSAR models like ComFa or ComSia use, for example, steric and electrostatic fields displayed around the superimposition of the known actives to make predictions for the activity of new compounds. Screening and selection of small molecules compatible with the corresponding fields can lead to novel small-molecules identification.[208] A combination of docking methods and 3D QSAR have also been performed, so-called receptor-based 3D QSAR.[209–213]

In the same idea, structure-based pharmacophore screening and particularly 3D-pharmacophores using descriptors derived from the structural and physicochemical properties of the interface can be used for PPI inhibitor screening.[214,215] Different pharmacophore modeling software exist such as Unity, Catalyst, Disco, or GASP[216] and are widely use for drug discovery projects.[214,217–219]

Unfortunately, in the case of PPIs, there are generally few reference inhibitors, but peptides, hot spots, or pharmacophores can be derived from the structure and properties of the protein partners involved in the interaction, and can be used as a reference. Combined with the knowledge of the structure of the interface, these methods can be very useful for finding new hits targeting PPIs. They can also be used when potential inhibitors have been identified in order to retrieve analogues and derive structure–activity relationships for lead optimization (Figure 14.2).

SUCCESSFUL APPLICATIONS ON PPI INHIBITORS IDENTIFICATION

Virtual screening procedures have already been successfully applied in numerous targets displaying "traditional" active sites, but even if it still seems to be challenging to target PPIs, several advances in the field are very encouraging and have been recently reviewed in Refs. 8 and 220–223. In general, when starting a project against a specific PPI, the structure of at least one protein partner (obtained via experimental or molecular modeling methods) is available but not the structure of any known inhibitor. That is why SBVLS methods are most often used, but if a hit is found, LBVLS can also be part of the drug design procedure to reenrich the hit list.

The p53 tumor suppressor is a transcription factor considered as the "guardian of the genome" as it reacts to cellular stress by activating transcription of numerous genes implied in the cell cycle arrest, DNA repair, and apoptosis.[224] Murine double minute 2 protein (MDM2) regulates p53 forming an inactive MDM2–p53 complex and addresses it for proteasomal degradation. MDM2 binds to a helical region of p53 close to the N-terminus (see Ref. 3 for a review of the interaction). Inactivation of p53 is common in about half of the tumors. Potential inhibitors of MDM2–p53 are thus seen as a promising path for cancer treatment.

Two applications of pharmacophore models on PPIs have been performed on the p53–MDM2 interface using pharmacophores derived from the structure of the MDM2 binding pocket (Figure 14.3). The first study was performed by Galatin and Abraham[225] using the program Unity. By screening the NCI database on a pharmacophore derived from the key binding residues of p53 (F19, L23, W26), they identified inhibitors that inhibit the interaction and activate p53-dependent transcription in MDM2 overexpressing cells. Bowman et al.[226] performed a similar study. They used their in-house method, MPS (multiple protein structure), to derive a receptor-based

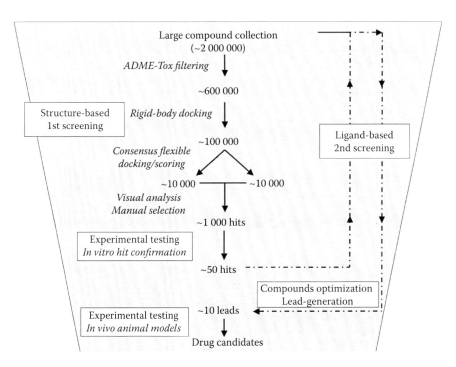

FIGURE 14.2 Example flowchart of a structure-based drug design protocol. The ADME-Tox filter is used to select druglike compounds in the initial compound collection. Once a binding pocket is identified in the protein–protein interface, the resulting compounds will be subjected to rigid-body docking followed by consensus flexible docking/scoring. The hits will then be validated *in vitro*. A similarity search will then be performed on the confirmed hits to find more active compounds that will be optimized using medicinal chemistry. The most promising compounds after optimization will then be evaluated *in vivo* to select drug candidates.

pharmacophore model, which identified the hot spots of binding. After screening their in-house 35,000 compound collection on the pharmacophore model, they identified five nonpeptide small-molecule inhibitors of the human p53–MDM2 interaction, the most potent displaying a Ki of 110 nM. Each inhibitor represented a novel scaffold on p53–MDM2 inhibition and their binding modes have been assessed using flexible induced-fit docking into the binding pocket showing that their structure mimics the key binding residues of p53.

A recent study of Weber, Holak, et al.[227] on the p53–MDM2 interaction led to an interesting protocol combining 2D similarity screening with the structure of Nutlin-3 (a known small-molecule inhibitor of the interaction).[228] They identified 278 compounds with various scaffolds from their vendors' database. Using the program LIGSITE,[48,49] they extracted the binding pocket available in the interface between p53 and MDM2. After generating different conformers for the 278 compounds identified through 2D similarity, they performed shape similarity with the shape of the pocket using the program M3dsml (Gerber molecular design). Using this protocol, they prioritized 131 compounds for experimental testing. By combining

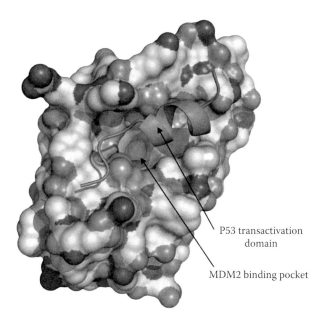

P53 transactivation
domain

MDM2 binding pocket

FIGURE 14.3 (SEE COLOR INSERT FOLLOWING PAGE 174.) X-ray structure of the MDM2–p53 fragment complex. MDM2 is displayed as a solid surface, whereas the p53 fragment is displayed as a cartoon colored in magenta. The MDM2 binding pocket is colored in orange. Figure generated with PyMol.

LBVLS and SBVLS, they identified novel scaffolds that could bind on the MDM2 pocket. After synthesis of analogues of the most promising compounds after experimental validation of the binding, they identified isoquinolin-1-one derivatives that can dissociate p53 and MDM2 *in vitro*, and thus provide a good starting point as potential inhibitors for the treatment of the numerous cancers associated with this interaction.[224]

During HIV-1 infection, there are interactions between auxiliary viral proteins and surface proteins of the hosts' CD4+ lymphocytes that impact directly on the efficiency of the infection and disease progression. HIV-1 Nef, an auxiliary viral protein, displays an SH3 domain binding surface that can be targeted to develop a new class of antiviral compounds.

Betzi et al.[229] performed a study on the SH3 domain–HIV-1 Nef interaction combining hierarchical SBVLS and LBVLS. They defined the binding site for docking with a hydrophobic groove on HIV-1 Nef bordered by key hydrophobic and charged residues (hot spots) involved in the interaction, so-called the RT loop binding region (RTBLR) cavity (Figure 14.4). They first performed an SBVLS step using FlexX on the National Cancer Institute diversity library filtered to contain only druglike compounds. A pharmacophoric filter based on the properties of the RTLBR cavity was then applied on the 335 compounds retained after SBVLS and led to 33 candidates that were visually inspected. The 10 most promising were tested experimentally and one compound, D1, was found active on the PPI in the micromolar range (Kd = 1.8 µM). Then, an LBVLS step by using a 2D-similarity screening based on D1

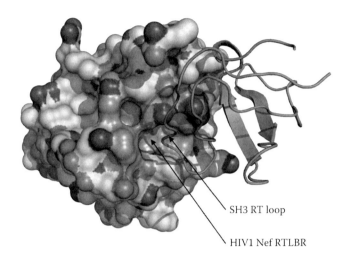

SH3 RT loop

HIV1 Nef RTLBR

FIGURE 14.4 X-ray structure of the HIV1 Nef/SH3 domain complex. HIV1 Nef is displayed as a solid surface, whereas SH3 domain is displayed as a cartoon colored in magenta. The RT loop binding region is colored in orange. Figure generated with PyMol.

substructure was performed on a 435,000-compound collection. The 70 most similar compounds were evaluated experimentally resulting in another promising candidate, DLC27, displaying a Kd of 0.98 µM on the interaction. These two compounds were highly original and were, to date, the first druglike molecules identified that bind HIV-1 Nef–SH3 binding surface in the micromolar range and are very promising for the development of new antiviral therapies.

CONCLUDING REMARKS

In the last few years, there has been major progress in the identification and development of PPI inhibitors. Until now, most of them have been retrieved using blind HTS or fragment-based screening, but recently, inhibitors issued from virtual screening protocols have arisen. The different successes presented are very promising for virtual screening methods that can constitute an interesting alternative to experimental screening. In particular, protocols using a wise combination of SBVLS and LBVLS methods have led to very interesting inhibitors displaying original scaffolds, which can be used as a basis to develop new compounds with therapeutical interest on challenging targets.

However, there are still several improvements to be made to routinely discover PPI inhibitors using virtual screening. The scoring functions used in virtual screening still need improvements, as in most of the cases there are still a lot of false positives. A better account for electrostatics and, in particular, solvation and accurate protonation and tautomeric states management can certainly improve the discriminating accuracy between actives and inactives. It also seems very important to take into account the particular physicochemical properties of the binding pocket defined in the interface in the parameters of the different tools used for screening. A better knowledge of the particular properties of PPI inhibitors will also certainly be of interest to optimize these parameters. Several other points need to be addressed including the treatment of flexibility on the receptor side and the evaluation of the potential of the hits retrieved after screening to be easily optimized.

It appears clearer with the recent successful applications that virtual screening methods can lead to promising compounds targeting PPIs and establish high expectations for future inhibitors targeting this challenging class of interactions.

REFERENCES

1. Toogood, P. L., Inhibition of protein–protein association by small molecules: Approaches and progress. *J Med Chem* 2002, 45(8), 1543–58.
2. Kortemme, T., Baker, D., Computational design of protein–protein interactions. *Curr Opin Chem Biol* 2004, 8(1), 91–97.
3. Murray, J. K., Gellman, S. H., Targeting protein–protein interactions: Lessons from p53/MDM2. *Biopolymers* 2007, 88(5), 657–86.
4. Sharma, S. K., Ramsey, T. M., Bair, K. W., Protein–protein interactions: Lessons learned. *Curr Med Chem Anticancer Agents* 2002, 2(2), 311–30.
5. Arkin, M., Randal, M., DeLano, W. L., Hyde, J., Luong, T. N., Oslob, J. D., Raphael, D. R., Taylor, J., Wang, J., McDowell, R. S., Wells, J. A., Braisted, A. C., Binding of small molecules to an adaptative protein–protein interface. *Proc Natl Acad Sci USA* 2003, 100, 1603–8.
6. Arkin, M. R., Wells, J. A., Small-molecule inhibitors of protein–protein interactions: Progressing towards the dream. *Nat Rev Drug Discov* 2004, 3(4), 301–17.
7. Lo Conte, L., Chothia, C., Janin, J., The atomic structure of protein–protein recognition sites. *J Mol Biol* 1999, 285(5), 2177–98.
8. Wells, J. A., McClendon, C. L., Reaching for high-hanging fruit in drug discovery at protein–protein interfaces. *Nature* 2007, 450(7172), 1001–9.
9. Shen, L., Villoutreix, B. O., Dahlback, B., Involvement of Lys 62(217) and Lys 63(218) of human anticoagulant protein C in heparin stimulation of inhibition by the protein C inhibitor. *Thromb Haemost* 1999, 82(1), 72–79.
10. Olson, S. T., Bjork, I., Sheffer, R., Craig, P. A., Shore, J. D., Choay, J., Role of the antithrombin-binding pentasaccharide in heparin acceleration of antithrombin-proteinase reactions. Resolution of the antithrombin conformational change contribution to heparin rate enhancement. *J Biol Chem* 1992, 267(18), 12528–38.
11. Villoutreix, B. O., Bastard, K., Sperandio, O., Fahraeus, R., Poyet, J. L., Calvo, F., Deprez, B., Miteva, M. A., In silico–in vitro screening of protein–protein interactions: Towards the next generation of therapeutics. *Curr Pharm Biotech* 2008, 9(1), 103–22.
12. Eyrisch, S., Helms, V., Transient pockets on protein surfaces involved in protein–protein interaction. *J Med Chem* 2007, 50(15), 3457–64.
13. Severson, W. E., Shindo, N., Sosa, M., Fletcher, T., III, White, E. L., Ananthan, S., Jonsson, C. B., Development and validation of a high-throughput screen for inhibitors of SARS CoV and its application in screening of a 100,000-compound library. *J Biomol Screen* 2007, 12(1), 33–40.
14. Doman, T. N., McGovern, S. L., Witherbee, B. J., Kasten, T. P., Kurumbail, R., Stallings, W. C., Connolly, D. T., Shoichet, B. K., Molecular docking and high-throughput screening for novel inhibitors of protein tyrosine phosphatase-1B. *J Med Chem* 2002, 45(11), 2213–21.

15. Alvesalo, J. K. O., Siiskonen, A., Vainio, M. J., Tammela, P. S. M., Vuorela, P. M., Similarity based virtual screening: A tool for targeted library design. *J Med Chem* 2006, 49, 2353–6.

16. Janin, J., Chothia, C., The structure of protein–protein recognition sites. *J Biol Chem* 1990, 265(27), 16027–30.

17. Chothia, C., Janin, J., Principles of protein–protein recognition. *Nature* 1975, 256(5520), 705–8.

18. Argos, P., An investigation of protein subunit and domain interfaces. *Protein Eng* 1988, 2(2), 101–13.

19. Jones, S., Thornton, J. M., Analysis of protein–protein interaction sites using surface patches. *J Mol Biol* 1997, 272(1), 121–32.

20. Jones, S., Thornton, J. M., Principles of protein–protein interactions. *Proc Natl Acad Sci USA* 1996, 93(1), 13–20.

21. Larsen, T. A., Olson, A. J., Goodsell, D. S., Morphology of protein–protein interfaces. *Structure* 1998, 6(4), 421–7.

22. Keskin, O., Ma, B., Rogale, K., Gunasekaran, K., Nussinov, R., Protein–protein interactions: Organization, cooperativity and mapping in a bottom-up Systems Biology approach. *Phys Biol* 2005, 2(2), S24–35.

23. Keskin, O., Tsai, C. J., Wolfson, H., Nussinov, R., A new, structurally nonredundant, diverse data set of protein–protein interfaces and its implications. *Protein Sci* 2004, 13(4), 1043–55.

24. Keskin, O., Gursoy, A., Ma, B., Nussinov, R., Towards drugs targeting multiple proteins in a systems biology approach. *Curr Top Med Chem* 2007, 7(10), 943–51.

25. Keskin, O., Nussinov, R., Similar binding sites and different partners: Implications to shared proteins in cellular pathways. *Structure* 2007, 15(3), 341–54.

26. Lawrence, M. C., Colman, P. M., Shape complementarity at protein/protein interfaces. *J Mol Biol* 1993, 234(4), 946–50.

27. Walls, P. H., Sternberg, M. J., New algorithm to model protein–protein recognition based on surface complementarity. Applications to antibody–antigen docking. *J Mol Biol* 1992, 228(1), 277–97.

28. Janin, J., Miller, S., Chothia, C., Surface, subunit interfaces and interior of oligomeric proteins. *J Mol Biol* 1988, 204(1), 155–64.

29. Miller, S., Lesk, A. M., Janin, J., Chothia, C., The accessible surface area and stability of oligomeric proteins. *Nature* 1987, 328(6133), 834–6.

30. Clackson, T., Wells, J. A., A hot spot of binding energy in a hormone-receptor interface. *Science* 1995, 267(5196), 383–6.

31. Fernandez-Recio, J., Totrov, M., Skorodumov, C., Abagyan, R., Optimal docking area: A new method for predicting protein–protein interaction sites. *Proteins* 2005, 58(1), 134–43.

32. Chakrabarti, P., Janin, J., Dissecting protein–protein recognition sites. *Proteins* 2002, 47(3), 334–43.

33. Caffrey, D. R., Somaroo, S., Hughes, J. D., Mintseris, J., Huang, E. S., Are protein–protein interfaces more conserved in sequence than the rest of the protein surface? *Protein Sci* 2004, 13(1), 190–202.

34. Valdar, W. S., Thornton, J. M., Protein–protein interfaces: Analysis of amino acid conservation in homodimers. *Proteins* 2001, 42(1), 108–24.

35. Keskin, O., Tuncbag, N., Gursoy, A., Characterization and prediction of protein interfaces to infer protein–protein interaction networks. *Curr Pharm Biotechnol* 2008, 9(2), 67–76.

36. Landau, M., Mayrose, I., Rosenberg, Y., Glaser, F., Martz, E., Pupko, T., Ben-Tal, N., ConSurf 2005: The projection of evolutionary conservation scores of residues on protein structures. *Nucleic Acids Res* 2005, 33(Web Server issue), W299–302.

37. Glaser, F., Rosenberg, Y., Kessel, A., Pupko, T., Ben-Tal, N., The ConSurf-HSSP database: The mapping of evolutionary conservation among homologs onto PDB structures. *Proteins* 2005, 58(3), 610–7.
38. Glaser, F., Pupko, T., Paz, I., Bell, R. E., Bechor-Shental, D., Martz, E., Ben-Tal, N., ConSurf: Identification of functional regions in proteins by surface-mapping of phylogenetic information. *Bioinformatics* 2003, 19(1), 163–4.
39. Nimrod, G., Glaser, F., Steinberg, D., Ben-Tal, N., Pupko, T., In silico identification of functional regions in proteins. *Bioinformatics* 2005, 21(Suppl 1), i328–37.
40. Sotriffer, C., Klebe, G., Identification and mapping of small-molecule binding sites in proteins: Computational tools for structure-based drug design. *Farmaco* 2002, 57(3), 243–51.
41. An, J., Totrov, M., Abagyan, R., Comprehensive identification of "druggable" protein ligand binding sites. *Genome Inform Ser Workshop Genome Inform* 2004, 15(2), 31–41.
42. Hajduk, P. J., Huth, J. R., Tse, C., Predicting protein druggability. *Drug Discov Today* 2005, 10(23-24), 1675–82.
43. Coleman, R. G., Salzberg, A. C., Cheng, A. C., Structure-based identification of small molecule binding sites using a free energy model. *J Chem Inf Model* 2006, 46(6), 2631–7.
44. Schulz-Gasch, T., Stahl, M., Binding site characteristics in structure-based virtual screening: Evaluation of current docking tools. *J Mol Model (Online)* 2003, 9(1), 47–57.
45. Halgren, T., New method for fast and accurate binding-site identification and analysis. *Chem Biol Drug Des* 2007, 69(2), 146–8.
46. Pettit, F. K., Bowie, J. U., Protein surface roughness and small molecular binding sites. *J Mol Biol* 1999, 285(4), 1377–82.
47. Dundas, J., Ouyang, Z., Tseng, J., Binkowski, A., Turpaz, Y., Liang, J., CASTp: Computed atlas of surface topography of proteins with structural and topographical mapping of functionally annotated residues. *Nucleic Acids Res* 2006, 34(Web Server issue), W116–8.
48. Huang, B., Schroeder, M., LIGSITEcsc: Predicting ligand binding sites using the Connolly surface and degree of conservation. *BMC Struct Biol* 2006, 6, 19.
49. Hendlich, M., Rippmann, F., Barnickel, G., LIGSITE: Automatic and efficient detection of potential small molecule-binding sites in proteins. *J Mol Graph Model* 1997, 15(6), 359–63, 389.
50. Brady, G. P., Jr., Stouten, P. F., Fast prediction and visualization of protein binding pockets with PASS. *J Comput Aided Mol Des* 2000, 14(4), 383–401.
51. An, J., Totrov, M., Abagyan, R., Pocketome via comprehensive identification and classification of ligand binding envelopes. *Mol Cell Proteomics* 2005, 4(6), 752–61.
52. Liang, J., Edelsbrunner, H., Woodward, C., Anatomy of protein pockets and cavities: Measurement of binding site geometry and implications for ligand design. *Protein Sci* 1998, 7(9), 1884–97.
53. Kastenholz, M. A., Pastor, M., Cruciani, G., Haaksma, E. E., Fox, T., GRID/CPCA: A new computational tool to design selective ligands. *J Med Chem* 2000, 43(16), 3033–44.
54. Bliznyuk, A. A., Gready, J. E., Identification and energetic ranking of possible docking sites for pterin on dihydrofolate reductase. *J Comput Aided Mol Des* 1998, 12(4), 325–33.
55. Kortvelyesi, T., Dennis, S., Silberstein, M., Brown, L., III, Vajda, S., Algorithms for computational solvent mapping of proteins. *Proteins* 2003, 51(3), 340–51.
56. Landon, M. R., Lancia, D. R., Jr., Yu, J., Thiel, S. C., Vajda, S., Identification of hot spots within druggable binding regions by computational solvent mapping of proteins. *J Med Chem* 2007, 50(6), 1231–40.

57. Laskowski, R. A., SURFNET: A program for visualizing molecular surfaces, cavities, and intermolecular interactions. *J Mol Graph* 1995, 13(5), 323–30, 307–8.

58. Laurie, A. T., Jackson, R. M., Q-SiteFinder: An energy-based method for the prediction of protein-ligand binding sites. *Bioinformatics* 2005, 21(9), 1908–16.

59. Jambon, M., Andrieu, O., Combet, C., Deleage, G., Delfaud, F., Geourjon, C., The SuMo server: 3D search for protein functional sites. *Bioinformatics* 2005, 21(20), 3929–30.

60. Schmitt, S., Kuhn, D., Klebe, G., A new method to detect related function among proteins independent of sequence and fold homology. *J Mol Biol* 2002, 323(2), 387–406.

61. Shulman-Peleg, A., Nussinov, R., Wolfson, H. J., Recognition of functional sites in protein structures. *J Mol Biol* 2004, 339(3), 607–33.

62. Gold, N. D., Jackson, R. M., SitesBase: A database for structure-based protein-ligand binding site comparisons. *Nucleic Acids Res* 2006, 34(Database issue), D231–4.

63. Dessailly, B. H., Lensink, M. F., Orengo, C. A., Wodak, S. J., LigASite—a database of biologically relevant binding sites in proteins with known apo-structures. *Nucleic Acids Res* 2008, 36(Database issue), D667–73.

64. Paul, N., Kellenberger, E., Bret, G., Muller, P., Rognan, D., Recovering the true targets of specific ligands by virtual screening of the protein data bank. *Proteins* 2004, 54(4), 671–80.

65. Brenk, R., Vetter, S. W., Boyce, S. E., Goodin, D. B., Shoichet, B. K., Probing molecular docking in a charged model binding site. *J Mol Biol* 2006, 357(5), 1449–70.

66. Armstrong, K. A., Tidor, B., Cheng, A. C., Optimal charges in lead progression: A structure-based neuraminidase case study. *J Med Chem* 2006, 49(8), 2470–7.

67. Krovat, E. M., Steindl, T., Langer, T., Recent advances in docking and scoring. *Curr Computer-Aided Drug Design* 2005, 1(1), 93–102.

68. Baurin, N., Baker, R., Richardson, C., Chen, I., Foloppe, N., Potter, A., Jordan, A., Roughley, S., Parratt, M., Greaney, P., Morley, D., Hubbard, R. E., Drug-like annotation and duplicate analysis of a 23-supplier chemical database totalling 2.7 million compounds. *J Chem Inf Comput Sci* 2004, 44(2), 643–51.

69. Krier, M., Bret, G., Rognan, D., Assessing the scaffold diversity of screening libraries. *J Chem Inf Model* 2006, 46(2), 512–24.

70. Lameijer, E. W., Kok, J. N., Back, T., Ijzerman, A. P., Mining a chemical database for fragment co-occurrence: Discovery of "chemical cliches." *J Chem Inf Model* 2006, 46(2), 553–62.

71. Bradley, M. P., An overview of the diversity represented in commercially-available databases. *J Comp Aid Mol Des* 2002, 16, 301–9.

72. Lameijer, E. W., Kok, J. N., Back, T., Ijzerman, A. P., The molecule evoluator. An interactive evolutionary algorithm for the design of drug-like molecules. *J Chem Inf Model* 2006, 46(2), 545–52.

73. Pirok, G., Mate, N., Varga, J., Szegezdi, J., Vargyas, M., Dorant, S., Csizmadia, F., Making "real" molecules in virtual space. *J Chem Inf Model* 2006, 46(2), 563–8.

74. Dalby, A., Nourse, J. G., Hounshell, W. D., Gushurst, A. K. I., Grier, D. L., Leland, B. A., Laufer, J., Description of several chemical structure file formats used by computer programs developed at Molecular Design Limited. *J Chem Inf Comput Sci* 1992, 32, 244–55.

75. Weininger, D., SMILES, a chemical language and information system. *J Chem Inf Comput Sci* 1988, 28, 31–36.

76. Berman, H. M., Westbrook, J., Feng, Z., Gilliland, G., Bhat, T. N., Weissig, H., Shindyalov, I. N., Bourne, P. E., The Protein Data Bank. *Nucleic Acids Res* 2000, 28(1), 235–42.

77. Bostrom, J., Greenwood, J. R., Gottfries, J., Assessing the performance of OMEGA with respect to retrieving bioactive conformations. *J Mol Graph Model* 2003, 21(5), 449–62.

78. Abagyan, R., Totrov, M., Kusnetsov, D., ICM (Internal Coordinate Mechanics). *J Comp Chem* 1994, 15, 488–506.

79. Kirchmair, J., Wolber, G., Laggner, C., Langer, T., Comparative performance assessment of the conformational model generators omega and catalyst: A large-scale survey on the retrieval of protein-bound ligand conformations. *J Chem Inf Model* 2006, 46(4), 1848–61.

80. Zou, J. W., Luo, C. C., Zhang, H. X., Liu, H. C., Jiang, Y. J., Yu, Q. S., Three-dimensional QSAR of HPPD inhibitors, PSA inhibitors, and anxiolytic agents: Effect of tautomerism on the CoMFA models. *J Mol Graph Model* 2007, 26(2), 494–504.

81. Oellien, F., Cramer, J., Beyer, C., Ihlenfeldt, W. D., Selzer, P. M., The impact of tautomer forms on pharmacophore-based virtual screening. *J Chem Inf Model* 2006, 46(6), 2342–54.

82. Kassel, D. B., Applications of high-throughput ADME in drug discovery. *Curr Opin Chem Biol* 2004, 8(3), 339–45.

83. Jenkins, K. M., Angeles, R., Quintos, M. T., Xu, R., Kassel, D. B., Rourick, R. A., Automated high throughput ADME assays for metabolic stability and cytochrome P450 inhibition profiling of combinatorial libraries. *J Pharm Biomed Anal* 2004, 34(5), 989–1004.

84. Rishton, G. M., Nonleadlikeness and leadlikeness in biochemical screening. *Drug Discov Today* 2003, 8(2), 86–96.

85. Martin, Y. C., A bioavailability score. *J Med Chem* 2005, 48(9), 3164–70.

86. Gasteiger, J., Physicochemical effects in the representation of molecular structures for drug designing. *Mini Rev Med Chem* 2003, 3(8), 789–96.

87. Li, A. P., Screening for human ADME/Tox drug properties in drug discovery. *Drug Discov Today* 2001, 6(7), 357–66.

88. Lombardo, F., Gifford, E., Shalaeva, M. Y., In silico ADME prediction: Data, models, facts and myths. *Mini Rev Med Chem* 2003, 3(8), 861–75.

89. Roche, O., Guba, W., Computational chemistry as an integral component of lead generation. *Mini Rev Med Chem* 2005, 5(7), 677–83.

90. Beresford, A. P., Segall, M., Tarbit, M. H., In silico prediction of ADME properties: Are we making progress? *Curr Opin Drug Discov Devel* 2004, 7(1), 36–42.

91. Muegge, I., Selection criteria for drug-like compounds. *Med Res Rev* 2003, 23(3), 302–21.

92. Lipinski, C. A., Drug-like properties and the causes of poor solubility and poor permeability. *J Pharmacol Toxicol Methods* 2000, 44(1), 235–49.

93. Lipinski, C. A., Lombardo, F., Dominy, B. W., Feeney, P. J., Experimental and computational approaches to estimate solbility and permeability in drug discovery. *Adv Drug Deliv Res* 1997, 23, 3–25.

94. Oprea, T. I., Current trends in lead discovery: Are we looking for the appropriate properties? *J Comput Aided Mol Des* 2002, 16(5-6), 325–34.

95. Oprea, T. I., Property distribution of drug-related chemical databases. *J Comput Aided Mol Des* 2000, 14(3), 251–64.

96. Veber, D. F., Johnson, S. R., Cheng, H. Y., Smith, B. R., Ward, K. W., Kopple, K. D., Molecular properties that influence the oral bioavailability of drug candidates. *J Med Chem* 2002, 45(12), 2615–23.

97. Caldwell, G. W., Yan, Z., Screening for reactive intermediates and toxicity assessment in drug discovery. *Curr Opin Drug Discov Devel* 2006, 9(1), 47–60.

98. Nassar, A. E., Talaat, R. E., Strategies for dealing with metabolite elucidation in drug discovery and development. *Drug Discov Today* 2004, 9(7), 317–27.

99. Socorro, I. M., Goodman, J. M., The ROBIA program for predicting organic reactivity. *J Chem Inf Model* 2006, 46(2), 606–14.

100. Nassar, A. E., Lopez-Anaya, A., Strategies for dealing with reactive intermediates in drug discovery and development. *Curr Opin Drug Discov Devel* 2004, 7(1), 126–36.

101. Feng, B. Y., Shelat, A., Doman, T. N., Guy, R. K., Shoichet, B. K., High-throughput assays for promiscuous inhibitors. *Nat Chem Biol* 2005, 1(3), 146–8.

102. McGovern, S. L., Helfand, B. T., Feng, B., Shoichet, B. K., A specific mechanism of nonspecific inhibition. *J Med Chem* 2003, 46(20), 4265–72.

103. Roche, O., Schneider, P., Zuegge, J., Guba, W., Kansy, M., Alanine, A., Bleicher, K., Danel, F., Gutknecht, E. M., Rogers-Evans, M., Neidhart, W., Stalder, H., Dillon, M., Sjogren, E., Fotouhi, N., Gillespie, P., Goodnow, R., Harris, W., Jones, P., Taniguchi, M., Tsujii, S., von der Saal, W., Zimmermann, G., Schneider, G., Development of a virtual screening method for identification of "frequent hitters" in compound libraries. *J Med Chem* 2002, 45(1), 137–42.

104. Nassar, A. E., Kamel, A. M., Clarimont, C., Improving the decision-making process in structural modification of drug candidates: Reducing toxicity. *Drug Discov Today* 2004, 9(24), 1055–64.

105. von Korff, M., Sander, T., Toxicity-indicating structural patterns. *J Chem Inf Model* 2006, 46(2), 536–44.

106. Gombar, V. K., Polli, J. W., Humphreys, J. E., Wring, S. A., Serabjit-Singh, C. S., Predicting P-glycoprotein substrates by a quantitative structure-activity relationship model. *J Pharm Sci* 2004, 93(4), 957–68.

107. Vermeulen, N. P., Prediction of drug metabolism: The case of cytochrome P450 2D6. *Curr Top Med Chem* 2003, 3(11), 1227–39.

108. Neugebauer, A., Hartmann, R. W., Klein, C. D., Prediction of protein–protein interaction inhibitors by chemoinformatics and machine learning methods. *J Med Chem* 2007, 50(19), 4665–8.

109. Kubinyi, H., Drug research: Myths, hype and reality. *Nat Rev Drug Discov* 2003, 2(8), 665–8.

110. Kuntz, I. D., Blaney, J. M., Oatley, S. J., Langridge, R., Ferrin, T. E., A geometric approach to macromolecule-ligand interactions. *J Mol Biol* 1982, 161(2), 269–88.

111. Miller, M. D., Kearsley, S. K., Underwood, D. J., Sheridan, R. P., FLOG: A system to select "quasi-flexible" ligands complementary to a receptor of known three-dimensional structure. *J Comput Aided Mol Des* 1994, 8(2), 153–74.

112. Pang, Y. P., Perola, E., Xu, K., Prendergast, F. G., EUDOC: A computer program for identification of drug interaction sites in macromolecules and drug leads from chemical databases. *J Comput Chem* 2001, 22(15), 1750–71.

113. Korb, O., Stützle, T., Exner, T. E., PLANTS: Application of ant colony optimization to structure-based drug design. *Lect Notes Compu Sci* 2006, 4150, 247–58.

114. Miteva, M. A., Lee, W. H., Montes, M. O., Villoutreix, B. O., Fast structure-based virtual ligand screening combining FRED, DOCK, and Surflex. *J Med Chem* 2005, 48(19), 6012–22.

115. Montes, M., Miteva, M. A., Villoutreix, B. O., Structure-based virtual ligand screening with LigandFit: Pose prediction and enrichment of compound collections. *Proteins* 2007, 68(3), 712–25.

116. Floriano, W. B., Vaidehi, N., Zamanakos, G., Goddard, W. A., III, HierVLS hierarchical docking protocol for virtual ligand screening of large-molecule databases. *J Med Chem* 2004, 47(1), 56–71.

117. Wang, J., Kang, X., Kuntz, I. D., Kollman, P. A., Hierarchical database screenings for HIV-1 reverse transcriptase using a pharmacophore model, rigid docking, solvation docking, and MM-PB/SA. *J Med Chem* 2005, 48(7), 2432–44.

118. Mozziconacci, J. C., Arnoult, E., Bernard, P., Do, Q. T., Marot, C., Morin-Allory, L., Optimization and validation of a docking-scoring protocol, application to virtual screening for COX-2 inhibitors. *J Med Chem* 2005, 48(4), 1055–68.

119. Brooijmans, N., Kuntz, I. D., Molecular recognition and docking algorithms. *Annu Rev Biophys Biomol Struct* 2003, 32, 335–73.

120. Jain, A. N., Surflex: Fully automatic flexible molecular docking using a molecular similarity-based search engine. *J Med Chem* 2003, 46(4), 499–511.

121. Rarey, M., Kramer, B., Lengauer, T., Time-efficient docking of flexible ligands into active sites of proteins. *Proc Int Conf Intell Syst Mol Biol* 1995, 3, 300–8.

122. Friesner, R. A., Banks, J. L., Murphy, R. B., Halgren, T. A., Klicic, J. J., Mainz, D. T., Repasky, M. P., Knoll, E. H., Shelley, M., Perry, J. K., Shaw, D. E., Francis, P., Shenkin, P. S., Glide: A new approach for rapid, accurate docking and scoring. 1. Method and assessment of docking accuracy. *J Med Chem* 2004, 47(7), 1739–49.

123. Schnecke, V., Kuhn, L. A., Database screening for HIV protease ligands: The influence of binding-site conformation and representation on ligand selectivity. *Proc Int Conf Intell Syst Mol Biol* 1999, 242–51.

124. Venkatachalam, C. M., Jiang, X., Oldfield, T., Waldman, M., LigandFit: A novel method for the shape-directed rapid docking of ligands to protein active sites. *J Mol Graph Model* 2003, 21(4), 289–307.

125. McMartin, C., Bohacek, R. S., QXP: Powerful, rapid computer algorithms for structure-based drug design. *J Comput Aided Mol Des* 1997, 11(4), 333–44.

126. Jenkins, J. L., Kao, R. Y., Shapiro, R., Virtual screening to enrich hit lists from high-throughput screening: A case study on small-molecule inhibitors of angiogenin. *Proteins* 2003, 50(1), 81–93.

127. Liu, M., Wang, S., MCDOCK: A Monte Carlo simulation approach to the molecular docking problem. *J Comput Aided Mol Des* 1999, 13(5), 435–51.

128. Trosset, J. Y., Scheraga, H. A., Reaching the global minimum in docking simulations: A Monte Carlo energy minimization approach using Bezier splines. *Proc Natl Acad Sci USA* 1998, 95(14), 8011–5.

129. Goodsell, D. S., Olson, A. J., Automated docking of substrates to proteins by simulated annealing. *Proteins* 1990, 8(3), 195–202.

130. Jones, G., Willett, P., Glen, R. C., Leach, A. R., Taylor, R., Development and validation of a genetic algorithm for flexible docking. *J Mol Biol* 1997, 267(3), 727–48.

131. Rosenfeld, R. J., Goodsell, D. S., Musah, R. A., Morris, G. M., Goodin, D. B., Olson, A. J., Automated docking of ligands to an artificial active site: Augmenting crystallographic analysis with computer modeling. *J Comput Aided Mol Des* 2003, 17(8), 525–36.

132. Kuhn, B., Gerber, P., Schulz-Gasch, T., Stahl, M., Validation and use of the MM-PBSA approach for drug discovery. *J Med Chem* 2005, 48(12), 4040–8.

133. Jiao, D., Golubkov, P. A., Darden, T. A., Ren, P., Calculation of protein-ligand binding free energy by using a polarizable potential. *Proc Natl Acad Sci USA* 2008, 105(17), 6290–5.

134. Huang, N., Kalyanaraman, C., Irwin, J. J., Jacobson, M. P., Physics-based scoring of protein-ligand complexes: Enrichment of known inhibitors in large-scale virtual screening. *J Chem Inf Model* 2006, 46(1), 243–53.

135. Kitchen, D. B., Decornez, H., Furr, J. R., Bajorath, J., Docking and scoring in virtual screening for drug discovery: Methods and applications. *Nat Rev Drug Discov* 2004, 3(11), 935–49.

136. Kramer, B., Rarey, M., Lengauer, T., Evaluation of the FLEXX incremental construction algorithm for protein-ligand docking. *Proteins* 1999, 37(2), 228–41.

137. Totrov, M., Abagyan, R., Rapid boundary element solvation electrostatics calculations in folding simulations: Successful folding of a 23-residue peptide. *Biopolymers* 2001, 60(2), 124–33.

138. Wei, B. Q., Baase, W. A., Weaver, L. H., Matthews, B. W., Shoichet, B. K., A model binding site for testing scoring functions in molecular docking. *J Mol Biol* 2002, 322(2), 339–55.

139. Zou, X., Sun, Y., Kuntz, I. D., Inclusion of solvation in ligand binding free energy calculations using the generalized-Born model. *J Am Chem Soc* 1999, 121(35), 8033–43.

140. Muegge, I., Martin, Y. C., A general and fast scoring function for protein-ligand interactions: A simplified potential approach. *J Med Chem* 1999, 42(5), 791–804.

141. Bohm, H. J., LUDI: Rule-based automatic design of new substituents for enzyme inhibitor leads. *J Comput Aided Mol Des* 1992, 6(6), 593–606.

142. Eldridge, M. D., Murray, C. W., Auton, T. R., Paolini, G. V., Mee, R. P., Empirical scoring functions: I. The development of a fast empirical scoring function to estimate the binding affinity of ligands in receptor complexes. *J Comput Aided Mol Des* 1997, 11(5), 425–45.

143. Gehlhaar, D. K., Verkhivker, G. M., Rejto, P. A., Sherman, C. J., Fogel, D. B., Fogel, L. J., Freer, S. T., Molecular recognition of the inhibitor AG-1343 by HIV-1 protease: Conformationally flexible docking by evolutionary programming. *Chem Biol* 1995, 2(5), 317–24.

144. Krammer, A., Kirchhoff, P. D., Jiang, X., Venkatachalam, C. M., Waldman, M., LigScore: A novel scoring function for predicting binding affinities. *J Mol Graph Model* 2005, 23(5), 395–407.

145. Bissantz, C., Folkers, G., Rognan, D., Protein-based virtual screening of chemical databases. 1. Evaluation of different docking/scoring combinations. *J Med Chem* 2000, 43(25), 4759–67.

146. Wang, R., Lu, Y., Wang, S., Comparative evaluation of 11 scoring functions for molecular docking. *J Med Chem* 2003, 46(12), 2287–303.

147. Kellenberger, E., Rodrigo, J., Muller, P., Rognan, D., Comparative evaluation of eight docking tools for docking and virtual screening accuracy. *Proteins* 2004, 57(2), 225–42.

148. Krovat, E. M., Langer, T., Impact of scoring functions on enrichment in docking-based virtual screening: An application study on renin inhibitors. *J Chem Inf Comput Sci* 2004, 44(3), 1123–9.

149. Kontoyianni, M., Sokol, G. S., McClellan, L. M., Evaluation of library ranking efficacy in virtual screening. *J Comput Chem* 2005, 26(1), 11–22.

150. Kontoyianni, M., McClellan, L. M., Sokol, G. S., Evaluation of docking performance: Comparative data on docking algorithms. *J Med Chem* 2004, 47(3), 558–65.

151. Xing, L., Hodgkin, E., Liu, Q., Sedlock, D., Evaluation and application of multiple scoring functions for a virtual screening experiment. *J Comput Aided Mol Des* 2004, 18(5), 333–44.

152. Chen, H., Lyne, P. D., Giordanetto, F., Lovell, T., Li, J., On evaluating molecular-docking methods for pose prediction and enrichment factors. *J Chem Inf Model* 2006, 46(1), 401–15.

153. Perola, E., Walters, W. P., Charifson, P. S., A detailed comparison of current docking and scoring methods on systems of pharmaceutical relevance. *Proteins* 2004, 56(2), 235–49.

154. Jain, A. N., Scoring functions for protein-ligand docking. *Curr Protein Pept Sci* 2006, 7(5), 407–20.

155. Yang, J. M., Chen, Y. F., Shen, T. W., Kristal, B. S., Hsu, D. F., Consensus scoring criteria for improving enrichment in virtual screening. *J Chem Inf Model* 2005, 45(4), 1134–46.

156. Feher, M., Consensus scoring for protein-ligand interactions. *Drug Discov Today* 2006, 11(9-10), 421–8.

157. Charifson, P. S., Corkery, J. J., Murcko, M. A., Walters, W. P., Consensus scoring: A method for obtaining improved hit rates from docking databases of three-dimensional structures into proteins. *J Med Chem* 1999, 42(25), 5100–9.

158. Clark, R. D., Strizhev, A., Leonard, J. M., Blake, J. F., Matthew, J. B., Consensus scoring for ligand/protein interactions. *J Mol Graph Model* 2002, 20(4), 281–95.

159. Paul, N., Rognan, D., ConsDock: A new program for the consensus analysis of protein-ligand interactions. *Proteins* 2002, 47(4), 521–33.

160. Helms, V., Protein dynamics tightly connected to the dynamics of surrounding and internal water molecules. *Chemphyschem* 2007, 8(1), 23–33.

161. Barril, X., Morley, S. D., Unveiling the full potential of flexible receptor docking using multiple crystallographic structures. *J Med Chem* 2005, 48(13), 4432–43.

162. Wei, B. Q., Weaver, L. H., Ferrari, A. M., Matthews, B. W., Shoichet, B. K., Testing a flexible-receptor docking algorithm in a model binding site. *J Mol Biol* 2004, 337(5), 1161–82.

163. Carlson, H. A., Protein flexibility and drug design: How to hit a moving target. *Curr Opin Chem Biol* 2002, 6(4), 447–52.

164. Carlson, H. A., Protein flexibility is an important component of structure-based drug discovery. *Curr Pharm Des* 2002, 8(17), 1571–8.

165. Carlson, H. A., McCammon, J. A., Accommodating protein flexibility in computational drug design. *Mol Pharmacol* 2000, 57(2), 213–8.

166. Ferrari, A. M., Wei, B. Q., Costantino, L., Shoichet, B. K., Soft docking and multiple receptor conformations in virtual screening. *J Med Chem* 2004, 47(21), 5076–84.

167. Kovacs, J. A., Chacon, P., Abagyan, R., Predictions of protein flexibility: First-order measures. *Proteins* 2004, 56(4), 661–8.

168. Cavasotto, C. N., Kovacs, J. A., Abagyan, R. A., Representing receptor flexibility in ligand docking through relevant normal modes. *J Am Chem Soc* 2005, 127(26), 9632–40.

169. Broughton, H. B., A method for including protein flexibility in protein-ligand docking: Improving tools for database mining and virtual screening. *J Mol Graph Model* 2000, 18(3), 247–57, 302–4.

170. Wong, C. F., Kua, J., Zhang, Y., Straatsma, T. P., McCammon, J. A., Molecular docking of balanol to dynamics snapshots of protein kinase A. *Proteins* 2005, 61(4), 850–8.

171. Zacharias, M., Rapid protein-ligand docking using soft modes from molecular dynamics simulations to account for protein deformability: Binding of FK506 to FKBP. *Proteins* 2004, 54(4), 759–67.

172. Meiler, J., Baker, D., ROSETTALIGAND: Protein-small molecule docking with full side-chain flexibility. *Proteins* 2006, 65(3), 538–48.

173. Pei, J., Wang, Q., Liu, Z., Li, Q., Yang, K., Lai, L., PSI-DOCK: Towards highly efficient and accurate flexible ligand docking. *Proteins* 2006, 62(4), 934–46.

174. Goodsell, D. S., Morris, G. M., Olson, A. J., Automated docking of flexible ligands: Applications of AutoDock. *J Mol Recognit* 1996, 9(1), 1–5.

175. Abagyan, R., Totrov, M., High-throughput docking for lead generation. *Curr Opin Chem Biol* 2001, 5(4), 375–82.

176. Teague, S. J., Implications of protein flexibility for drug discovery. *Nat Rev Drug Discov* 2003, 2(7), 527–41.

177. Merlot, C., Domine, D., Cleva, C., Church, D. J., Chemical substructures in drug discovery. *Drug Discov Today* 2003, 8(13), 594–602.

178. Guner, O. F., Hughes, D. W., Dumont, L. M., An integrated approach to three-dimensional information management with MACCS-3D. *J Chem Inf Comput Sci* 1991, 31(3), 408–14.

179. Xue, L., Godden, J. W., Bajorath, J., Mini-fingerprints for virtual screening: Design principles and generation of novel prototypes based on information theory. *SAR QSAR Environ Res* 2003, 14(1), 27–40.

180. Xue, L., Godden, J. W., Stahura, F. L., Bajorath, J., Design and evaluation of a molecular fingerprint involving the transformation of property descriptor values into a binary classification scheme. *J Chem Inf Comput Sci* 2003, 43(4), 1151–7.

181. Labute, P., A widely applicable set of descriptors. *J Mol Graph Model* 2000, 18(4-5), 464–77.

182. Mazza, C. B., Rege, K., Breneman, C. M., Sukumar, N., Dordick, J. S., Cramer, S. M., High-throughput screening and quantitative structure-efficacy relationship models of potential displacer molecules for ion-exchange systems. *Biotechnol Bioeng* 2002, 80(1), 60–72.

183. Fechner, U., Franke, L., Renner, S., Schneider, P., Schneider, G., Comparison of correlation vector methods for ligand-based similarity searching. *J Comput Aided Mol Des* 2003, 17(10), 687–98.

184. Marrero Ponce, Y., Cabrera Perez, M. A., Romero Zaldivar, V., Gonzalez Diaz, H., Torrens, F., A new topological descriptors based model for predicting intestinal epithelial transport of drugs in Caco-2 cell culture. *J Pharm Pharm Sci* 2004, 7(2), 186–99.

185. Moro, S., Bacilieri, M., Ferrari, C., Spalluto, G., Autocorrelation of molecular electrostatic potential surface properties combined with partial least squares analysis as alternative attractive tool to generate ligand-based 3D-QSARs. *Curr Drug Discov Technol* 2005, 2(1), 13–21.

186. Diaz, H. G., Bastida, I., Castanedo, N., Nasco, O., Olazabal, E., Morales, A., Serrano, H. S., de Armas, R. R., Simple stochastic fingerprints towards mathematical modelling in biology and medicine. 1. The treatment of coccidiosis. *Bull Math Biol* 2004, 66(5), 1285–311.

187. Gao, H., Shanmugasundaram, V., Lee, P., Estimation of aqueous solubility of organic compounds with QSPR approach. *Pharm Res* 2002, 19(4), 497–503.

188. Glem, R. C., Bender, A., Arnby, C. H., Carlsson, L., Boyer, S., Smith, J., Circular fingerprints: Flexible molecular descriptors with applications from physical chemistry to ADME. *IDrugs* 2006, 9(3), 199–204.

189. Mason, J. S., Beno, B. R., Library design using BCUT chemistry-space descriptors and multiple four-point pharmacophore fingerprints: Simultaneous optimization and structure-based diversity. *J Mol Graph Model* 2000, 18(4-5), 438–51, 538.

190. Mason, J. S., Good, A. C., Martin, E. J., 3-D pharmacophores in drug discovery. *Curr Pharm Des* 2001, 7(7), 567–97.

191. Xue, L., Bajorath, J., Molecular descriptors in chemoinformatics, computational combinatorial chemistry, and virtual screening. *Comb Chem High Throughput Screen* 2000, 3(5), 363–72.

192. Xue, L., Godden, J. W., Bajorath, J., Evaluation of descriptors and mini-fingerprints for the identification of molecules with similar activity. *J Chem Inf Comput Sci* 2000, 40(5), 1227–34.

193. Cramer, R. D., Jilek, R. J., Guessregen, S., Clark, S. J., Wendt, B., Clark, R. D., "Lead hopping." Validation of topomer similarity as a superior predictor of similar biological activities. *J Med Chem* 2004, 47(27), 6777–91.

194. Labute, P., Derivation and applications of molecular descriptors based on approximate surface area. *Methods Mol Biol* 2004, 275, 261–78.

195. Bajorath, J., Selected concepts and investigations in compound classification, molecular descriptor analysis, and virtual screening. *J Chem Inf Comput Sci* 2001, 41(2), 233–45.

196. Sheridan, R. P., Kearsley, S. K., Why do we need so many chemical similarity search methods? *Drug Discov Today* 2002, 7(17), 903–11.

197. Rush, T. S., III, Grant, J. A., Mosyak, L., Nicholls, A., A shape-based 3-D scaffold hopping method and its application to a bacterial protein–protein interaction. *J Med Chem* 2005, 48(5), 1489–95.

198. Reid, D., Sadjad, B. S., Zsoldos, Z., Simon, A., LASSO-ligand activity by surface similarity order: A new tool for ligand based virtual screening. *J Comput Aided Mol Des* 2008, 22(6-7), 479–87.

199. Wale, N., Watson, I. A., Karypis, G., Indirect similarity based methods for effective scaffold-hopping in chemical compounds. *J Chem Inf Model* 2008, 48(4), 730–41.

200. Tsunoyama, K., Amini, A., Sternberg, M. J., Muggleton, S. H., Scaffold hopping in drug discovery using inductive logic programming. *J Chem Inf Model* 2008, 48(5), 949–57.

201. Jakobi, A. J., Mauser, H., Clark, T., ParaFrag—an approach for surface-based similarity comparison of molecular fragments. *J Mol Model* 2008, 14(7), 547–58.

202. Wale, N., Karypis, G., Watson, I. A., Method for effective virtual screening and scaffold-hopping in chemical compounds. *Comput Syst Bioinformatics Conf* 2007, 6, 403–14.

203. Tanrikulu, Y., Nietert, M., Scheffer, U., Proschak, E., Grabowski, K., Schneider, P., Weidlich, M., Karas, M., Gobel, M., Schneider, G., Scaffold hopping by "fuzzy" pharmacophores and its application to RNA targets. *Chembiochem* 2007, 8(16), 1932–6.

204. Sperandio, O., Andrieu, O., Miteva, M. A., Vo, M. Q., Souaille, M., Delfaud, F., Villoutreix, B. O., MED-SuMoLig: A new ligand-based screening tool for efficient scaffold hopping. *J Chem Inf Model* 2007, 47(3), 1097–110.

205. Zhao, H., Scaffold selection and scaffold hopping in lead generation: A medicinal chemistry perspective. *Drug Discov Today* 2007, 12(3-4), 149–55.

206. Brown, N., Jacoby, E., On scaffolds and hopping in medicinal chemistry. *Mini Rev Med Chem* 2006, 6(11), 1217–29.

207. Mauser, H., Guba, W., Recent developments in de novo design and scaffold hopping. *Curr Opin Drug Discov Devel* 2008, 11(3), 365–74.

208. Myung, P. K., Sung, N. D., 2D-QSAR and HQSAR of the inhibition of calcineurin-NFAT signaling by blocking protein–protein interaction with N-(4-oxo-1(4H)-naphthalenylidene)benzenesulfonamide analogues. *Arch Pharm Res* 2007, 30(8), 976–83.

209. Sippl, W., Development of biologically active compounds by combining 3D QSAR and structure-based design methods. *J Comput Aided Mol Des* 2002, 16(11), 825–30.

210. Sippl, W., Contreras, J. M., Parrot, I., Rival, Y. M., Wermuth, C. G., Structure-based 3D QSAR and design of novel acetylcholinesterase inhibitors. *J Comput Aided Mol Des* 2001, 15(5), 395–410.

211. Sippl, W., Receptor-based 3D QSAR analysis of estrogen receptor ligands—merging the accuracy of receptor-based alignments with the computational efficiency of ligand-based methods. *J Comput Aided Mol Des* 2000, 14(6), 559–72.

212. Lozano, J. J., Pastor, M., Cruciani, G., Gaedt, K., Centeno, N. B., Gago, F., Sanz, F., 3D-QSAR methods on the basis of ligand-receptor complexes. Application of COMBINE and GRID/GOLPE methodologies to a series of CYP1A2 ligands. *J Comput Aided Mol Des* 2000, 14(4), 341–53.

213. Tervo, A. J., Nyronen, T. H., Ronkko, T., Poso, A., A structure-activity relationship study of catechol-O-methyltransferase inhibitors combining molecular docking and 3D QSAR methods. *J Comput Aided Mol Des* 2003, 17(12), 797–810.

214. Dror, O., Shulman-Peleg, A., Nussinov, R., Wolfson, H. J., Predicting molecular interactions in silico: I. A guide to pharmacophore identification and its applications to drug design. *Curr Med Chem* 2004, 11(1), 71–90.

215. Bowman, A. L., Lerner, M. G., Carlson, H. A., Protein flexibility and species specificity in structure-based drug discovery: Dihydrofolate reductase as a test system. *J Am Chem Soc* 2007, 129(12), 3634–40.

216. Patel, Y., Gillet, V. J., Bravi, G., Leach, A. R., A comparison of the pharmacophore identification programs: Catalyst, DISCO and GASP. *J Comput Aided Mol Des* 2002, 16(8-9), 653–81.

217. Langer, T., Krovat, E. M., Chemical feature-based pharmacophores and virtual library screening for discovery of new leads. *Curr Opin Drug Discov Devel* 2003, 6(3), 370–6.

218. van Drie, J. H., Pharmacophore discovery—lessons learned. *Curr Pharm Des* 2003, 9(20), 1649–64.

219. Kirchmair, J., Ristic, S., Eder, K., Markt, P., Wolber, G., Laggner, C., Langer, T., Fast and efficient in silico 3D screening: Toward maximum computational efficiency of pharmacophore-based and shape-based approaches. *J Chem Inf Model* 2007, 47(6), 2182–96.

220. Arkin, M., Protein–protein interactions and cancer: Small molecules going in for the kill. *Curr Opin Chem Biol* 2005, 9(3), 317–24.

221. Yin, H., Hamilton, A. D., Strategies for targeting protein–protein interactions with synthetic agents. *Angew Chem Int Ed Engl* 2005, 44(27), 4130–63.

222. Fletcher, S., Hamilton, A. D., Protein–protein interaction inhibitors: Small molecules from screening techniques. *Curr Top Med Chem* 2007, 7(10), 922–7.

223. Fletcher, S., Hamilton, A. D., Targeting protein–protein interactions by rational design: Mimicry of protein surfaces. *J R Soc Interface* 2006, 3(7), 215–33.

224. Lane, D. P., Cancer. p53, guardian of the genome. *Nature* 1992, 358(6381), 15–16.

225. Galatin, P. S., Abraham, D. J., A nonpeptidic sulfonamide inhibits the p53-mdm2 interaction and activates p53-dependent transcription in mdm2-overexpressing cells. *J Med Chem* 2004, 47(17), 4163–5.

226. Bowman, A. L., Nikolovska-Coleska, Z., Zhong, H., Wang, S., Carlson, H. A., Small molecule inhibitors of the MDM2-p53 interaction discovered by ensemble-based receptor models. *J Am Chem Soc* 2007, 129(42), 12809–14.

227. Rothweiler, U., Czarna, A., Krajewski, M., Ciombor, J., Kalinski, C., Khazak, V., Ross, G., Skobeleva, N., Weber, L., Holak, T. A., Isoquinolin-1-one inhibitors of the MDM2-p53 interaction. *ChemMedChem* 2008, 3(7), 1118–28.

228. Vassilev, L. T., Vu, B. T., Graves, B., Carvajal, D., Podlaski, F., Filipovic, Z., Kong, N., Kammlott, U., Lukacs, C., Klein, C., Fotouhi, N., Liu, E. A., In vivo activation of the p53 pathway by small-molecule antagonists of MDM2. *Science* 2004, 303(5659), 844–8.

229. Betzi, S., Restouin, A., Opi, S., Arold, S. T., Parrot, I., Guerlesquin, F., Morelli, X., Collette, Y., Protein protein interaction inhibition (2P2I) combining high throughput and virtual screening: Application to the HIV-1 Nef protein. *Proc Natl Acad Sci USA* 2007, 104(49), 19256–61.

Index